TELECOMMUNICATIONS SYSTEMS AND TECHNOLOGY

TELECOMMUNICATIONS SYSTEMS AND TECHNOLOGY

Michael Khader
William E. Barnes

NEW JERSEY INSTITUTE
OF TECHNOLOGY

Prentice Hall

Upper Saddle River, New Jersey   *Columbus, Ohio*

Library of Congress Cataloging-in-Publication Data

Khader, Michael.
    Telecommunications systems and technology / Michael Khader,
    William E. Barnes.
        p.   cm.
    ISBN 0-13-660705-5
    1. Telecommunication systems.   I. Barnes, William E.
    II. Title.
    TK5105.K48   2000
    621.382—DC21

98-56382
CIP

Publisher: Charles E. Stewart, Jr.
Associate Editor: Kate Linsner
Production Editor: Alexandrina Benedicto Wolf
Production Coordination: Custom Editorial Productions, Inc.
Cover Photo: SuperStock
Cover Design Coordinator: Karrie Converse-Jones
Cover Designer: Mark Shumaker
Production Manager: Deidra M. Schwartz
Marketing Manager: Ben Leonard

This book was set in Palatino by Custom Editorial Productions, Inc. and was printed and bound by
Courier/Westford. The cover was printed by Phoenix Color Corp.

© 2000 by Prentice-Hall, Inc.
Pearson Education
Upper Saddle River, New Jersey 07458

Printed in the United States of America

10 9 8 7 6 5 4 3 2 1

ISBN   0-13-660705-5

Prentice-Hall International (UK) Limited, *London*
Prentice-Hall of Australia Pty. Limited, *Sydney*
Prentice-Hall Canada Inc., *Toronto*
Prentice-Hall Hispanoamericana, S. A., *Mexico*
Prentice-Hall of India Private Limited, *New Delhi*
Prentice-Hall of Japan, Inc., *Tokyo*
Prentice-Hall (Singapore) Pte. Ltd., *Singapore*
Editora Prentice-Hall do Brasil, Ltda., *Rio de Janeiro*

To my mother, Zahia, from Mike
To my parents, Wallace and Charlotte, from Bill

Telecommunications is one of the fastest growing areas in the information technology field. When you answer the telephone and hear a computer talking to you on the other end, or when you attempt to reach someone via the telephone and are greeted with a voice response system telling you what button to push, you benefit from the advances in telecommunications. The Internet is another example of how telecommunications is changing the way we do business and live our lives. In the last five years, large corporations have grown with the Internet, e-mail traffic now exceeds that of the U.S. Postal Service, and the Internet is competing with television for our attention.

We had three major criteria to follow in preparing this text: be as state of the art as possible, slant the book with an industry perspective, and include the fundamental concepts of telecommunications that do not change. This text provides the reader with an insight into how the systems used in telecommunications are put together and work. The book attempts, in a straightforward manner, to introduce the user to concepts necessary to a good understanding of the operations common to many telecommunications systems.

Chapter 1 is devoted to fundamental concepts necessary to gain an insight into the overall operations of communication devices. This chapter introduces the communication system model, with a particular focus on its components. The types of communications media and their characteristics, such as twisted pair, optical fibers, radio waves, and coaxial cabling, are discussed. The history and operations of the public telephone network, and the types of switching techniques used in today's modern networks, are presented. The chapter closes with an introduction to protocols and layering, including a generic example of machines using protocols to communicate.

Chapter 2 covers the techniques of digital communications over the public telephone networks, specifically the pulse code modulation technique. We discuss the digital transmission facilities and the time-division multiplexing techniques that are used often, including the T1 and T3 facilities. We also present the formats used in the synchronous optical networks (SONET). Chapter 3 is devoted to applications of T1 and T3, subscriber's digital loop carrier systems (SDLC), and the technique of wave-division multiplexing.

Chapter 4 includes switching systems and the techniques used in switching digital information, specifically packet and circuit switching.

We also present a case study that focuses on the architecture of the Lucent Technologies' Definity private branch exchange (PBX), as well as computer telephony applications in switching systems. In Chapter 5, we discuss implementations of computer telephony and the concept of standard messaging. Then we devote a substantial part of the chapter to a case study that focuses on the design of a small-size switch using a personal computer for a platform.

Chapter 6 presents the topic of modems and modulation techniques used in modem communications. More than half of this chapter is devoted to compression techniques and entropy encoding algorithms. The Huffman code, run length, and the modified READ algorithms are also presented.

Chapter 7 is devoted to protocols and standards. We discuss the open system interconnection (OSI) reference model and then present typical layers as specified in the OSI reference model. We discuss the physical layer, the data link layer, and the network layer in great detail. High-level data link controller (HDLC) specifications and operations are emphasized, as well as uses in layer 2 of the OSI reference model. We then present the X.25 and ISDN standards, including all three layers as mapped into the OSI reference model.

In Chapter 8, we discuss broadband ISDN, and the asynchronous transfer mode (ATM) including the format of the ATM cell, the layers of ATM, and the adaptation layer of ATM known as AAL5. Concepts that are important to the understanding of ATM switching and communications, including the virtual circuit and the virtual path, are discussed. Chapter 8 concludes with a discussion of local area network (LAN) operations over an ATM network.

Chapter 9 presents the basics of and the need for local area networks. We present the media used in local area networks, and the media access techniques including ALOHA, CS/CD/MA, and token passing. Chapter 9 also discusses bridges and routers, as well as typical examples of local area networks.

In Chapter 10, we present Internet*working*, including the techniques and protocols used, the TCP/IP protocol suite, and networking management protocols. We present a detailed description of IPv4 and IPv6, their different formats, and the addressing techniques used in both protocols. We then present the UNIX implementation of sockets and their use in Internetworking.

Chapter 11 delves into the burgeoning world of wireless communications. A general discussion of wireless is followed by a discussion of pagers, the cellular concept, cell design issues, and power considerations. We also introduce multiple access and relate the specifics of FDMA, TDMA, and CDMA systems. PCS systems are introduced in this chapter as well.

No book on telecommunications systems and technology is complete without a chapter on multimedia communications. In Chapter 12, we present an introduction to multimedia communications, bandwidth requirements for multimedia communications, Joint Photographic Expert Group (JPEG) standards, Motion Picture Expert Group (MPEG) standards, and the discrete cosine transform used in compressing images in both the JPEG and MPEG.

This book is designed for undergraduates and professionals in the telecommunications field who want to learn more about any of the areas just described. The undergraduate audience may include students in the following disciplines:

- Electrical engineering
- Electrical/Computer engineering technology
- Computer science
- Computer engineering
- Telecommunication and information networks
- Telecommunication management

The math prerequisites for this text do not exceed those of a first-year college student. We attempted to avoid mathematical calculations that may obscure the intent of this text, which is to present the reader with a description of how things work and why they work the way they do.

For the instructors using this book, an instructor's manual is available.

We have established a Web page for this text. Anyone wishing to submit comments, corrections, or suggestions is encouraged to do so through this site at www-ec.njit.edu/~khader/telecomtext.

ACKNOWLEDGMENTS

Mike Khader would like to thank his family (Josephine, Daniela, Nadia, and Amanda) for their patience and unqualified support. Bill Barnes also thanks his family (Colleen, Myles, Jason, Nicole, Meghan, and Erika) for their help and forbearance during the long months of this project.

We thank Charles Stewart for his contagious enthusiam and support for this project. We owe a very special thanks to our copy editor, Ginjer Clarke, and our project editor, Kevin Walzer, for their meticulous and insightful contributions to this project. Among our colleagues at New Jersey Institute of Technology, we would particularly like to acknowledge Bob English, Edna Randolph, and Marcia Eddings. The students on whom the material was tested also deserve our gratitude. Finally, we would like to thank the following reviewers for their valuable feedback: Robert E. Morris, DeVry Institute of Technology; Jeffrey L. Rankinen, Pennsylvania College of Technology; and Edward Peterson, Arizona State University.

BRIEF CONTENTS

CONTENTS

Basic Concepts

OBJECTIVES
Specific topics in this chapter include the following:

- The components of a communication system.
- The types of communications based on the flow of information among the communicating parties and the configuration of the parties in a communication situation.
- An overview of transforming information into data and then into signals as well the type of transmission media.
- An historical perspective of telecommunication networks.
- The concept of switching and the idea of protocols.

1.1 INTRODUCTION

All activities of production and consumption, whether in business or government, in work or play, involve the process of material transformation—the storage, manipulation, and transport of physical matter—and the process of information transformation—the storage, processing, and communication of information.

Technological advances of the last one hundred years have steadily reduced the effort that society devotes to material transformation. The reduction is seen, for instance, in the fact that one hundred years ago nearly all U.S. workers were engaged in manual labor in agriculture, mining, and manufacturing. Today those sectors employ less than one-quarter of the workforce.

More than three-quarters of workers today are engaged in the process of information transformation and could be called *information workers.* We are now in the midst of another wave of change—the information revolution. This revolution consists of a core of advances in information technologies and their impact on ever-widening circles of social activities. We can model these activities as a flow process that has three phases. The first phase is initiated by the production of the information technologies, the second is the restructuring of the information technologies, and the last phase—changes in civic and personal relations—is a result of the preceding two.

1

The outer circles of the information revolution are still obscure, but some changes are visible. One widely observed trend is globalization. Corporations are becoming global as telecommunication and computer networks permit the collection of data and the coordination of corporate decisions around the world. You are probably aware of the term *global village*. The foundation of the global village may be seen in the homogenous culture resulting from the worldwide broadcast of the same TV entertainment and news programs. Government and business may become more intrusive as databases of information about citizens, hitherto isolated by agency or distance, now become interconnected, centralized, and accessible to all.

However, other trends run counter to globalization and centralization. Because access to information is much cheaper than before, citizens have better means to counter the propaganda from powerful organizations. By communicating over the Internet, geographically isolated, like-minded individuals can come together in small groups to pursue their common interests and to form new cultural islands outside mass culture. (Unfortunately, much of the information on the Internet should be viewed skeptically because sources are not always available or verifiable.) A movement such as this would have been impossible a decade ago. Further in the future lies the possibility that education will become a truly lifelong process, enabling a much fuller development of our potential.

Peering successfully into the future and the outer reaches of the information revolution requires the imagination of a poet and the vision of the seer. Yet the technologies at the core of the revolution are now visible and can be described using the language and concepts of the communication engineer and technologist, the computer scientist, and the economist. These technologies comprise advances in computers, telecommunications, signal processing, and their application in diverse domains. A subset of these advances relates to telecommunication technologies and their applications.

Advances in telecommunications technologies, as well as the core principles of telecommunications, are the substance of this book. We begin this chapter with a formulation of the basic communication concepts necessary for the understanding of these exciting technologies. We first introduce the simple models of communications (e.g., two people talking) and we advance this concept and its applications into the telecommunications field. Then we briefly examine the concept of telecommunication networks as they exist today contrasted with the way they were one hundred years ago. We also introduce *switching*, a concept that is necessary for an understanding of networking. We model a switch as a black box and give its functions and mode of operations later on in the chapter.

Protocols are the rules that govern communications among different entities within a telecommunication network. These rules and the framework within which they are constructed are briefly introduced to give the reader an overview of the intrinsic marriage between telecommunication networks and computers.

1.2 BASIC COMPONENTS OF A COMMUNICATION SYSTEM

Communication is an important part of our lives, because we are almost always involved in some form of it. Some examples of everyday communication, as shown in Figure 1–1, are face-to-face conversations, sending or receiving a letter, telephone conversations, watching a film at the movies or on television, using a computer to send E-mail, attending a lecture, and reading a book.

In the examples just cited, we observe that each communication system has its own characteristics; however, several properties are common to all of those systems. The main common attribute is that the aim of communication is to transfer information from one point to another. In data communication systems, we generally call this information *data* or *message.*

The message may take a number of different forms. It may be composed of factual information such as a well-prepared lecture. It may be composed of emotional information such as a heated argument in a telephone conversation or the arrangement of a piece of music. The message, however, is extremely important, and all of the processes of communication have been devised because there is a message to send.

In order to send a message from one point to another, three system components must be present. We need (1) a source, which generates a message and

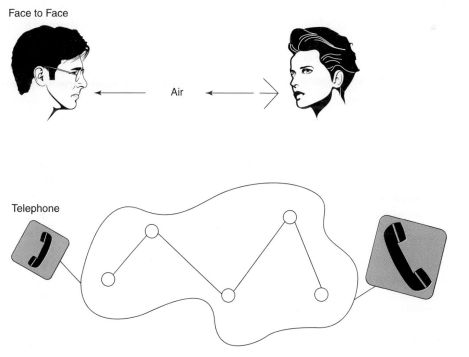

Figure 1–1
Examples of day-to-day communications.

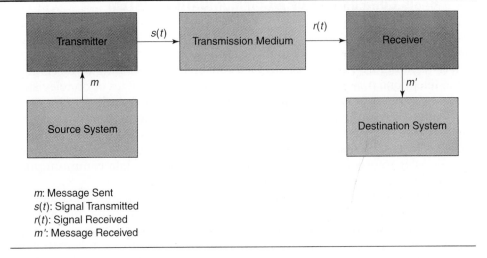

m: Message Sent
s(t): Signal Transmitted
r(t): Signal Received
m': Message Received

places it on (2) a transmission medium, which carries the message to (3) the receiver. These elements are the minimum requirements in any communication
process, and the absence of any one of them makes the communication process incomplete, and in many cases useless. The elements of a common communication system are illustrated in Figure 1–2.

The fundamental elements can be presented in many forms depending on
the particular communication system. Now let us examine some everyday examples to determine the source, medium, and receiver in each case:

- In a telephone conversation, the source might be you in New York City;
  the medium is the telephone network; and the receiver is your sister in
  San Francisco.
- In a face-to-face conversation, taking place around the kitchen table, the
  source may be Mrs. Jones; the medium is the air, which carries the sound
  waves of her voice; and her daughter is the receiver of the message.
- When an instructor gives a lecture in the classroom, she sends a message through the medium of air to a number of individual receivers,
  who are the members of the audience.

For communication to be effective, the message must be understood. If, while
talking to a friend, you use words that he or she does not know, then you have
not communicated meaningfully. If you pick up a Chinese version of this
book, you probably would not understand it. No matter how well presented
the material is, if you cannot read Chinese, the message is meaningless to you.
Similarly, if a computer is expecting information to come along a data line at a
particular speed and in a particular code, and the information comes in at a
different speed or in a different code, then effective communication has not
taken place.

The overall characteristics of a communications system are defined and limited by the individual characteristics of the source, medium, and receiver. The kind of information to be conveyed often dictates the type of source, medium, and receiver that will be used in a communication system. A color film of a horse race can convey visual and sound information about the race. We can see whether the sky is blue or overcast; we can see what color the jockeys are wearing; we can see who wins the race; but we cannot smell the training padlocks due to the limitation of the particular communication system involved.

In London, computers are used to handle betting on horse races. By looking at the information coming over the communication lines, the computer can tell how much money is being bet on which horse so that it can compute the odds and determine how much money to pay the winner. Due to limitations in the communication system, however, the computer cannot sense the feeling of anticipation of the jockeys and horses at the starting gate. Later you will see how a single component in a communication system can limit the performance of the entire system—the well-known weakest link of a chain principle.

In a communication system, interference can occur during the transmission process, and the message may get corrupted. Any such undesired disturbance in the system is called *noise*. A low-flying jet plane is a source of noise that interferes with a conversation.

1.3 TYPES OF COMMUNICATIONS

Using the examples we have seen thus far, we can classify the types of communications in a few different ways. One method of classification is to consider the flow of information and determine if it is one-way communication or two-way communication. In certain types of communication, information flows only in one direction, for example, radio/TV stations and receivers at home. This type is called *one-way* or *simplex communication*. If the two parties involved in communication can both send and receive information, the communication is *two-way* or *duplex communication*.

Duplex communication can be further divided into two types: *full-duplex communication* and *half-duplex communication*. Full-duplex or two-way simultaneous communication occurs when both parties in a communication situation can simultaneously send and receive messages. Ordinary face-to-face and telephone conversations are typically in full-duplex mode, although politeness dictates that it operates in half-duplex mode. Full-duplex communication is analogous to an interstate highway that runs north and south. On the other hand, half-duplex or two-way alternate communication refers to the mode in which one party sends a message, but the other party is disabled from sending while receiving. Although both parties can send and receive messages, they cannot do so simultaneously and they must take turns in sending (and receiving). Telegraph communications used to be of this type. Half-duplex communication

simplex communication, duplex communication, full-duplex communication, and half-duplex communication

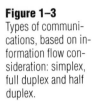

Figure 1–3
Types of communications, based on information flow consideration: simplex, full duplex and half duplex.

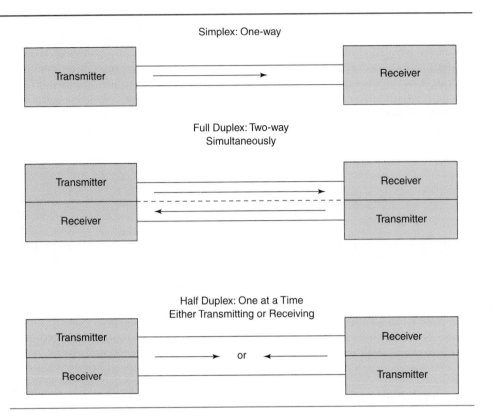

works like one-lane roads or bridges where traffic is regulated by stoplights or police. The flow alternates in the two directions. In computer-to-computer communication, the two computers often have an option of negotiating whether they will use full-duplex or half-duplex mode. Figure 1–3 illustrates these types of communication based on the information flow considerations.

A second way of classifying communication systems is by considering the number of devices that share the physical communication link. If only two devices share the link, we refer to the link as a *point-to-point link*. If more than two devices share the link, the link is said to be a *multipoint link*. Note that in a multipoint link, the sender of a message has to identify the intended recipient of the message. This step is not necessary in a point-to-point link. Figure 1–4 depicts these two types of communication scenarios based on the number of devices on a link.

point-to-point
link, and
multipoint link

Classification of communication modes helps us describe a particular communication system. However, certain concepts must be understood for one to discern the operation and use of modern communication systems. Analog and digital messages, transmission of messages, encoding of messages, routing and switching of messages, the network of highways and roadways that messages travel over, and regulation of the flow of messages within

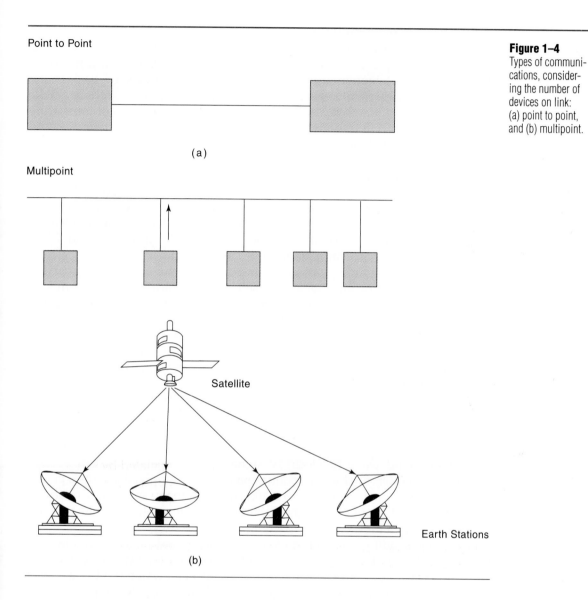

Point to Point

(a)

Multipoint

Satellite

Earth Stations

(b)

Figure 1–4
Types of communi-
cations, consider-
ing the number of
devices on link:
(a) point to point,
and (b) multipoint.

that network are a few of the themes that are necessary to an understanding of the particulars of a communication system. We will describe some of these themes in the remaining sections of this chapter.

1.4 TRANSMISSION CONCEPTS

In this section, we examine principles related to allowing a message to reach its destination once it has left the source. A message may take different shapes and forms before arriving at the receiver. This is acceptable, of course, as long as the intended message is understood by the receiving entity. For example, when you talk on the telephone, the sound waves you produce are converted to an electrical signal that is capable of traveling over the telephone network. At the receiving telephone, that electrical signal is reconverted back to the sound wave that originally was produced at the transmitting telephone. In the following subsection, we deal with these issues and techniques of translating a message. *Message*, in this context, means a sound wave, a computer file, and other entities that belong in these categories.

1.4.1   Analog and Digital

Analog and *digital* are terms that have entered the mainstream and are heard in everyday discussions about cell phones and televisions. These words, however, have specific meanings and implications as to how a communication system operates. The term *analog* can be equated with continuous and *digital* with discrete. Common examples abound, such as digital versus analog clocks, digital versus analog tuning on radios, and digital versus analog thermostats.

These terms, *analog* and *digital*, are used in communication literature mainly in three different contexts: (1) information or data, (2) signals and signaling, and (3) transmission.

In the context of information or data, the term *analog data* refers to data, such as voice, video, or pictures, in which the intensity or energy levels can vary continuously over time taking on all possible values over an interval range. The term *digital data* refers to data, such as alphanumeric characters, that can only take on a specific finite set of values, for example, values from the character set (A–Z, a–z, 0–9, special symbols such as backslash, blank, tab, etc.) available on a keyboard.

signal

In the context of signaling, the term *analog signal* refers to a physical *signal* (such as a sound wave or electrical signal) whose energy level or voltage varies over an interval of values continuously. For example, the signals generated by a traditional telephone set are analog signals. A *discrete*, or *digital signal*, in contrast, would be a signal wherein only a few selected voltage levels are possible, thereby leading to discontinuous step functions of time. For in-

bit

stance, when sending a sequence of *bits*, such as 0s and 1s (discussed later in this chapter), one may use a signal wherein each 1 bit is sent as a signal of –5

bit time

volts for a certain duration of time (called *bit time*) and each 0 bit is sent as a signal of 0 volts. Many other signal encoding schemes are used in practice. We will discuss them throughout the book as needed.

It should be noted that analog data might be transmitted in the form of analog or digital signals. For instance, speech data are carried in the form of electrical analog signals in most of the analog telephone loops that connect the home telephone set to the central office switch. On the other hand, speech data can also be digitized and carried in the form of digital signals within a digital network. Similarly, while digital data may be carried in the form of digital signals, it can also be sent in the form of analog signals. For example, a 0 bit may be sent as an analog waveform of one type, and a 1 bit as a waveform of another type. This type of conversion from digital to analog (and reverse) is usually performed by a device known as a *modem* (modulator-demodulator). The modem will convert the digital signals of a home personal computer (PC) into analog signals for transmission over the local loop and will then eventually convert them back to digital signals for a remote computer.

A signal loses energy and becomes distorted in shape as it travels through a medium such as air or cable. You do not expect your friend to hear you across a football field as well as if he were standing next to you during a conversation. This energy loss is referred to as *attenuation*. To compensate for attenuation, which can cause loss or errors in information transfer, analog transmission resorts to boosting, or amplification, of the signals. The amplification increases the energy level of the signals to help raise the signal level above the noise in the system (as discussed later in this chapter).

The signal is used as a vehicle to carry information or data from one point to another within a communication system. Information can be organized in many forms, including computer files. In the next section, we discuss the relationships among computer files, the data representation of such files, and the signals that may be used to transmit them.

1.4.2 Files, Data, and Signals

Information—or user data—is generally organized and stored in computing machines as files. File examples include a computer program, a document, or a numerical reading from an experiment. One of the principal goals of data communication is to allow users to exchange data in a reliable, efficient, and understandable manner. Consider the file HAMLET.TXT, which contains line 98 from Act I, Scene 5 of *Hamlet:* "Yea, from the table of my memory I will wipe away all trivial fond records."

The data are represented internally in the machine by a unique bit pattern for each character. Eight-bit entities are usually used to represent each character and the mapping between a character and its 8-bit representation is performed by a character code. It is important to note that bit patterns depend entirely on the choice of the character code, and the choice of a character code is typically independent of the characters themselves. The data in Figure 1–5 are shown with their binary representation using American Standard Code for Information Interchange (ASCII). Any character code could have been chosen.

To transmit the file HAMLET.TXT, the user data are converted into a signal. A signal is the representation of the data external to the machine. This external representation permits the exchange of data between machines. One example is a signal that takes on discrete voltage levels for both a 1 and a 0 bit. After the data are converted to a signal, the signal is placed on a physical medium for transmission. Some examples of a physical medium (described in later sections) include twisted-pair wire, coaxial cable, optical fiber, and the atmosphere. Figure 1–5 shows the bits being transmitted serially (i.e., one bit at a time), where a low signal represents a 0 bit and a high signal represents a 1 bit. Serial transmission may be contrasted with parallel transmission, such as

Figure 1–5
Files, data, and
signals.

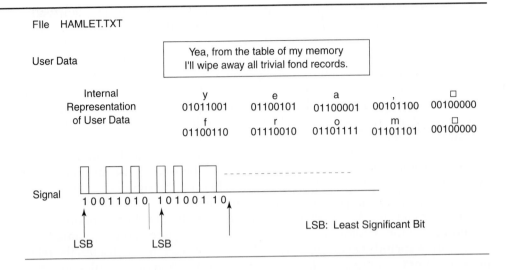

inside a computer, where 8, 16, 32, or 64 bits are transmitted at the same time but require 8, 16, 32, or 64 wires or lines. Of course, this method is not practical for transmitting over any distance. In our example, the least significant bit of each character is transmitted first. This convention is typical, but the most significant bit is transmitted first in some cases. Sending the HAMLET.TXT file can be accomplished using one of two transmission modes, synchronous or asynchronous, described in the following section.

1.4.3   Asynchronous and Synchronous Transmission

Serial transmission may be accomplished using either synchronous or asynchronous techniques. The use of these terms refers strictly to the relative timing of characters within a transmission block or a message. *Asynchronous transmission* means that each character in a message is transmitted as an individual entity, without regard to when the previous character was transmitted. Each character must have at least two framing bits associated with it; one framing bit indicates the beginning of the character and at least one other framing bit indicates the end of the character.

asynchronous
transmission

Synchronous transmission

Synchronous transmission means that all characters in a message are sent contiguously, one after the other. Framing characters indicate the beginning and end of the entire message block. Figure 1–6 depicts the synchronous and asynchronous transmission of the HAMLET.TXT file.

frame

A synchronous message is typically referred to as a *block* or a *frame*. It is sent as a contiguous bit stream, with no break between characters. Typically a synchronous message comprises the following fields (as shown in Figure 1–7):

- SYNC-CHARACTERS—These characters indicate to the receiver that a message is arriving. In other words, the message will immediately follow the SYNC-characters.

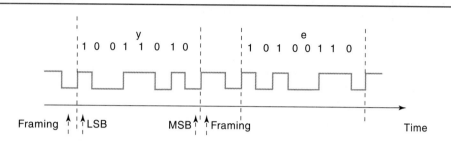

Figure 1–6
Asynchronous
and synchronous
transmission.

Asynchronous Transmission
*Timing of each bit is specified within a character.
*Inter-character time is nonuniform.
*Each character must be framed.

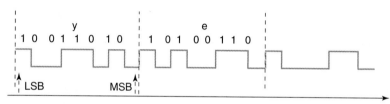

Synchronous Transmission
*Characters, within a block, are
 sent contiguously.
*Each block is framed.

LSB: Least Significant Bit
MSB: Most Significant Bit

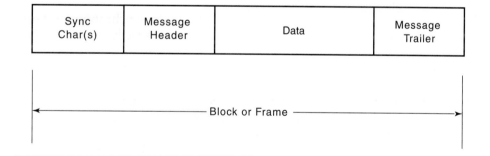

Figure 1–7
Synchronous
message format.

Sync Char(s)	Message Header	Data	Message Trailer

Block or Frame

- MESSAGE HEADER—Beginning-of-message framing, which includes the sequence number of the block (or frame). Messages have to be numbered so that they can be reordered according to their sequence number if they arrive at the receiver out of order.
- DATA—User information. In the example, data that will be part or all of the contents of the file HAMLET.TXT depending on the size of the data field.
- MESSAGE TRAILER—End-of-message framing, which may include error detection information.

1.4.4   Theoretical Aspects of Data Transmission

In this section, we will examine the limits imposed on transmission speed by noise and the bandwidth of a channel. We will also briefly discuss topics that we consider relevant to understanding a communication system, including frequency representation of signals, passband and bandwidth of a channel, signaling rate and bit rate, noise, the Shannon capacity theorem, and transmission media.

The Sine and Cosine Waves   Figure 1–8 shows a drawing of both a sine and a cosine wave. Both are examples of periodic analog functions; the sine and cosine repeat their basic pattern over and over again. The basic pattern, repeated after a

Figure 1–8
Sine and cosine
waveforms.

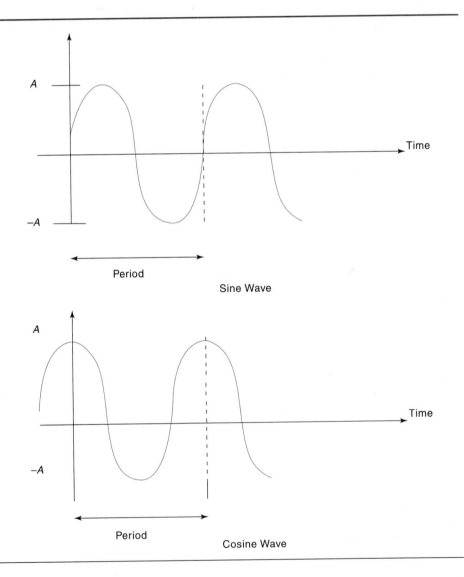

Sine Wave

Cosine Wave

length of time referred to as a *period,* is shown for both functions. Sine and cosine are examples of analog waveforms since they are smooth and continuously varying functions. The governing mathematical representations of both the sine and cosine wave are given by

Sine wave: $A \sin (2\pi ft + \theta)$
Cosine wave: $A \cos (2\pi ft + \theta)$

where A is amplitude
          f is frequency
          t is time
          θ is phase.

In general, the amplitude A, the frequency f, and the phase θ may be constant or time-varying functions. The reciprocal of the period is the frequency and is measured in cycles per second or equivalently Hertz (Hz). For example, if the period of a sine wave is 0.001 seconds, it has a frequency of 1,000 Hz.

Sine waves contain three important properties: amplitude, frequency, and phase shift. Consider a person who is playing the guitar. As a string is plucked, it vibrates and sends out a certain sound. The movement of the string from its neutral position over time is plotted as a time wave. The *amplitude* of the sine wave, then, is the magnitude of this movement and is detected by the ear as the loudness of the sound. Figure 1–9(a) shows two sine waves that differ only in amplitude. amplitude

The *frequency,* in our example, is the number of vibrations made each second, detected as the musical note. Figure 1–9(b) shows two sine waves that differ only in frequency. As discussed earlier, the frequency is measured in cycles per second or Hz. frequency

Finally, consider two guitars that play the same note at slightly different times. The *phase shift* will be the difference in time between the two events. This phenomenon might be observed in a two-propeller plane; a low-frequency vibrating hum is heard when the propellers are out of phase with each other. Figure 1–9(c) shows two sine waves that are of the same amplitude and frequency, but that are slightly out of phase. The difference in phase between the two sine waves can be measured in degrees and is referred to as the *phase angle*. To measure this angle, one period is defined as 360 degrees. Thus, as an example, if two sine waves were a quarter of a period apart, the case shown in Figure 1–9(c), the phase angle would be 360/4, or 90 degrees. phase shift

Fourier Series Expansion   Almost all time-varying periodic waveforms (they repeat at regular intervals) can be decomposed into a sum (of a possibly infinite set) of sine functions. The French physicist Fourier discovered this property of sine functions. The *Fourier series* for a function is its decomposition into sine waves. Figure 1–10 shows a train of square waves together with the sum of the first few terms of its Fourier series. Fourier series

As shown in Figure 1–10, some of the frequencies for the sine waves required to represent the train of square pulses will be high. To reproduce the

Figure 1–9
Sine wave character-
istics: (a) different
amplitudes, (b) dif-
ferent frequencies,
and (c) different
phase angles.

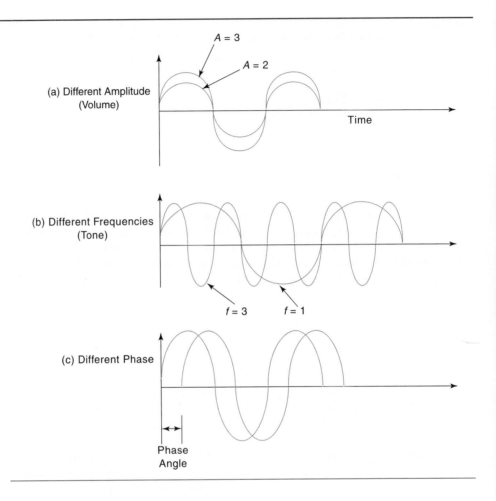

(a) Different Amplitude (Volume)

A = 3

A = 2

Time

(b) Different Frequencies (Tone)

f = 3   f = 1

(c) Different Phase

Phase
Angle

sharp corner of the square signal, you must have a sine wave that rises sharply. Since low-frequency sine waves rise slowly, some of the components of the Fourier expansion must be high-frequency signals. When we speak of signal, such as voice, containing frequencies from 50 to 15,000 Hz (one Hz is one cycle per second), we are describing the frequencies of the sine waves in the Fourier series for the function. Fourier series expansion and its twin, Fourier transform, are useful tools for analyzing the frequency behavior of signals in communication systems. In the following sections, we present concepts used to characterize a communication system.

Passband and Bandwidth of a Channel   If you examine the title of this section, you will find that we introduced three terms: channel, passband, and bandwidth. First, a *channel* is a conduit through which information passes. A communication channel can exist on many types of transmission media, for example, coaxial cable, copper wire, optical fiber, air, and water.

channel

Almost any periodic time-varying signal may be expressed as the sum of a possibly infinite set of sine waves, e.g., Fourier series of a periodic square wave.

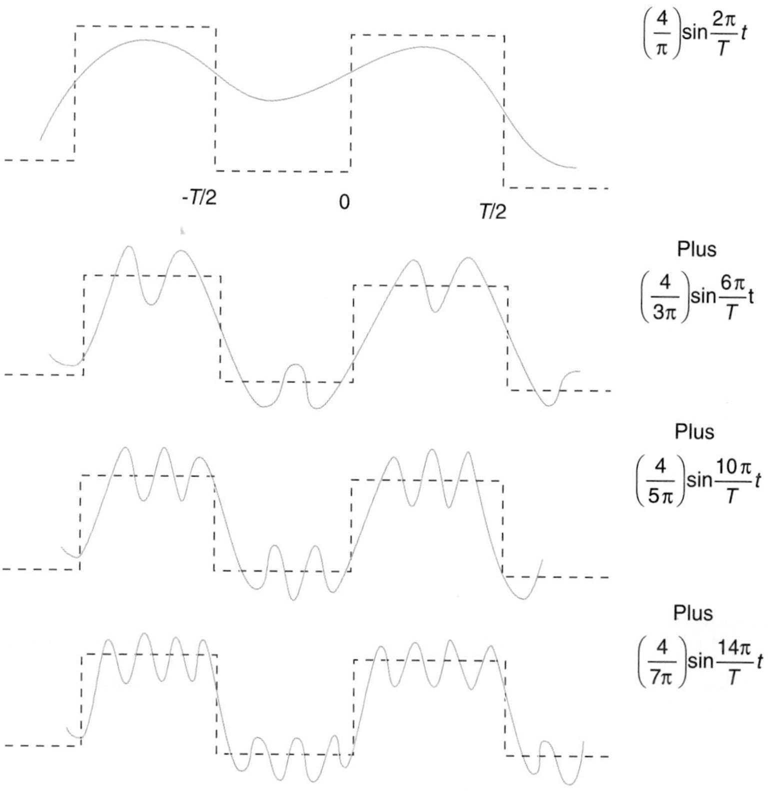

Figure 1–10
Fourier series expansion.

$$F(t) = \left(\frac{4}{\pi}\right)\sin\frac{2\pi}{T}t + \left(\frac{4}{3\pi}\right)\sin\frac{6\pi}{T}t + \cdots + \left(\frac{4}{\pi}\right)\left(\frac{1}{2n-1}\right)\sin\frac{2p(2n-1)}{T}t$$

$$\left(\frac{4}{\pi}\right)\sin\frac{2\pi}{T}t$$

-T/2   0   T/2

Plus

$$\left(\frac{4}{3\pi}\right)\sin\frac{6\pi}{T}t$$

Plus

$$\left(\frac{4}{5\pi}\right)\sin\frac{10\pi}{T}t$$

Plus

$$\left(\frac{4}{7\pi}\right)\sin\frac{14\pi}{T}t$$

Not all frequencies can pass with equal ease through a communication channel. This may be due to the characteristics of the channel itself or due to a *filter* that is placed on the channel for certain applications. (A filter is a device that allows only certain frequency signals to pass through it while severely attenuating other signals at different frequencies.) One of the most widely used filters in the phone network allows only signals with frequencies between 300 Hz and 3300 Hz to pass through. All other frequencies are severely attenuated. Thus, we have our second definition: the *passband* is the range of frequencies that will pass through a given communication channel.

Finally, the *bandwidth* of a channel is the width of the passband. For example, the local loop in the telephone system may be restricted to carrying signals

filter

passband

bandwidth

Figure 1–11
Passband and
bandwidth of a
channel.

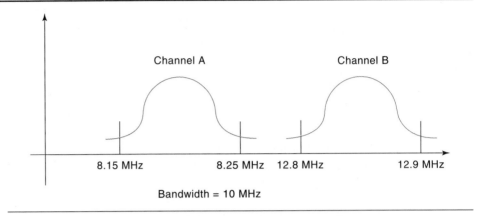

between 300 and 3300 Hz. Thus the passband is 300 to 3300 Hz, and the bandwidth is 3000 Hz (or 3 kilohertz—kHz). It is important to note that two channels with different passbands may have the same bandwidth. For example, if channel A has a passband of 76 to 80 kHz and channel B has a passband of 90.25 to 94.25 kHz, both have a bandwidth of 4 kHz. You can see that we can easily get the bandwidth from the passband but the bandwidth, alone, doesn't give us enough information to determine the passband. Figure 1–11 shows two channels with a bandwidth of 100 kHz.

In an imaginary and perfect communication world, an infinite number of signaling elements per second could be transmitted through a communication channel. In fact, however, there is a theoretical limit on the information-carrying capacity of a transmission channel because of the finite bandwidth that any channel has available. A physical channel, due to limitations of the transmission medium and/or the possible presence of filters, supplies the user with a bandwidth-limited channel.

Noise   *Noise* is a phenomenon that occurs on almost all communication channels. It is a random event, typically described by using statistical terms such as *mean* and *standard deviation*. Noise on a telephone line usually occurs in bursts, which may be audible as a click. A burst of noise with a duration of 0.1 seconds is barely noticeable to a human. However, during 2400 bits per second transmission, 0.1 seconds of noise can destroy 240 bits (about 24 characters).

There are three primary types of noise: thermal noise, crosstalk, and impulse noise. *Thermal noise* occurs due to thermal agitation of electrons, causing random motion. It is present in all electronic devices and transmission media. Thermal noise is uniformly distributed across the frequency spectrum and is referred to as *white noise*. It is often characterized by a hissing sound. *Crosstalk* occurs when signals from one communication path are heard on another path. Nearly everyone who uses a telephone has heard another conversation on the line; this is crosstalk. Crosstalk occurs when two wires are electrically coupled. That is, an electrical current on one wire will produce an electromagnetic field.

noise

thermal noise

crosstalk

If a second wire picks up the electromagnetic field, it will result in an unwanted current on the second wire. *Impulse noise* is a noncontinuous noise, consisting of irregular pulses or noise spikes with a short duration and large amplitude. It can be generated from a variety of sources, such as the movement of contacts in mechanical switches, power hits, lightning, and even animals chewing on the wire. Impulse noise is characterized by clicks.

impulse noise

    The study of noise is important because of the limitation it imposes on the amount of information that can be transmitted through a communication channel. The number of bits that can be pushed over the channel is impacted by noise. In particular, the limiting quantity is the ratio of the signal power to the noise power, generally written as the signal-to-noise ratio *(S/N)*.

    The maximum transmission rate, or capacity, of a channel, as a function of the *S/N* ratio and the bandwidth, is given by the *Shannon capacity theorem*. Figure 1–12 demonstrates the effect of signal-to-noise ratio on transmission speed. Suppose that levels *K* and (*K* + 1) represent two different signal levels on the channel. The actual transmitted signal will be the data signal level plus

Shannon capacity theorem

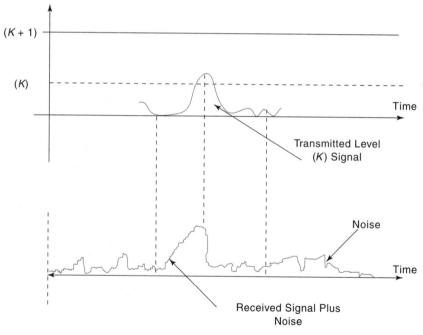

Figure 1–12
The concept of signal-to-noise-ratio.

* To reduce effect of noise:
A. Increase separation between levels.
B. Slow rate of transmission.
C. Add error-correction information.

Result: Decrease the effective data transfer rate.

background noise. If the noise level is too large and/or the signal levels are too close together, the transmitted data may be corrupted and interpreted by the receiver incorrectly.

In 1948, Claude Shannon of Bell Telephone Laboratories described the maximum number of bits per second that can be transmitted on a channel as a relationship between bandwidth and signal-to-noise ratio. Shannon showed that:

$$C = B \times \log_2 (1 + S/N)$$

where C is the channel capacity (bits per second)
          B is the bandwidth of the channel (in Hertz)
          S/N is the signal to noise ratio.

Applying the Shannon theorem to the local loop assuming a signal-to-noise ratio of 1000 and a bandwidth of 3 kHz, the maximum bit rate is

$$C = B \times \log_2 (1 + S/N) = 3000 \times \log_2 (1 + 1000)$$
$$C < = 30{,}000 \text{ bits per second.}$$

1.4.5 Transmission Media

Transmission media can be generally categorized into two types: guided and unguided media. *Guided media* include wire media such as *twisted-pair wire*, *coaxial cable*, and *fiber optic cable*. *Unguided* or *wireless media* include satellite, microwave radio, and cellular transmission. In guided media, the transmission characteristics determine their limitations, such as bit rate, bandwidth, and the repeater spacing. In unguided media, the transmitting antenna is very important in determining the limitations. In the following subsections, we examine the attributes of those media.

Twisted Pair Twisted pair is the most common transmission medium for both analog and digital data. It simply consists of two insulated copper wires arranged in a spiral pattern. A wire pair acts as a transmission link. Twisted wire pairs come in a range of gauges based on their diameters. Each gauge is designated with an American Wire Gauge (AWG) number, where a higher number designates a smaller diameter. AWG 22 and AWG 24 wires are two of the common types of twisted pair cables used for data and voice (telephone) transmission. Twisted-pair cable is commonly bundled in 2 to 3000 pairs per group.

Wire pairs can be used to carry both analog and digital signals. For analog signals, amplification of the signal will be needed about every 5 to 6 kilometers (km). For digital signals, repeaters will be used about every 2 or 3 km.

Twisted pair is limited in distance, data rate, and bandwidth as compared to some of the other transmission media. Attenuation (or power loss) in twisted pair is a strongly increasing function of the frequency. It is susceptible to thermal noise, impulse noise, and interference. Several techniques are available to reduce such impairments. Twisting of the wire reduces low-frequency interference. Crosstalk among adjacent wire pairs in the same cable can be re-

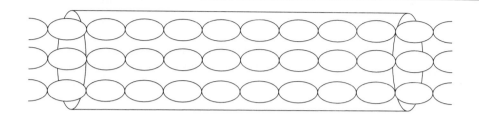

Figure 1–13
Twisted-pair wire.

duced by different lengths of twists of these pairs. Shielding the wire with metal braids can also reduce interference.

Usually one wire of a wire pair provides the ground potential. This is referred to as an *unbalanced line*. A *balanced line* is one in which both wires of a pair are above ground potential, carrying signals with equal amplitude but opposite phase.

balanced line and unbalanced line

Twisted-wire pairs are commonly used to connect residential phones to central office equipment, called the *local loop*. They are also used to connect office phones to office telephone equipment and to connect printers and terminals to local area networks. Figure 1–13 shows a diagram of a bundle of three twisted-wire pairs.

local loop

Coaxial Cables   Coaxial cables have better shielding than twisted pairs, so signals can travel longer distances at higher transmission speed. Two kinds of coaxial cables are widely in use—a 50-ohm cable usually used in digital transmission and a 75-ohm cable used in analog transmission. A single copper wire covered by insulating material is at the center of the coaxial cable. The insulated material is encased in a cylindrical conductor usually woven as a braided mesh. This outer conductor is enclosed in a protective plastic sheath.

The construction and shielding of the coaxial cable give it a good combination of high bandwidth and excellent noise immunity. The possible bandwidth depends on the cable length. For 1-mile cables, a data rate of 1 to 1.5 gigabits (1 billion or 10^9)per second of digital transmission is feasible. Longer cables can be used but at lower data rates or with repeated signal amplification. Coaxial cables used to be widely deployed within the telephone system but have largely been replaced by fiber optics on long-haul routes. The cable companies use a coaxial cable system known as *broadband coaxial. Broadband* is a term used by the telephone company for bandwidths higher than 4 kHz. Broadband networks use standard cable television technology. The cable can be used for analog transmission of up to 300 MHz and higher (450 MHz). Figure 1–14 shows the construction of the coaxial cable.

Optical Fiber   An optical transmission system has three components: a light source, a transmission medium, and a detector. A pulse of light indicates a 1 bit and the absence of light indicates a 0 bit. The optical transmission medium is an ultra-thin (hair-thin) fiber of glass. The detector generates an electrical

Figure 1–14
Coaxial cable.

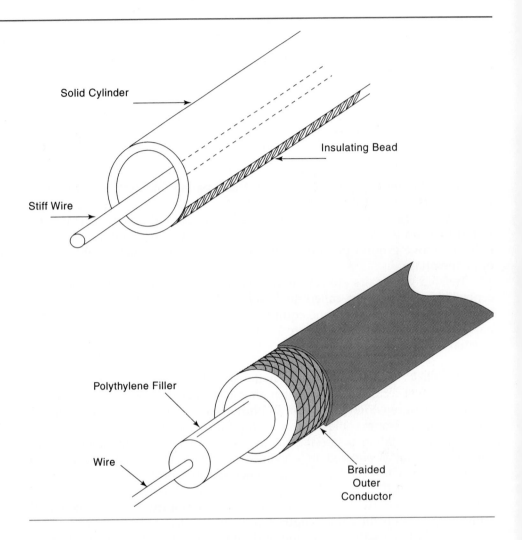

Solid Cylinder

Insulating Bead

Stiff Wire

Polythylene Filler

Wire

Braided
Outer
Conductor

pulse when light falls on it. Attaching a light source to one end of an optical fiber and a detector to the other, we obtain a unidirectional transmission system that accepts an electrical signal, converts and transmits that signal by light pulses, and then reconverts the output to an electrical signal at the receiving end.

This transmission system would leak light and would be useless in practice except for a very important principle of physics. When a light beam passes from one medium to another, for example from air to water, the beam is refracted (bent) at the air-water boundary as shown in Figure 1–15(a). Here we see the light beam incident on the boundary at an angle a1 and emerging at an angle b1. The amount of *refraction* depends on the properties of the two media

refraction

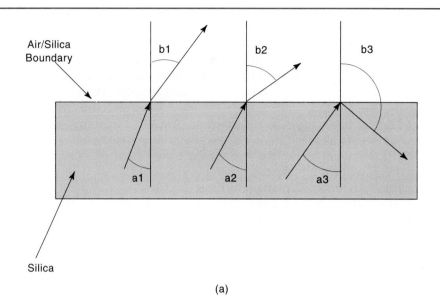

Air/Silica
Boundary

b1   b2   b3

a1   a2   a3

Silica

(a)

Figure 1–15
(a) Three examples of a light ray from inside a silica fiber impinging on the air/silica boundary at different angles, and (b) light trapped by total internal reflection.

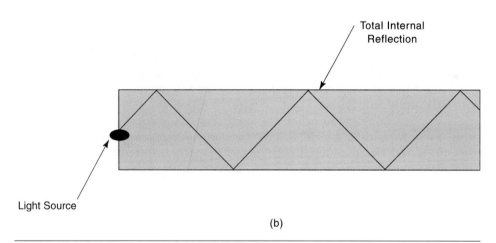

Total Internal
Reflection

Light Source

(b)

(in particular, their indices of refraction). For an angle of incidence greater than certain critical value, the light is reflected back into the air; none of it escapes into the water. Thus if we have a fiber and the light ray incident at or above a critical angle, the light ray is trapped inside the fiber as shown in Figure 1–15(b). The ray can propagate for many miles with virtually no losses.

Figure 1–15(b) shows only one trapped ray, but since any light ray incident on the boundary above the critical angle will be reflected internally, many different rays will be traveling inside the fiber at different angles. Each ray is said to have a different *mode.* A fiber having this property is called a *multimode fiber.*   multimode fiber

If the fiber's diameter is reduced to a few wavelengths of light, the fiber acts like a waveguide. In this case the light can travel only in a straight line without bouncing. This type of fiber is called a *single-mode fiber*. Single-mode fibers are more expensive since it is harder to make a fiber with a diameter in the order of light wavelengths. Their advantage, however, is that they can travel a long distance (about 22 miles) and transmit data at a rate of several gigabits per second.

Optical fibers are made from glass. Ancient Egyptians made glass thousands of years ago, but their glass was not as transparent as that used today. Optical fibers are made from extremely transparent glass, such that if we fill the ocean with it, the bottom of the ocean would be viewed similar to seeing the ground from an airplane on a clear day.

The attenuation of light through glass depends on the wavelength of light. For the kind of glass used in fibers, the attenuation is given by the following equation

$$\text{Attenuation in (dB)} = \frac{10 \log_{10} \text{transmitted-power}}{\text{received-power}}$$

Attenuation of 3 dB gives a factor of two loss (attenuation loss = $10 \log_{10} [2/1]$ = 3 dB), and attenuation of 6 dB gives a factor of four loss (attenuation loss = $10 \log_{10} [4/1]$ = 6 dB).

The Fiber Cable   Construction of fiber optic cable is similar to coaxial cable construction (Figure 1–16(a)). At the core is a single glass fiber through which the light propagates. The core is surrounded by glass cladding that has a lower index of refraction than the core. The cladding is encased within a protective thin plastic jacket. Fibers are typically grouped together in bundles, protected by an outer sheath (Figure 1–16(b)). Terrestrial fiber sheaths are normally laid in the ground within a meter of the surface, where they are occasionally subject to attacks by backhoes or gophers. In deep water, they are simply laid on the bottom, where they can be snagged by fishing trawlers or even eaten by sharks.

Connecting Fiber Cables   Fibers can be connected in three different ways. First, we can terminate the fiber with connectors that plug into fiber sockets. Light traveling through connectors will lose about 10 to 20 percent. However, connectors make fiber configuration easy.

Second, they can be spliced mechanically. The two fibers are laid carefully, and then the ends are cut next to each other in a special sleeve and clamped in place. Alignment can be improved by passing light through the junction and then making small adjustments to improve the signal passing between the two fibers. Mechanical splices produce about 10 percent loss in light. They require trained personnel to perform them (a trained person takes about five minutes to make a mechanical splice).

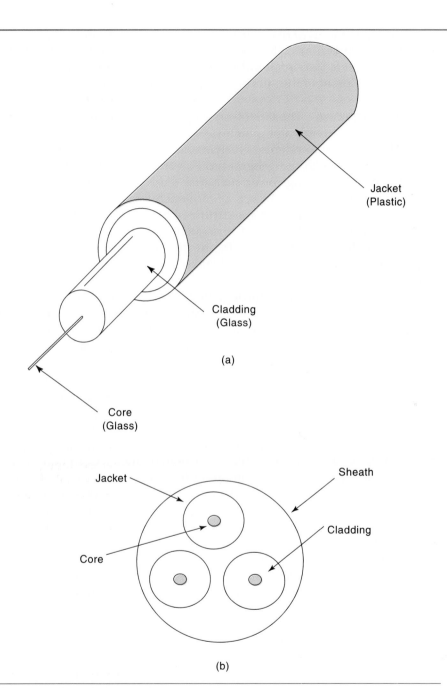

Figure 1–16
(a) Side view of a single fiber, and (b) end view of a sheath with three fibers.

Jacket
(Plastic)

Cladding
(Glass)

(a)

Core
(Glass)

Jacket

Sheath

Core

Cladding

(b)

Third, two pieces of fiber can be fused (melted) to form a solid connection. A fusion splice is almost as good as a single drawn fiber, but still a small amount of light is lost. In all three methods, the light traveling encounters refraction at the point of the splice.

line-of-sight transmission

Terrestrial Microwave   The tall towers, large horns, and dish antennas that you often see while driving along highways are the relay or repeater stations for terrestrial microwave systems, which operate by using *line-of-sight transmission.* Line-of-sight refers to the geographical arrangement of the transmitting antennas and the receiving tower such that the wave travels in a straight line from the transmitter to the receiver. Since microwaves travel in a straight line, if the towers are too far apart, the earth will get in the way (think about a San Francisco to Amsterdam link). Consequently, repeaters are needed periodically. The higher the towers are, the further apart they can be. For a 100 m high tower, repeaters can be spaced 80 km apart (assuming no large hills are in between).

multipath fading

Microwaves do not pass through buildings well. In addition, even though the beam may be well focused at the transmitter, some divergence occurs in space. Some waves may be refracted off low-lying atmospheric layers and may take slightly longer to arrive than direct waves. The delayed waves may arrive out of phase with the direct wave and thus cancel the signal. This effect is called *multipath fading.* Some operators keep 10 percent of their channel capacity idle as spare to switch on in case multipath fading temporarily wipes out some frequency band.

Figure 1–17 shows an example of a microwave transmission configuration. The most common type of microwave antenna is a rigidly fixed parabolic dish–shaped antenna approximately 10 feet in diameter. Common frequencies used for microwave transmission are in the range of 2 GHz to 40 GHz. The higher the frequency employed, the higher the bandwidth and data rate. Although microwave transmission requires line-of-sight transmission, it needs far fewer amplifiers and repeaters than does coaxial cable for the same distance. Microwave is used for both voice and television/cable transmission and for short point-to-point links between buildings in cities.

Microwave bands are regulated by the Federal Communication Commission (FCC). The 4 GHz, 6 GHz, and 11 GHz bands are commonly used for long-haul communications. The 12 GHz band is used to provide TV signals to local cable TV installations and the 22 GHz band is commonly used for point-to-point links between buildings.

uplink
downlink

Satellite Microwave   A communication satellite is essentially a microwave relay station. It links two or more ground-based transmitters/receivers called *earth stations.* The satellite receives the analog or digital signals transmitted by an earth station on one frequency band called the *uplink.* It then repeats or amplifies the signal on another frequency band known as the *downlink.*

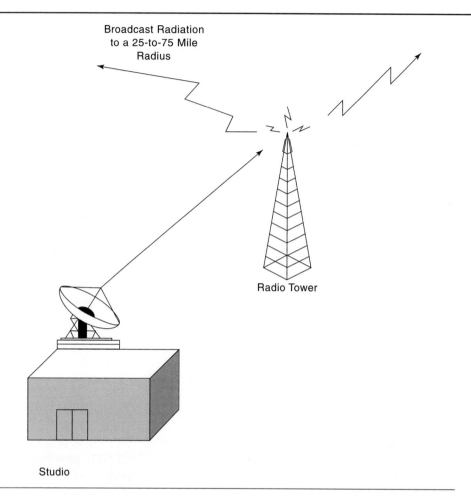

Broadcast Radiation
to a 25-to-75 Mile
Radius

Radio Tower

Studio

Figure 1–17
An example of a
microwave
configuration.

A single satellite can operate on a number of frequency bands and is called a *transponder.*

To ensure that the satellite is in the line-of-sight of earth stations it communicates with despite the rotation of the earth, it is easier when the satellite is made to rotate at a period equal of that of the earth. This is possible when the satellite is at a distance of approximately 35,800 km. A satellite of this type is referred to as being in a *geostationary,* or *geosynchronous, orbit* since it is then in a fixed position relative to the earth station.

One geostationary satellite can cover approximately 120 degrees east to west on the earth's surface. This area is called the *footprint* of the satellite. A   footprint
satellite is therefore ideal for broadcast over a large area. It also has a large bandwidth. Figure 1–18 shows a satellite used in a point-to-point link (a), and as a broadcast link (b).

Figure 1–18
(a) Point-to-point link via satellite microwave, and (b) broadcast link via satellite microwave.

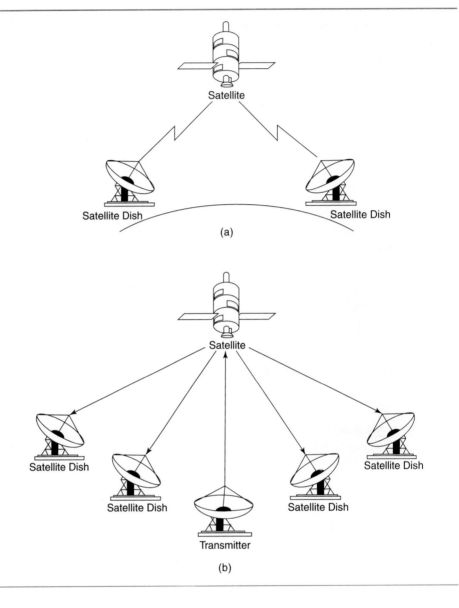

1.5   NETWORKING CONCEPTS

A communication infrastructure encompasses several transmission links, switches that those links carry messages into and out of, and equipment that allows the user to utilize the infrastructure. When these elements are interconnected to provide a particular communication service, we call the resultant structure a *network*. In this section, we explore the concept of networking. We examine the oldest and the most famous network of all—the telephone network. Then we present a brief discussion of computer networks that grew out of business needs.

network

1.5.1   The Telephone Network: An Historical Perspective

In 1876, if Alexander Graham Bell tripped over a pile of horse manure on his way to the patent office, Elizabeth Gray might have beat him to it, and today we would be talking about the Gray system. Instead, we talk about the Bell system, which has been in existence for more than one hundred years. Alexander Graham Bell invented a pair of telephones. Around 1890, simple networks connected telephones by manually operated switches. In this network, as shown in Figure 1–19, the signal is analog, as indicated by the letter *A* on the links. To call another telephone, a customer first rings the operator and provides the phone number of the intended party. The operator then determines which line goes either directly to the other party or to another operator along a path to the other party. In the latter case, the operators talk to one another, decide how to handle the call, and the procedure of constructing the path continues, possibly involving other operators. Eventually, one operator rings the destination, and if the telephone is picked up, the two parties are connected. The parties remain connected for the duration of the conversation and are disconnected by the operator at the end of the call.

Now let's look at how the transmission lines are allocated to the phone conversation. This is accomplished by means of *circuit switching*, where *circuit*

circuit switching

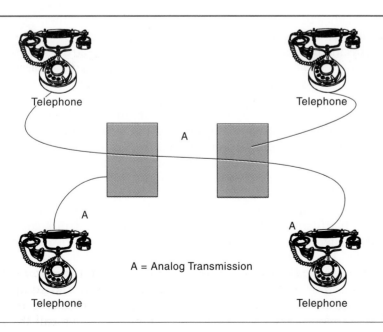

Figure 1–19
The telephone network as it existed around 1890.

Figure 1–20
The telephone net-
work around 1988.
The transmissions
are analog (A) or dig-
ital (D). The switches
are electronic and ex-
change control infor-
mation by using a
data network called
common channel
signaling (CCS).

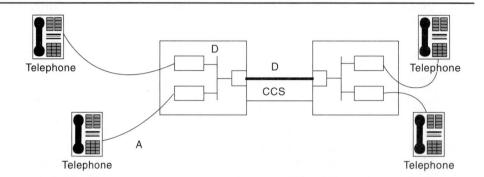

Figure 1–20
The telephone net-
work around 1988.
The transmissions
are analog (A) or dig-
ital (D). The switches
are electronic and ex-
change control infor-
mation by using a
data network called
common channel
signaling (CCS).

refers to the capability of transmitting one telephone conversation along one link. To set up a call, a set of circuits has to be connected, joining the two telephone sets. By modifying the connection, the operators can switch the circuits. Circuit switching occurs at the beginning of a telephone call. Operators were later replaced by mechanical switches and eventually by electronic switches.

Figure 1–20 illustrates the telephone network around 1988. One major development at this stage is that the transmission of the voice signal between switches is digital, as indicated by the letter *D*, instead of analog.

bit stream

An electronic interface in the switch converts the analog signal traveling on the link from the telephone set to the switch into a digital signal, called a *bit stream*. The same interface converts the digital signal that travels between the switches into an analog signal before sending it from the switch to the telephone.

CCS

The switches themselves are computers, which makes them very flexible. This flexibility allows the telephone company to modify connections by sending specific instructions to the computer. Figure 1–20 also shows a major development—*common channel signaling (CCS)*. CCS is a data communication network that the switches use to exchange control information among themselves. This conversation between switches serves the same function as the conversation that took place between operators in the manual network. CCS separates the functions of call control from the transfer of voice. Combined with the flexible computerized switches, this separation of function facilitates new services such as call waiting, call forwarding, and call back.

trunks

In current telephone networks, the bit stream in the *trunks* (lines connecting switches) and access links (lines connecting subscribers' telephones to the switch) are organized in the digital signal (DS) hierarchy. The links themselves—the hardware—are called *digital carrier systems*. Trunk capacity is divided into a hierarchy of logical channels. In North America these channels are called DS-1, . . . ,DS-4 and have rates ranging from 1.544 to 274.176 MBPS (megabits per second). The basic unit is set by the DS-0 channel, which carries 64 KBPS (kilobits per second) and accommodates one voice circuit. Since the 1980s, the transmission links of the telephone network have been changing to optical fibers using the SONET, or Synchronous Optical Network, standard (described in detail in Chapter 2).

The Telephone Network Structure   Currently, the telephone system is organized as a highly redundant, multilevel hierarchy. The following, although highly simplified, describes the present configuration. Each telephone has a pair of wires that goes directly to the telephone company's nearest end office, which is also called the *local central office*. The distance is typically 1 to 10 km, being smaller in cities than urban areas. The two-wire connection between each subscriber's telephone and the end office is known as the local loop. At one time, 80 percent of AT&T's capital value was the copper in the local loops. AT&T was then, in effect, the world's largest copper mine.

In the United States alone, there are approximately 20,000 local central offices. The concentration of the area code and the first three digits of the telephone number uniquely specify a local central office, which is why the rate structure uses this information. Customers are referred to as *subscribers*. If a subscriber attached to a particular end office calls a subscriber attached to the same end office, the switching mechanism within the office sets up a direct electrical connection between the two local loops. This connection remains intact for the duration of the call.

If the called telephone is attached to another end office, a different procedure has to be used. Each end office (local switch) has a number of outgoing lines to one or more nearby switching centers, called *toll offices* (or if they are within the same local area, *tandem switches*). These lines are called *toll-connecting trunks*. If both the calling party and the called party end offices happen to have a toll-connecting trunk to the same toll office (a likely occurrence if they are close by), the connection may be established within the toll office.

If the calling party and the called party do not have a toll office in common, the path will have to be established somewhere higher up in the hierarchy. Figure 1–21 shows the hierarchy as it currently exists. There are primary, sectional, and regional offices that form a network by which the toll offices are connected. The toll, primary, sectional, and regional exchanges communicate with

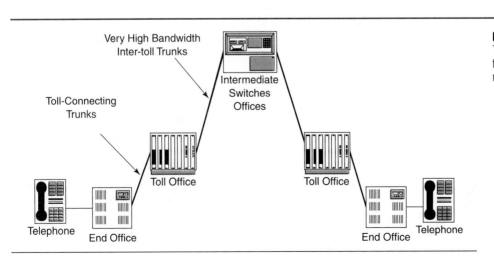

Figure 1–21
Typical circuit route for a call of a medium distance.

Very High Bandwidth Inter-toll Trunks

Intermediate Switches Offices

Toll-Connecting Trunks

Toll Office

Toll Office

Telephone

End Office

End Office

Telephone

each other via high bandwidth intertoll trunks (also called *interoffice trunks*). The number of different kinds of switching centers and their topologies (e.g., two sectional offices may have a direct connection or may go through a regional office) vary from country to country depending on telephone densities.

Telephone Call Scenario   The telephone set in our homes is connected to the local central office via a pair of wires called *tip and ring* (the twisted pair). Telephone calls are initiated by subscribers (users) by going off-hook; that is, by lifting the telephone's handset off its cradle or hook. The act of lifting the handset causes a switch to close. This action completes a direct current path in the local loop between the telephone and the local switch. The loop closure represents an off-hook signal between the subscriber's telephone and the local switch. This signal from the subscriber's telephone set is an act of requesting service.

To acknowledge this service request, the local switch station returns a dial tone to the caller. Essentially, the local switch is saying, "I received your request and I'm ready to process it if I can." The subscriber then proceeds to dial the desired number either by using rotary-type dialing (rarely used today) or a touch tone dialer. If the number dialed belongs to a local subscriber, then the local switch senses whether the dialed telephone is busy; if it is, then a busy tone is sent back to the originating subscriber. The busy tone is sent half a second on and half a second off.

If the dialed telephone is idle, the local switch sends an alerting (ringing) signal toward the dialed telephone and simultaneously sends a ring-back tone to the originating telephone. The ringing signal and the ring-back tone are sent at the same frequency—on for 2 seconds and off for 4 seconds. If all goes well and someone picks up the ringing telephone, the switch cuts off the ringing signal and the ring-back tone. Then the local switch connects a talk path between the two parties. The local switch continues to monitor the status of the telephone call until one of the parties on the call decides to hang up. This act causes a switch to open and thus the DC current is discontinued. The switch detects the on-hook condition and releases the talk path and the call is discontinued. Figure 1–22 shows a local telephone call scenario where both the calling and the called parties are connected to the same local switch.

For long-distance or message-unit calls, the call is routed through trunk lines and via toll stations. Toll stations are switch stations used to select long-distance trunk lines to route the call. The ends of the trunk lines are connected via other toll stations to local switch stations. Longer distances between parties may require longer trunk lines and additional intermediate switching stations. The hierarchy of the telephone switching stations, shown in Figure 1–23, begins with the local switch station, which has the direct lines to the end user—the subscriber. Connecting subscribers at the local switching station completes many calls; others require connections through high levels of switching. Local switch stations are classified as class 5 stations. They are connected to each other and in clusters to a tandem switch, which is also classified as a class 5. Table 1–1 outlines the structure of the public telephone network as it exists today.

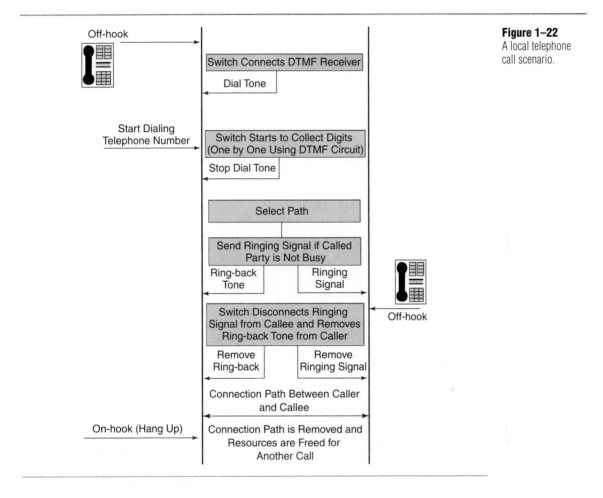

Figure 1–22
A local telephone call scenario.

Tandem switches, in turn, may connect the incoming call to other tandem switch stations, which pass the call to a local station to be connected to the called subscriber. Tandem stations are usually the beginning of the long-distance or toll or trunk line network. They are connected through toll stations (class 4), which attempt to route the call to another toll station. In the event that the call is not within a specific primary area, it is routed up through the class 3 *primary* station. Several of these primary stations are further interconnected through class 2 *sectional* stations and finally to the class 1 *regional* stations. The hierarchy takes the form of a switch tree. There are 12 regional centers in the United States (10) and Canada (2). The number of stations for each level in descending order increases until the bottom of the tree is reached at the local office level. There are approximately 20,000 local switch stations. It should be noted that the switching procedures are designed so that the minimal number of connections is made to complete a call. As shown in Figure 1–23, parallel-level connections as well as cross connections exist between levels to meet this purpose.

Figure 1–23
Telephone network
hierarchy.

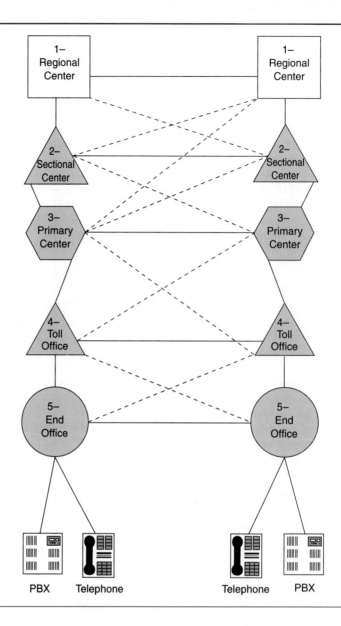

Table 1–1
The telephone
system hierarchy as
it exists today.

Order	Name	Comments
Class 1	Regional Center	Top of the hierarchy
Class 2	Sectional Center	
Class 3	Primary Center	
Class 4	Toll Center	Now "point of presence" (POP) where local exchange meets IEX
Class 5	End Office	In the local exchange carrier (LEC) area

Most of the upper-level switching stations incorporate a *repeater,* which is an amplifier and signal shaper, to regenerate information that may have deteriorated as it traveled along the long trunk line. Note that analog signals are amplified and digital signals are regenerated (repeated). repeater

Two-wire cabling is used for local loops to carry voice and data between two stations. These two wires are the active line and the signal return line required to complete an electrical circuit. Conversations using the local loop are conducted in a full-duplex manner; that is, the parties at both ends of the line can talk to each other at the same time. Long-distance trunk lines (which are carrying digital signals) employ four-wire cables to reduce signal degradation. One pair of wires carries signals in one direction. A second pair of lines carries signals back in the other direction. In essence, each pair of lines is operating in a simplex manner; that is, allowing communication in one direction. However, the use of both pairs allows communication to occur in a duplex (two-way) manner.

Data Communication over the Telephone Network   Initially, the telephone system was designed to carry electrical signals representing voice waves. With advances in the workplace, it became apparent that other modes of communication were required. Computers became an essential part of the work environment, and methods to interconnect them and have them communicate with each other were developed. The method used to allow computers to exchange information is referred to as *data communications.* Since the analog telephone system with its vast (but limited bandwidth) network was already in place, it became obvious that this network can be used to carry computer information if the computer information could be converted into a signal suitable for transport over such a network. The early device that made this conversion possible is called the *modem,* short for modulator/demodulator. However, before we examine this device in more detail, let us look at a general overview of data communications from rudimentary to highly advanced methods. data communications

For data communication to take place, it is not necessary to use telephone lines or special-purpose data lines. In fact, many organizations use data communications right now. If you want to send information from one point to another, various media are available to perform this task. You can use the mail for transmitting documents, cassettes, floppy discs, or CD-ROMs. You can use air express, you can transmit the data along a telephone line, or you can physically carry a box of documents from one point to another. In each case, the aim is the same—to get the information from one point to another and to get it there in one piece. The particular medium that you use is determined by examining a number of factors, such as the cost, the speed, the reliability, and the availability of the medium, as well as the urgency of your own requirements. You should go through the exercise of evaluating the various media that are available for two reasons: (1) to determine which particular medium or combination of media is optimal for your purposes, and (2) to select a suitable back-up in case the prime system fails. For example, if you rely on using air express to carry CD-ROMs from one city to another, what do you do if an airline goes on strike?

The parameters affecting which method to use to deliver information can still apply to the more complex methods used in today's advancing communication technology. Methods that are available today range from faxes to the more advanced instantaneous and simultaneous transfer of documents, images, and voice. Today you are able to watch a football game on your computer monitor while you discuss a homework assignment with a friend; both the assignment and the football game are instantaneously displayed on your monitor and your colleague's monitor at the same time.

What makes this multitask environment possible is the continuous and rapid evolution of data communications technologies. The term *data communications* deserves some elaboration. As mentioned in Section 1.2, data are defined as the representation of information in a manner that is understood by both human and machine. For example, to fax a document, the facsimile machine scans the document. Once scanned, the document is represented internally by sequences of 1s and 0s. The fax machine then, via a modem, converts those 1s and 0s into successions of analog signals suitable for transport over the local telephone loop. The receiving modem recreates from the analog signals the sequences of 1s and 0s that were originally transmitted. The facsimile machine then prints the pages corresponding to the information.

Data communications may involve a computer with one or more terminals connected by communications lines, video terminals connected to other video terminals by special-purpose communication lines (or by phone lines), and a number of computers interconnected by the Internet. The examples are endless in today's information technology. The communication lines can be ISDN | standard telephone lines, *integrated services digital networks (ISDN)*, or dedicated high-speed data communication lines. High-speed lines may include Ethernet, T1, T3, and fiber optics.

When two computers owned by the same company or organization and located close to each other need to communicate, it is often easier to run a cable between them. LANs work this way. However, when the distances are large, many computers are involved, or the cables would have to pass through a public road or other public right of way, the costs of running private cables are usually prohibitive. Furthermore, in almost every country in the world, stringing private transmission lines across (or underneath) public property is also illegal. Consequently, the network designers must rely upon the existing telecommunication infrastructure.

PSTN | These facilities, especially the *public switched telephone network (PSTN)*, were usually designed many years ago, with a completely different goal in mind—transmitting the human voice in a more or less recognizable form. Their suitability for use in computer-to-computer communication is often marginal at best, but the situation is rapidly changing with the introduction of fiber optics and digital technology. In any event, the telephone system is so tightly intertwined with (wide area) computer networks, that it is worth devoting considerable time studying the PSTN. Today the delineation between computers and computerized switches within the networks is blurred. The substance of this book is really devoted to computer communication in many of its forms.

1.6   SWITCHING CONCEPTS

Switches are essential components in a network, regardless of whether the network is public or private. As shown in Figure 1–24, any switching system can be modeled as a black box that performs the function of routing signals over transmission routes. As mentioned earlier, a network consists of a number of switches of various types and transmission links interconnecting them. This configuration coupled with a user's equipment provides the information technologies most commonly used.

From a conceptual point of view, a switch can be modeled as a box with a number of transmission links connecting to it and a control mechanism that allows interconnection among a subset of these links as required. The examples we presented earlier regarding the scenario of making a telephone call illustrate this point.

Two predominant types of switching mechanism exist, depending, in general, on the network employed. The type that is commonly used in the telephone network is called *circuit switching*. Local area networks and other high-speed digital data communication networks use *packet switching*.

packet switching

In a telephone conversation, once a number of circuits are allocated for a particular conversation, this allocation stays in place, unchanged, until the telephone

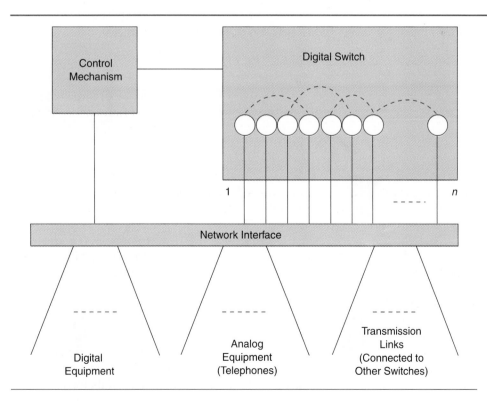

Figure 1–24
Elements of a one-node switch.

call is ended. The switching system in this case does not concern itself so much with the contents of the conversation. You can imagine this process as stretching a pipe between two telephones, along many paths, and plumbing joints. What goes into the pipe on one side comes out of it in the same order at the other side. For two-way conversation you may need two pipes, one for each direction. Those pipes have to exist until the call is completed.

As depicted in Figure 1–25, a path is established for a conversation between telephone A and telephone B. This path goes through nodes (switches) 1, 3, and 5. The allocation by the switches of the transmission media and the resources required to establish this call will stay in place until one of the parties decide to end the conversation.

In contrast, packet switching is concerned with messages being transmitted. The message is fragmented into smaller chunks and routed along different paths as available. At the receiving end, the smaller chunks (called *packets* as we will see later in the book) are reassembled and the original message is formed. Figure 1–26 demonstrates packet switching. You may ask this question: What if one of the smaller chunks is lost? Well, if that happens (and it does), the data that were transmitted will get corrupted. Therefore, some form of error detection and retransmission must be implemented to guard against this situation.

At the beginning of the Cold War, the military was looking for ways to protect its communication networks against enemies spying on sensitive conversations.

Figure 1–25
The concept of
circuit switching.

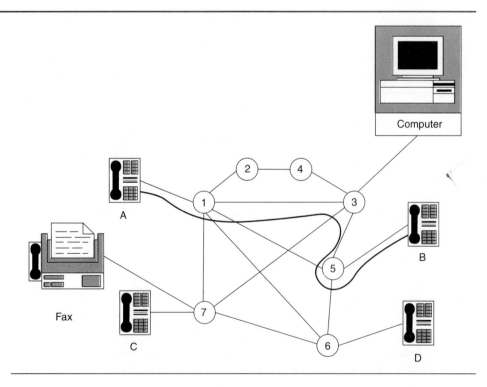

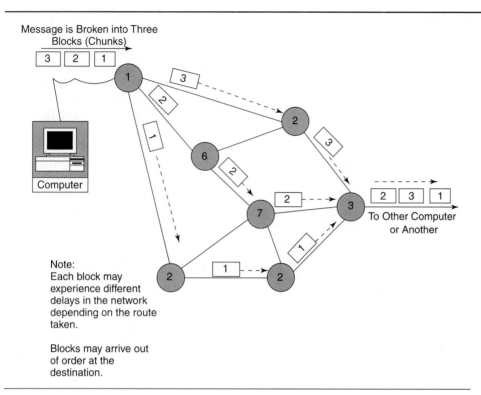

Message is Broken into Three Blocks (Chunks)

Computer

Note:
Each block may experience different delays in the network depending on the route taken.

Blocks may arrive out of order at the destination.

To Other Computer or Another

Figure 1–26
The concept of packet switching.

You can see that, in the circuit switching scheme, one needs only to place a tap along any point on the path of the telephone call, and all phone calls along that path can be heard. In packet switching, however, this task is not so easy. The digitized voice message is broken down into smaller chunks; each chunk on its own does not convey intelligent information. And since each chunk can take a different route while traveling from point to point along the network, listening in by tapping a line becomes very difficult. The technology that gave us the Internet was initiated in this manner.

When equipment and switches in a network need to communicate, they have to adhere to a set of rules depending on their level of importance in the network. Rules dictating when to ask, when to listen, when to wait, and when to ask for help are in place to help orchestrate the smooth operations of the network. These rules are called *protocols*. In the next subsection, we will examine this concept. We also dedicate Chapter 7 to this topic.

1.7 LAYERING AND PROTOCOLS

To reduce their design complexity, most networks are organized as a series of layers or levels, each one built upon the one below it. The number of layers, the name of each layer, the contents of each layer, and the function of each layer differ from network to network. However, in full networks, the purpose of each layer is to offer certain services to the higher layers, shielding those layers from the details of how the offered services are actually implemented.

Why use layers? Because people build layered networks. Most networks humans build sit on top of each other. The *road network* is a meshed communications network linking crossroads—that is, communication nodes—villages, and towns, which serve as communication nodes but are also where the final network users reside.

Another communication network is the *postal network.* It is made up of post offices where the mail traffic is switched in a packet switching–type mechanism. The postal road network runs on top of the road network. But a one-to-one correspondence does not occur between the road switching points and the postal switching points; not all villages have post offices. The post offices form a plane of logically interconnected nodes that are physically interconnected by roads. We also know that the postal network does not use roads only. It also uses the railroad network and the airline network.

If you need a piece of mail sent, you may simply drop it in a mailbox. You do not worry about whether this piece of mail will be sent by the airline network or the railroad network as long as it gets to its destination within the expected time.

Another example of layering and protocols is two diplomats who do not speak a common language but who need to communicate in writing. Let us say that one diplomat speaks French and the other speaks Chinese. Let us assume the language used for communication is English. The Chinese diplomat gives the document to his or her secretary (the layer below), and the secretary gives it to the translator (a lower layer). The document is translated into English and sent by fax (a further lower layer) to the office of the diplomat in France. A clerical staff person (the lowest layer in the office) picks up the fax. He or she realizes that this message needs translation and passes it to the translator (English to French), who then hands the document to the diplomat's assistant (the layer just below the diplomat), who in turn hands the document to the diplomat.

The communicating parties in this case did not need to know how the message was transferred from one functional office to the next. All they were concerned with was the information within the document. Many layers of supporting function were utilized to complete the transaction. Similarly, if you send an E-mail, you are concerned only with the message's content, as is your recipient. Again, though, several layers of support were needed in the delivery of that message.

Protocol layering in communication networks is not much different. Computers and switching nodes communicate with each other by the use of similar layering techniques to that in the preceding example. If we say the French diplomat and the Chinese diplomat are a level (layer) N, both communicate within the primitives of that layer and are unconcerned with the primitives of the layers beneath them. In the Chinese diplomat's office, let us say we have four layers: the diplomat (layer N), the assistant (layer $N-1$), the translator (layer $N-2$), and the clerical staff person (layer 1), which is the lowest layer in this case. The lowest layer is concerned only with the transfer of the document from the Chinese office to the French office. This layer treats the document as a group of papers that is to be transferred in one form or another to the corresponding office. This layer is uninterested in the contents of the document. A similar structure exists in the French office.

If we view each machine communicating within a network as a diplomat's office, each machine will house a number of layers necessary to establish communications with other machines. Layer N on a machine carries on a conversation with layer N on another machine. The rules and conventions used in this conversation are collectively known as the *layer protocol*. Basically, a protocol is an agreement between the communicating parties on how communication is to proceed. As an analogy, when a woman is introduced to a man, she may choose to offer her hand. The man in turn may decide to shake it or kiss it, depending on whether she is an American lawyer at a business meeting or a European princess at a ball. Violating the protocol will make communication more difficult, if not impossible.

A five-layer network is shown in Figure 1–27. The entities composing the corresponding layers on different machines are called *peers*. In other words, the peers communicate using the protocol.

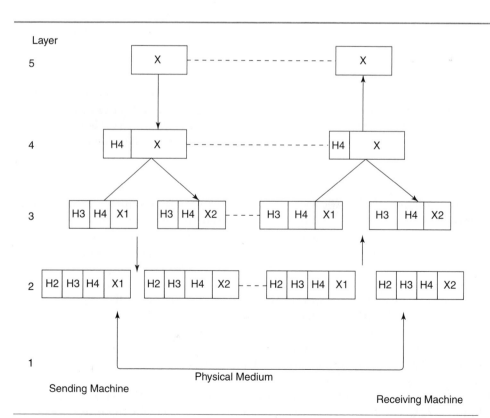

Figure 1–27
Communication using five-layer protocol stack.

In reality no data are directly transferred from layer N on one machine to layer N on another machine. Instead, each layer passes data and control information to the layer immediately below it, until the lowest layer is reached. Below layer 1 is the physical medium through which actual communication occurs (in the diplomat example this was the facsimile machine). In Figure 1–27 virtual communication is shown by dotted lines and physical communication by solid lines.

Between each pair of vertically adjacent layers, there is an interface. The interface defines which primitive operations and services the lower layer offers to the upper one. When network designers decide how many layers to include in a network and what each one must do, one of the most important considerations is defining a clean interface between the layers. Doing so, in turn, requires that each layer perform a specific collection of well-understood functions. To minimize the amount of information that must be passed between layers, clean-cut interfaces make it simpler to replace the implementation of one layer with a completely different implementation (e.g., all the telephone lines are replaced by satellite channels). All that is required of the new implementation is that it offers exactly the same set of services to its upstairs neighbor as in the old implementation.

architecture   A set of layers and protocols is called *network architecture.* The specification of an architecture must contain enough information to allow the implementers to write the program or build the hardware for each layer so that it will correctly obey the appropriate protocol. Neither the details of the implementation nor the specification of the interfaces is part of the architecture, because these are hidden away inside the machine and are not visible from the outside. It is not even necessary that all machines in a network be the same, provided that each machine can correctly use all the protocols. A list of protocols used by a certain system—one protocol per layer—is called a *protocol stack.* The subjects of network architectures, protocol stacks, and protocols themselves are essential to an understanding of telecommunications. These topics are featured prominently in this book.

To further explain this concept, let us consider a technical example: How do we provide communication to the top layer of the five-layer network shown in Figure 1–27? A message, X, is produced by an application process running in layer 5 and given to layer 4 for transmission. Layer 4 puts a *header* in front of the message to identify the message and passes the message with the header to layer 3. The header includes control information, such as sequence numbers, to allow layer 4 on the destination machine to deliver the message in the right order if the lower layer did not maintain the sequence. In some layers, headers also contain sizes, times, and other control fields.

In many networks, the size of messages transmitted in the layer 4 protocol is not limited, but a limit is nearly always imposed by layer 3 protocol. Consequently, layer 3 breaks up the incoming (from layer 4) message into smaller units, called *packets,* inserting layer 3 header to each packet. In this example, X is split into two parts, X1 and X2.

Layer 3 decides which of the outgoing lines to use and passes the packets to layer 2. Layer 2 adds not only a header to each piece but also a trailer and gives the resulting unit to layer 1 for physical transmission. At the receiving machine, the message moves upward, from layer to layer, with headers being stripped off as it progresses. None of the headers of layer N is passed to layer $N + 1$ when the message moves upward.

The important point, as seen in Figure 1–28, is the relation between virtual and actual communications and the difference between protocols and interfaces. The peer process in layer 4, for example, consequently "thinks" of their communication as being "horizontal," using the layer 4 protocol. Each one is

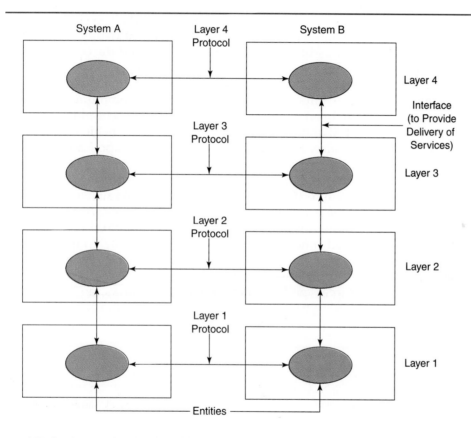

Figure 1–28
Protocol stack.

* Each subsystem is composed of a number of entities.
* An entity is a hardware and/or software module.
* Two forms of interactions among entities:
  – Access services across an interface (within the same system, but across layers).
  – Exchange data and control information according to a protocol (within a layer but across systems).

likely to have a procedure called something like *send to other side* and *get from other side*, even though these procedures actually communicate with lower layers across the 3–4 interface, not with the other side.

1.8 SUMMARY

In this chapter, we presented a survey of some of the basic concepts necessary for a good understanding of the telecommunication technologies and their impact on the information technology production and structure. We have discussed the essential components of a typical communication system and the effect of the absence of any one component. We presented two types of communications based on the directional information flow and the configuration of devices and links.

Although a strong background in math is not required to read this book, a very high-level view of important theories used in telecommunication systems design was introduced. We briefly outlined the idea behind the Fourier series expansion and Shannon's theorem of the maximum bit rate over a channel experiencing noise. These theories have important applications in the field of telecommunications transmission, as we will see in Chapter 2.

We have also examined in this chapter the networking and switching concepts and how they are intertwined. The concept of layered networks, protocols, and switching is the basis of modern telecommunications. As we will see later in the book, a minimum of three protocol layers must reside in a given node in a network at all times. Protocol stacks, including up to the seventh layer, can reside in the communicating equipment, whether this equipment is a computer running sophisticated multimedia applications or a mainframe that supports remote access and Internet services.

REVIEW QUESTIONS
AND PROBLEMS

1. List the necessary three elements in a communication system.

2. List the types of communications based on
   a. The directional flow of information.
   b. The configuration of devices and links.

3. Describe what the terms analog and digital mean. Classify each of the following signals (analog or digital):
   a. Electrical voltage representing a sound wave.
   b. Sampled point of the electrical voltage in a sound wave.

4. What are the contents of a computer file in terms of coded information? For the text in Figure 1–29, give the equivalent ASCII representation:

5. Give the data representation of the phrase "I'm sorry" in:
   a. Synchronous transmission.
   b. Asynchronous transmission.

6. What is the amplitude, frequency, period, and phase for each of the following:
   a. $f(t) = 20 \sin (1000t + 30)$
   b. $f(t) = M \cos (50t - 60)$
   c. $f(t) = 30 \sin 2t$

I'm sorry I could not get back to you earlier. I will be able to make it to the meeting today.

Sincerely,
Mike

Figure 1–29
Text from computer file.

7. State the Fourier expansion theorem.

8. What is a first harmonic in a Fourier series expansion?

9. What is a fourth harmonic in a Fourier series expansion?

10. Why is the Fourier series expansion important in the design of communication systems? Give a typical example of how it may be used.

11. **a.** What is the bandwidth of a channel?
    b. Are there any differences between bandwidth and passband of a channel?

12. Using the Shannon capacity theorem, find the maximum bit rate on a channel that has a bandwidth of 4000 Hz and a SNR of 30 dB.

13. What is the maximum bit rate that can be transmitted over twisted pair type wires in the local loop?

14. Describe the physics behind fiber optic transmission. Give three reasons why fiber optic transmission is becoming very popular in telecommunication networks.

15. Compare coaxial cable transmission with fiber optic transmission and list where each would be used.

16. Give the two types of microwave communications.

17. What is meant by "line-of-sight?"

18. List the three methods used to connect fiber cables.

19. List the major elements in a networking infrastructure.

20. Describe the public telephone network hierarchy.

21. What is the voltage that is present on the twisted pair connecting a telephone to the central office?

22. What are the minimum functions required from a central office to perform its operation?

23. Describe the function of the CCS. Why is it considered one of the major innovations in signaling between switches?

24. What is the role of modems in data communication?

25. What are the two major types of switching techniques and in what type networking may each be used?

26. Why are protocols essential for smooth communication over a network?

27. Show the steps that may take place when two computers attempt to communicate. Assume that computer A talks to computer C via computer B and that each of the three computers houses a three-layer protocol stack.

Digital Transmission

2.1 INTRODUCTION

In this chapter, we introduce the reader to the techniques of digital transmission in telecommunication systems. We begin with an historical overview of digital signaling leading to the current technology used in digitizing the voice signal. Sections 2.4 and 2.5 treat three widely used digitization techniques: pulse code modulation, delta modulation, and adaptive delta modulation. A practical functional representation of a device that implements pulse code modulation for voice signals is given. We follow the digitization techniques with the concept of multiplexing, with specific emphasis on time-division multiplexing. The discussion of time-division multiplexing leads into a detailed description of T1 transmission techniques and characteristics.

Digital transmission systems have evolved in speed and sophistication to handle thousands of voice channels on a single transmission facility. T3 is one of those systems that is able to bundle several hundred voice conversations, data exchanges, video, and other applications. We present a detailed description of T3 structure, operations, and maintenance as the last part of this chapter.

2.2   HISTORICAL OVERVIEW

It might be surprising to know that the earliest telecommunication devices were digital in nature. Smoke signals, drum beats, and the telegraph—these devices created messages coded by the presence or absence of an event, rather than by continually changing events. A puff of smoke was sent into the air or not, the drum was struck or not, the telegraph key was tapped or not. All these digital devices produce signals that were discretely variable—on or off, present or absent.

It is perhaps ironic that Alexander Graham Bell was trying to improve upon a digital device—the telegraph—when he accidentally discovered the telephone. The telegraph had been around for approximately 42 years, without great advances in technology, when Bell came up with his idea of a harmonic telegraph. Bell reasoned that since telegraph lines were already strung across the country, it would make a lot of sense to find a way for these wires to carry more than one message at a time. The harmonic telegraph was an attempt to do just that, but while Bell was working on the harmonic telegraph (a digital device in nature), an accident occurred. While adjusting the equipment, a misaligned transmit reed relay inadvertently created an electrical signal that covered the human voice frequency (VF) range. When all the relays in the far end vibrated in response to the incoming electrical signal, voice frequency was reproduced. This led Bell to search for a practical telephone device and eventually to the famous event where he uttered the famous words, "Come here, Watson, I need you."

For the next 80 years, the telephone network remained essentially analog. The digital telegraph with its obscure codes was replaced by the analog transmission of voice that anyone with access to a telephone could easily use.

As the population density increased, so did the need for more cables in the analog telephone network to support the added demands. The question brings us back to Bell's experiment in 1876—how to carry more than one message at a time over the same transmission facility, a concept we have already seen, known as *multiplexing*. If Bell could (almost) do it with telegraph signals in the late-nineteenth century, certainly we could do it in the mid-twentieth century, and that's exactly what happened.

multiplexing

Bell was experimenting with frequency-division multiplexing, dividing the frequency range of a transmission facility into separate transmission paths. Frequency-division multiplexing is a good technique, but it has inherent limitations. Any transmission medium can only transmit or receive a given range of frequencies. This range, as we saw in Chapter 1, is called the bandwidth.

Due to the nature of the transmission media, there is an inverse relationship between bandwidth and the distance over which the media can carry the signal accurately. The higher the bandwidth the shorter the distance and vice versa. As mentioned earlier, 80 percent of all voice information is contained between 300 Hz and 3000 Hz, and about 99 percent falls within the range of 300 Hz and 3400 Hz.

Based on the fact that a range of 300 Hz to 3400 Hz is more than adequate for voice communications, the analog voice network has standardized on that bandwidth for voice circuits. In order to multiplex voice circuits then, we must find a way to send the entire range of frequencies to the other end, along with other voice circuits of the same frequency range.

Two primary techniques have evolved: frequency-division multiplexing *(FDM)* and time-division multiplexing (TDM). *Frequency-division multiplexing* divides up the bandwidth of the transmission facility into frequency segments that can carry separate signals. Frequency-division multiplexing lends itself well to analog radio transmission because of the wide bandwidth available. Over copper wires, however, FDM is limited because of the rapid attenuation (loss) of the higher frequency signals due to the inverse relationship between bandwidth and distance.

FDM

Time-division multiplexing divides the capacity of the transmission facility into time slots and assigns each channel the use of a slot. Time-division multiplexing works over any transmission medium but is not well suited to analog signals.

TDM

The telephone system that evolved from the days of Alexander Graham Bell was designed to provide analog dial-up telephone service. Everything was based on voice communications services on a switched (non-dedicated) basis. Since voice was the primary service provided, the telephone set evolved into a device that took the sound wave from the human vocal cords and converted that sound into an analog (equivalent) electrical current. The human voice produces constantly changing variables of both amplitude (the height of the wave) and frequency (the number of cycles per second). As these constantly changing variables of amplitude and frequency are generated, an electrical wave is produced.

2.2.1   Reasons to Convert to Digital Signals

The telephone set converts the sound wave into electrical energy that will be carried down the twisted pairs. As the electrical energy is introduced to the wires (twisted pairs for this reference), certain characteristics begin to work on the energy. First, resistance exists in the wires. This resistance starts to impede the flow of electricity drawing away some of the signal strength. Second, the wires act as an antenna, drawing in noise, such as static, or electricity from other conversations taking place on adjacent pairs. The loss of power, coupled with the added noise, continues to distort and attenuate the signal.

To overcome this problem, the telephone company used analog amplifiers at various points along the line to boost the signal strength. The amplifiers are normally placed on circuits longer than 18,000 feet. The amplifiers are used to ensure that the signal gets to the other end; however, amplifiers were prone to failure and had other undesirable qualities in magnifying the signal.

Unfortunately, as the signal is amplified, the noise on the wires is amplified as well. Therefore, noise that has been amplified accumulates on the circuit, preventing comprehension of the original voice over the total noise and eventually rendering the transmission worthless. Much of the problem stems from the fact that amplifiers cannot distinguish the analog signal from the noise.

T-carrier system

The *T-carrier system* was born when the telephone company wanted to overcome this problem and enhance the quality of calls by making better use of the cabling plant it had. The T-carriers are telecommunication systems that digitize many voice signals and bundle the digitized signals into a single digital bit stream that travels to the destination. Once received, this stream of bits is unbundled, and an inverse conversion from digital to analog takes place to reproduce the original voice (or what sounds like the original voice).

The bundling of the digitized stream is accomplished using the time-division multiplexing technique. This process allowed the telephone company to increase its call-carrying capacity by taking advantage of the unused transmission capacity of its existing wire-pair facilities—back to what Alexander Graham Bell was trying to do in 1876. Also, this digital technique of communication improved the quality of transmission by migrating away from the analog signals and the noise problem associated with them.

PCM

The T-carrier was the first successful system designed to use digitized voice transmission. It identified many of the standards used today in digital switching and transmission, including a modulation technique known as *pulse code modulation (PCM)*. The transmission rate was established at 1.544 megabits per second (1,544,000 bits per second), which became the building block for the North American digital hierarchy and the AT&T digital standard. Figure 2–1 illustrates how the T-carrier system is used.

The digital switching systems introduced into the telephone network were designed to be compatible with the T-carrier system. The digital switching matrix of the 4 and 5 electronic switching systems (ESS) was developed by Western Electric and the digital multiplexing systems (DMS 10, 100, and 250) were developed by Northern Telecom Inc. The pulse code modulation technique was used in these systems to provide lower cost and better quality dial-up telephone services. However, this same technology underlies the idea of a full end-to-end digital network service that is the basis of the integrated service digital networks (ISDN).

Figure 2–1
Digital transmission of the voice signal along the public network.

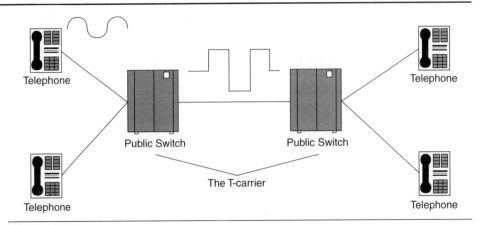

Telephone

Telephone

Public Switch

Public Switch

Telephone

Telephone

The T-carrier

2.3 ANALOG-TO-DIGITAL CONVERSION

The nature of the voice signal is analog. To use a digital facility to transmit such a signal, a form of conversion from analog to digital and back to analog needs to take place (often referred to as A/D and D/A). Since digital transmission dictates the use of a digital bit stream, the analog wave must be converted into a usable format—digital pulses represented by 1s and 0s. As the analog signal is transmitted to the network, it must be converted from a wave of varying amplitude and frequency to a digital format that is represented in the form of 1s and 0s (or the presence and absence of a voltage).

The technique used for converting the voice, video, and other analog signals into a digital format suitable for travel over the digital network is known as *pulse code modulation* (PCM). In this modulation technique, the analog signal is translated into a pulse train, which in turn is coded into a bit stream of 1s and 0s representing the original signal. When the distant equipment on the digital network receives this bit stream, it recreates what sounds or looks like the original signal being transmitted. In the next section, we discuss the process by which a bit stream is created from analog signals.

2.4 PULSE CODE MODULATION

Developing a bit stream using pulse code modulation involves three equally important steps: sampling, quantization, and coding. Figure 2–2 shows the steps required in PCM communication. The result is a serial binary signal, which may (or may not) go through additional modulation steps and conditioning. *Sampling* consists of periodically measuring the values of the analog signal, $v(t)$, a continuous (as opposed to being made up of discrete values) real-valued function of time. These values are called *samples*. *Quantization* consists of representing the samples by rounding each true sample value to a near value from a finite set of discrete quantities. Those quantities cover the entire range of the signal being sampled. *Coding* entails taking the quantized values and representing them, usually as 8-bit binary numbers. In the following sections we will examine, in more detail, these three steps.

sampling

quantization

coding

2.4.1 Sampling

As explained in the previous section, sampling involves taking periodic snapshot measurements of the input analog signal. It starts by dividing the original signal into uniformly spaced pulses as shown in Figure 2–3 (a, b, c). The amplitude of the samples (pulses) rises and falls with the amplitude of the original signal. The resulting wave is a pulse amplitude modulation (PAM) signal.

As illustrated in Figure 2–3, if a continuous signal is multiplied into a pulse train with a fixed amplitude using point-by-point multiplication, another pulse train, with amplitudes following the original analog signal will result. This pulse train is called the pulse amplitude modulation (PAM) signal.

Figure 2–2
Simplified block
diagram of PCM
communication.

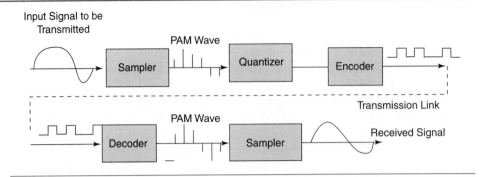

Input Signal to be
Transmitted

One question arises, however: How many times should we measure the values of the analog signal in a given interval of time? The minimum number of times the values of the analog signal is measured in a given interval, which is known as the sampling rate, is the subject of the *Nyquist sampling theorem.*

Nyquist sampling theorem

According to the Nyquist rule, for a valid digital representation to be created, the analog wave must be sampled at a rate that is at least twice the highest frequency component of that signal. For an analog channel delivered by the local telephone company, the frequency on a twisted pair can range from 300 Hz to approximately 3400 Hz. According to the Nyquist theorem, we

Figure 2–3
Creating a PAM
wave for a single si-
nusoid: (a) a sinu-
soid signal, (b) a
pulse train, and
(c) the result of
passing (a) and (b)
through a point-by-
point multiplier.

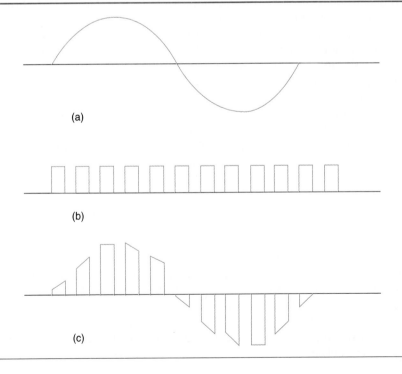

will have to sample the line at a rate of twice the 3400 Hz—the highest frequency component—which is 6800 samples per second. In practice, however, a sampling rate of 8000 samples per second is used. Therefore, taking one sample of the line every 125 microseconds (1/8000) to digitize the analog voice signal generated by the analog telephone set is sufficient for the reconstruction of that signal by the receiving destination.

In practice, the systems used for digital transmission require sampling several channels. The resulting pulse modulation signal from each voice channel undergoes *interleaving* with other PAM signals to form a single serial stream of pulses that represents the digitized signals of all channels. North American systems sample 24 voice channels in sequence. The European PCM systems sample 30 voice channels.

interleaving

2.4.2 Quantization

The objective of the quantization step in the PCM development process is to represent each sample by a fixed number of bits. For example, if the amplitude of the PAM wave resulting from the sampling process ranges between +1 and −1 volts, there can be infinite values of voltage between +1 and −1 volts. For instance, one value can be −0.276898799 volts. To assign a different binary sequence to each voltage value, we would have to construct a code of infinite length. Therefore, we must take a limited number of voltage values between +1 and −1 volts to represent the original signal, and these values must be discrete. We can set 20 discrete values between +1 and −1 volts, in increments of .1V. Because we are converting to a binary world, we select the total number of discrete values to be a binary number multiple (i.e., 2, 4, 8, 16, 32, 64, 128, 256, and so on). This system facilitates binary coding. For instance, if there were four values, they would be as follows: 00, 01, 10, and 11. This is a 2-bit code. A 3-bit code will yield eight different binary numbers. We find then that a code with length n-bits yields 2^n total possible binary combinations. A 7-bit code gives 128 (2^7) binary combinations.

In the quantization process we want to present the *coder* with discrete voltage values. We assume that our quantization steps were in 0.1-volt increments and that our voltage measure for one sample is 0.58 volts. That would have to be rounded off to 0.6 volts, the nearest discrete value. Note that there is a 0.02-volt error, the difference between 0.58 and 0.60 volts. Figure 2–4 shows one cycle of the PAM wave where we use possible codes or levels between +1 and −1 volts. Thus we can assign eight possibilities above the origin and eight possibilities below the origin. The 16 quantum steps are coded as in Table 2–1.

Examination of Figure 2–4 shows that step 12 occurs twice. Neither time it is used is it the true value of the impinging sinusoid. It is a round-off value. These round-off values are shown with dashed lines, which follow the general outline of the sinusoid. The horizontal dashed lines show the point where the quantum changes to the next higher or the next lower level if the sinusoid curve is above or below that level. Take step 12 in the curve, for example. The curve is passing through a maximum and is given two values of 12. For the first value, the actual curve is above 12 and for the second value, below. That

Table 2–1

Sixteen coded quantum steps ranging from +1 volt to −1 volts.

Step Number	Code	Step Number	Code
0	0000	8	1000
1	0001	9	1001
2	0010	10	1010
3	0011	11	1011
4	0100	12	1100
5	0101	13	1101
6	0110	14	1110
7	0111	15	1111

quantizing
distortion

error, in the case of 12, from the quantum value to the true value, is called *quantizing distortion*. This distortion is the major source of imperfection in PCM systems. The more quantizing steps, the better quality the system will deliver. However, increasing the number of quantizing steps has two major costs: one is the cost of designing a system with large binary code size needed and the other is the time it takes to process this large number of quantizing steps by the coder. A very large number of quantizing steps may induce unwanted delays in the system.

Consider Figure 2–4, maintaining the −1, 0, and +1 volts relationship, let us double the number of quantum steps from 16 to 32. What improvement would be gained in quantization distortion? First, determine the step increment in millivolts (mV) in each case. In the first case, the total range of 2000 mV would be divided into 16 steps, or 125 mV per step. The second case would have 2000 ÷ 32 or 62.5 mV per step. For the first case, the worst possible quantization error (distortion) occurs when an input to be quantized is at half-step level, or in this case 125 ÷ 2 or 62.5 mV above or below the nearest quan-

Figure 2–4

Quantization and resulting coding using 16 quantizing steps.

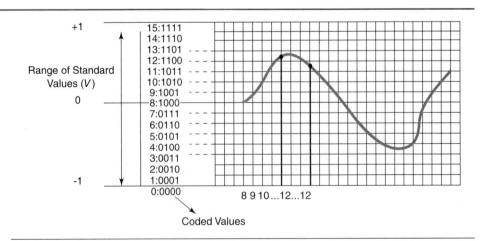

tization step. For the 32-step case, the worst quantizing error, at the half-step level, is 62.5 ÷ 2, or 31.25 mV. Thus, the improvement in decibels for doubling the number of quantizing steps is

$$20\log 62.5/31.25 = 20\log 2 \text{ or } 6 \text{ dB (approximately)}.$$

This is valid for linear quantization only. Thus, doubling the number of quantizing steps for a fixed range of input values reduces the error by 6 dB.

    Voice transmission presents a problem because of its wide dynamic range, which can be defined as the level range from the loudest syllable of the loudest talker to the lowest syllable of the quietest talker. Using linear quantization, we find we need 2048 discrete steps to provide any fidelity at all. The 2048 steps require 11-bit binary coded words. That means increasing the transmission bandwidth requirement; that is, to maintain the same data rate with an increased number of bits per value would require faster transmission and, therefore, an increased bandwidth. Increasing the bandwidth in order to increase the number of quantizing steps may not be practical considering the cost of the circuitry and the processing time.

    Nonlinear techniques are used to deal with quantization distortion in a more effective way. An old analog technique called *companding* is used to achieve the improvement sought in quantizing error. Companding stands for *comp*ression/exp*anding*. Compression takes place at the transmitting circuit side and expanding occurs at the receiving end. Compression reduces the dynamic range with little loss of fidelity, and expansion returns the signal to its normal condition.

companding

    In this technique, the voltage range between the lowest and highest level is segmented in a nonlinear manner. Instead of dividing the voltage range into a fixed number of steps, we divide the entire range into segments. Within each segment, there will be a fixed number of steps. The voltage range of each segment, however, will vary according to the levels of the voltages. The lower the voltage levels, the lower the range of a segment will be, so the segments will be relatively small, around 0V. As shown in Figure 2–5, there are eight segments for the entire range between +5 and –5 volts, but the individual range for each segment varies as follows

Segment #	Assigned voltage range
4, 5	0 to 0.5, 0 to –0.5
3, 6	0.5 to 1.5, –0.5 to –1.5
2, 7	1.5 to 3.0, –1.5 to –3.0
1, 8	3.0 to 5.0, –3.0 to –5.0

Observe that the number of steps is the same for all segments. Thus, the low voltage levels, where the segments are smaller, have less voltage per step than the higher levels. This form of compression is shown in Figure 2–6.

    By guaranteeing more quantization steps per volt for low-level speech than higher levels we ensure low quantization distortion at the lower level. Since the low levels are more susceptible to noise, the overall quantization distortion is

Figure 2–5
Nonlinear quantiza-
tion using eight seg-
ments with each
segment assigned
two steps (two coded
words).

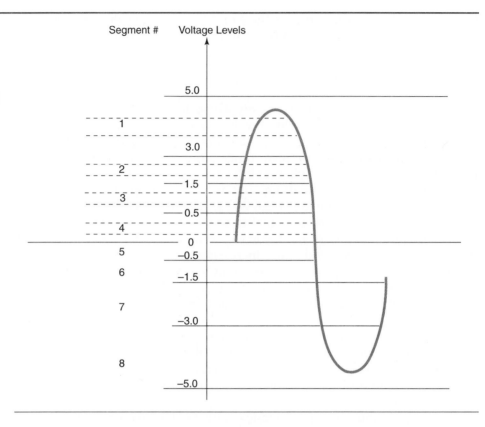

minimized. Figure 2–7 shows eight coded sequences are assigned to each level grouping (segment). The smallest range rises only 0.0666 volts from the origin (0 volts). The largest range extends over 0.5 volts, and it is assigned only eight coded sequences (quantizing steps). Although the voltage ranges get progressively smaller, they all are assigned eight steps each. There are 8 steps for the ranges 0–1/64, 1/64–1/32, 1/32–1/16, 1/16–1/8, 1/8–1/4, 1/4–1/2, and 1/2–1. Thus, the quantizing distortion is minimized at the low level.

2.4.3 Coding for Modern PCM Systems

Modern PCM systems use an 8-bit code with improved quantizing distortion performance. The companding and coding are carried out simultaneously. The compression and later expanding functions are logarithmic. A pseudo-logarithmic curve made up of linear segments imparts finer granularity to low-level signals and less granularity for the higher-level signals. The logarithmic curve follows one of two laws: the *A-law* and the *μ-law*. The curve for the A-law may be plotted from the following formula

A-law, and μ-law

$$Y = \frac{AX}{1 + \log A} \text{ for } 0 \le v \le \frac{V}{A}$$

$$Y = \frac{1 + \log (AX)}{1 + \log A} \quad \text{for } \frac{V}{A} \leq v \leq V$$

where $A = 87.6$. The curve for the μ-law is plotted following the formula

$$|Y| = \frac{\log (1 + \mu \; |X|)}{\log (1 + \mu)}$$

where $\mu = 100$ for the original North American system and 255 for later North American (DS-1) systems. In these formulae

$$X = \frac{v}{V}$$

$$Y = \frac{i}{B}$$

where v = instantaneous input voltage.
   V = maximum input voltage for which peak limitation is absent.
   i = number of quantization steps starting from the center of the range.
   B = number of quantization steps on each side of the center of the range.

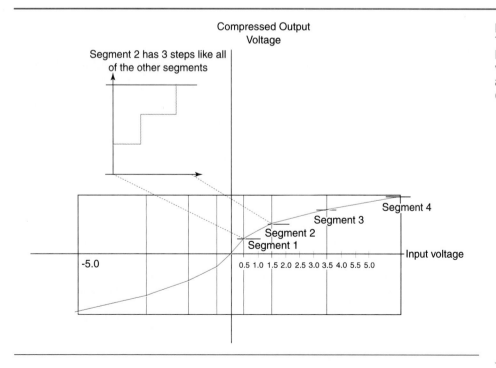

Compressed Output
Voltage

Segment 2 has 3 steps like all of the other segments

-5.0

0.5 1.0 1.5 2.0 2.5 3.0 3.5 4.0 5.5 5.0

Segment 4
Segment 3
Segment 2
Segment 1
Input voltage

Figure 2–6
The relationship between the input voltage (−5 to +5) and the compressed output voltage.

The signal-to-distortion ratio in PCM is given by the parameters A and μ for the companding laws. A and μ determine the range over which a signal-to-distortion ratio is comparatively constant, about 26 dB. For A-law companding, $S/D = 37.5$ dB can be expected ($A = 87.6$) and for μ-law companding $S/D = 37$ dB (μ = 255).

Let us examine Figure 2–7 depicting the A-law used in the E1 European multiplex system. The curve consists of linear piecewise segments, seven above and seven below the origin. Counting the 2 segments by the origin, there are 16 segments. Each segment has 16 quantization steps represented in 8-bit PCM-coded words.

As shown in Figure 2–8, the first bit in a coded word (most significant bit) tells the distant receiver if the sample is a positive or a negative voltage. (In

Figure 2–7
Thirteen-segment approximation of the A-law curve used with E1 PCM equipment.

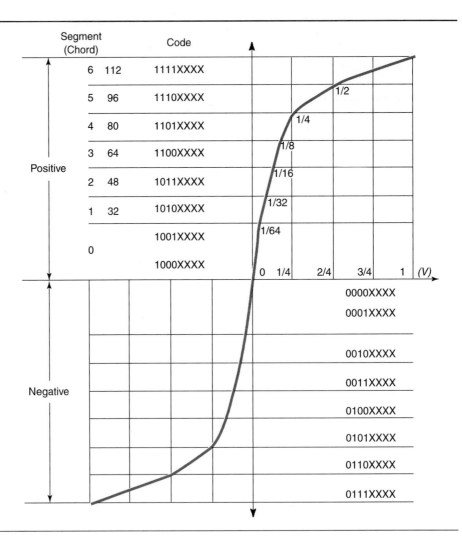

Figure 2–8
PCM code example.

Figure 2–7, all positive samples, above the origin, start with a binary 1, and those below the origin start with a binary 0.) The next three bits identify the segment. The last four bits, shown in the Figure 2–7 as *XXXX*, identify where in the segment that voltage line is located.

Suppose the far end received the binary sequence 11010100 in an E1 system. The first bit indicates that the voltage is positive (i.e., above the origin in Figure 2–7). The next three bits, 101, indicate that the sample is in segment 4. The last four bits, 0100, tell us where it is in the segment, as further illustrated in Figure 2–9.

Figure 2–10 shows the equivalent logarithmic curve for the North American DS-1 system. It uses a 15-segment approximation of the logarithmic μ-law curve ($\mu = 255$). The segments cutting the origin are collinear and are counted as one. So again, we have a total of 16 segments.

The coding process in the North American PCM system uses the straightforward binary codes we have just seen in the above example. Shown in Table 2–2 are the code levels for the North American PCM system.

2.4.4   Quantization Distortion

Quantization distortion has been defined as the difference between the original signal waveform as presented to the PCM system and the equivalent quantized value. For a linear system with *n* binary digits per sample, the ratio

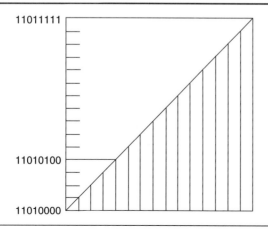

Figure 2–9
The European E1 system, coding of segment (chord) 4 (positive). Details of segment #4 (101).

Figure 2–10
Piecewise linear approximation of the μ-law logarithmic curve coding.

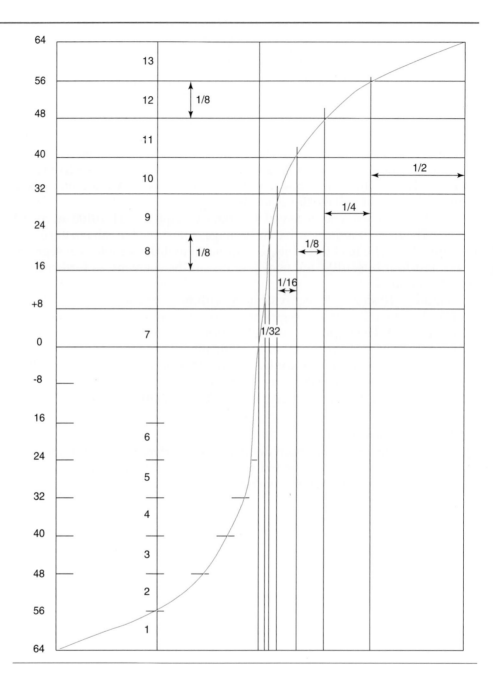

Code Level		1	2	3	4	5	6	7	8
					Digit Number				
255	(Peak positive)	1	0	0	0	0	0	0	0
239		1	0	0	1	0	0	0	0
223		1	0	1	0	0	0	0	0
207		1	0	1	1	0	0	0	0
191		1	1	0	0	0	0	0	0
175		1	1	0	1	0	0	0	0
159		1	1	1	0	0	0	0	0
143		1	1	1	1	0	0	0	0
127	(Center level)	1	1	1	1	1	1	1	1
126	(Nominal zero)	0	1	1	1	1	1	1	1
		0	1	1	1	0	0	1	0
95		0	1	1	0	0	0	0	0
79		0	1	0	1	0	0	0	0
63		0	1	0	0	0	0	0	0
47		0	0	1	1	0	0	0	0
31		0	0	1	0	0	0	0	0
15		0	0	0	1	0	0	0	0
2		0	0	0	0	0	0	1	1
1		0	0	0	0	0	0	1	0
0	(Peak negative level)	0	0	0	0	0	0	1*	0

*One digit added to ensure that timing content of transmitted pattern is maintained.

Table 2–2
Code levels for the North American PCM system.

of the full-load sine wave power to quantizing distortion power, S/D, is given by

$$S/D = 6n + 1.8 \text{ dB}$$

where n is the number of bits used for each PCM code word, expressing the sample.

Example

The older AT&T D1 system uses a 7-bit word to express a sample (level). What is the signal to quantizing distortion ratio for that system?

Solution

7-bit word with uniform quantization yields
Signal-to-quantization distortion ratio

$$S/D = 6 \times 7 + 1.8 = 43.8 \text{ dB}$$

With each binary digit added to the PCM code word, the signal-to-distortion ratio is increased by 6 dB for linear quantization. In practical PCM systems using 8-bit PCM code words, the signal-to-distortion ratio ranges from 33 dB to 38 dB, depending largely on the talker sound signal levels.

In the older PCM system we just discussed, samples are measured periodically and are then quantized and coded (for instance, using an 8-bit PCM word, and then transmitted in the resulting bit stream). Using a sampling rate of 8000 samples per second (recall that this is the minimum sample rate required for a voice signal), we need a communication channel with 64,000 bits per second bandwidth to carry a single voice conversation. Enhanced PCM encoding techniques are used to make better utilization of the channel bandwidth, yielding several voice connections over the 64,000 bit per second channel. In the next sections, we present two of these techniques: delta modulation and adaptive delta modulation.

2.5 · DELTA MODULATION

delta modulation

The purpose of *delta modulation* is to minimize the effects of noise without increasing the number of bits being sent. This increases the signal-to-noise ratio, improving system performance. The idea behind delta modulation is to take samples close enough to each other so that each sample's amplitude does not vary by more than a single quantizing step size. Then instead of sending a binary code representing the step size, a single bit is sent, signifying whether the sample size has increased or decreased (from the previous sample) by a single step. This process is illustrated in Figure 2–11. We first sample and quantize the

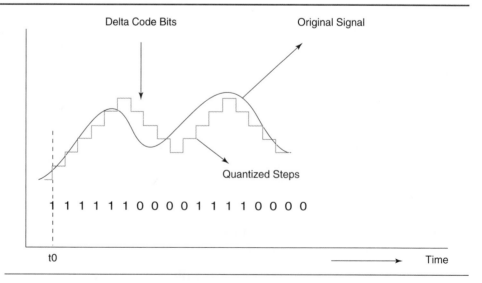

Figure 2–11
Delta encoding of a quantized waveform.

Delta Code Bits

Original Signal

Quantized Steps

1 1 1 1 1 0 0 0 0 1 1 1 1 0 0 0 0

t0

Time

original signal as with pulse code modulation. If the sample currently being coded is above the previous sample, then the current binary bit is set to a logic-1. If the sample is lower than the previous sample, then the bit is set to a 0.

Example

What is the binary code sent for the original signal in Figure 2–11, using delta modulation?

Solution

As shown in Figure 2–11, the original signal is quantized at a given step rate. A sample whose amplitude is one step size above the previous sample generates a logic-1 bit. And a sample that is lower generates a 0-logic level. The bit pattern generated by the signal starting with the first sample on the left therefore is:

111111000011110000

The functional block diagram for a delta modulator is shown in Figure 2–12. Samples from the original signal are compared to the output of a staircase generator. If the result of the comparison shows the original signal to be larger than the staircase voltage, the comparator is set high. This is sent as a logic-1 and causes the staircase generator to increase by a step. If the comparator indicates that the staircase voltage is greater than the original signal, then the comparator goes low and causes the staircase generator to decrease by one step.

Delta modulation systems are useful with signals that change slowly. A problem occurs when the original signal changes its amplitude up and down at a fast rate (Figure 2–13). In other words, the signal has a high frequency Fs. At slower slope changes, the replicated staircase tracks fairly close to the original

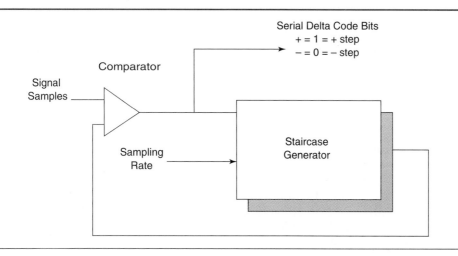

Figure 2–12
Delta modulator.

Figure 2–13
Delta modulation
slope overload.

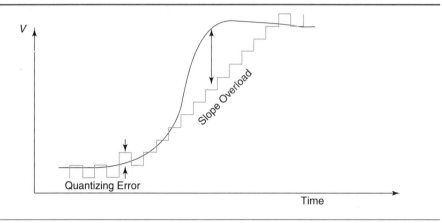

Fs. A rapid increase in slope causes a widening gap between the original signal and that of the reconstructed waveform at the receiver. This difference is known as *slope overload.*

slope overload

2.5.1   Adaptive Delta Modulation

As we just saw, delta modulation works well as long as the original signal does not have an abrupt or large variation in frequency, which causes slope overload. However, analog signals that are driven from audio or video information (rather than from voice) vary in frequency. For a delta modulation system using fixed step sizes, this presents a problem. Increasing the sampling rate can minimize the slope overload problem, but this is not desirable because it increases the channel bandwidth requirements. *Adaptive delta modulation*, an alternative solution, deals with the slope overload problem by varying the step size, so that the quantized signal more closely follows the original signal. The dotted staircase in Figure 2–14 shows the objective of adaptive delta modulation. To make the staircase voltage track the original signal more accurately, the step size is increased on the leading edge and decreased on the trailing edge.

adaptive delta
modulation

When using adaptive delta modulation, a long string of 1s or 0s indicates that the staircase generator is continually trying to play catch up with the original signal; that is, it is trying to continually add or subtract step voltages from the output. The more frequently the digital bits alternate between 0 and 1, the more closely the staircase voltage is tracking the original waveform. An adaptive modulator causes the step to increase whenever it detects many consecutive 1s or 0s and causes the size to decrease when it detects alternation in the logic level of the digital data stream. The result of varying the step size is shown in Figure 2–14. The rapidly changing leading edge begins to generate a set of consecutive 1s. Immediately the adaptive delta modulator increases the step size. As the peak is reached, alternating 1s and 0s occur and the modulator reduces the step size. It is increased on the down slope side after a string of 0s is detected. The second slope, rising slowly, begins to cause alternating 1s and 0s

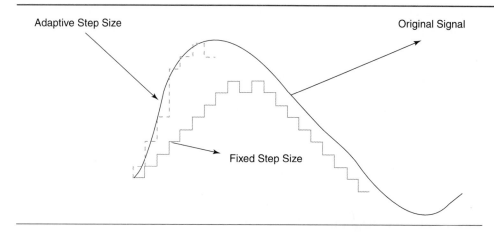

Figure 2–14
Fixed and adaptive
step sizing.

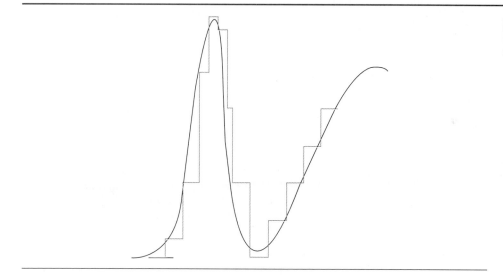

Figure 2–15
Example of adaptive
delta modulation.

to appear because of the large step size from the previous down slope. The adaptive delta modulator decreases the step size, causing the staircase output to once again follow the original signal more closely. See Figure 2–15 for an example of adaptive modulation.

2.6 CODECS

A codec is a device designed to convert analog signals, such as voice communications, into PCM-compressed samples to be sent onto the digital network. On the receiving end, the codec replicates the original analog signal from the

coder/decoder

incoming PCM stream. The term *codec* is an acronym for *coder/decoder*, signifying the pulse coding and decoding functions of the device. In the early days of PCM, the coding and decoding functions of PCM were managed by several individual devices, each performing one of the tasks necessary for PCM communications. Individual devices were needed to perform PCM functions such as sampling, quantizing, encoding, decoding, filtering, and companding. With the advances in semiconductor manufacturing technologies, a single codec IC chip offered by one of many semiconductor manufacturers (e.g., Intel, MITEL, Texas Instruments, AT&T, and Motorola) performs all of these tasks.

In general, a codec chip can operate at different data rates ranging from 64 kHz to 2048 MHz, including the standard T1 rate of 1.544 MHz. The chip usually allows the selection of the μ-law or the A-law—the companding techniques discussed earlier—for the compression function. Figure 2–16 depicts a functional block diagram for the transmit side of a typical codec chip. The receiving side of the chip is shown in Figure 2–17.

The following technical description of the circuitry in a codec chip is helpful for understanding the chip's internal operation but is not essential for understanding the codec's use.

As seen in Figure 2–16, the input signal is fed through an operational amplifier that conforms to the standard noninverting rule of amplification with a gain of $Av = 1 + R2/R1$. If no amplification is desired, the op-amp can act as a buffer by making $R2 = 0$. The op-amp's output is fed through a low-pass filter followed by a band-pass filter and then to a sample-and-hold circuit. The band-pass filter is designed to produce a flat response from 300 Hz to 3 kHz. Additional high-pass filtering is used to reject power line frequencies (50 and 60 Hz).

Samples are held for encoding and μ-law or A-law compression in an internal sample-and-hold circuit. An additional auto-zero unit is used to correct for any DC offset that may be introduced up to this point. The compressed PCM-serialized data are sent out through the transmit-data line.

Figure 2–16
Functional block diagram of the transmit side of a typical codec chip.

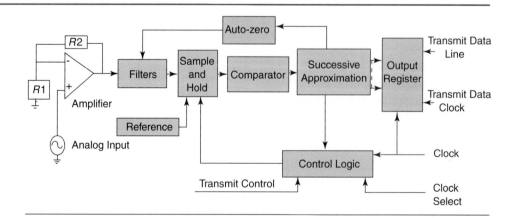

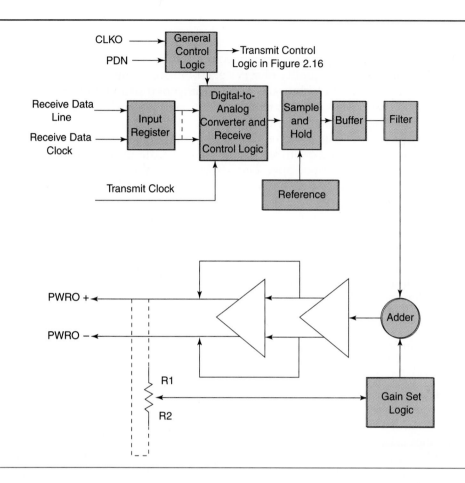

Figure 2–17
Functional block
diagram of a typical
codec chip receive
block.

On the receive side of the codec (Figure 2–17), the serial data stream is shifted into the input register and then output in parallel to the digital-to-analog converter. The resulting analog signal is held in a sample-and-hold circuit until the next coded sample is shifted into the input register. The decoded samples are then shaped and filtered before being sent through the power amplifier. The filter is a passband filter with a sharp roll-off at 4000 Hz. It has a flat response, like the transmit filter, between 300 Hz and 3000 Hz. The output level is determined by the voltage level applied to the gain set circuitry.

Typical PCM systems handle sampling, quantizing of samples, and coding for several voice channels, and transmit the resulting single bit stream over the appropriate transmission facility. What enables the PCM system to accomplish the task of digitizing several voice signals into a single bit stream is the time-division multiplexing technique previously discussed. In the following sections, we examine the idea of multiplexing with particular focus on the time division technique.

2.7   MULTIPLEXING

The fundamental telephone company objectives have been increasing profitability, offering better quality of service, and introducing new services. Multiplexing is one of the techniques used to improve profitability by packing more voice channels over the existing transmission wires. Transmission media usually have capacities great enough to carry more than one voice channel. In other words, their bandwidths are considerably greater than the 3.1 kHz needed for transmitting a single human voice. At the upper end of the spectrum, microwave and fiber optic circuits can carry thousands of voice channels; at the lower end of the spectrum, each voice channel may be split into 12 or 24 telegraph channels.

When a facility is set up, such as a chain of microwave links, which has a broad bandwidth, the goal is to efficiently use the available bandwidth by making it carry as many channels as possible. It is often desirable, from a design and economic point of view, to construct a communication link with as wide a bandwidth as possible and then divide the bandwidth among as many users as possible. Many separate signals are *multiplexed* together so that they can travel as one signal over a high bandwidth path.

In a multiplexed system, two or more signals are combined so that they can be transmitted together over one physical cable or radio link. The original signals may be voice, video, data, or other types of signals. The resulting combined signal is transmitted over a system with a suitably high bandwidth. The received signal must be split up into the original separate signals of which it was composed. The process of combining two or more signals into one is called *multiplexing*. The process of splitting up the composite signal into separate ones is called *demultiplexing*.

multiplexing
demultiplexing

The idea of multiplexing is also used in data processing. For example, a multiplexing channel on a computer is one on which several devices can operate at the same time. Several printers, or other input/output devices, operate simultaneously, and the bits sent to them or received from them are multiplexed as they travel along the signal channel.

Multiplexing, in general, means the use of one facility to handle several separate but similar operations simultaneously. In telecommunications language it means the use of one telecommunication link to handle several channels of voice, video, or data. Multiplexing is possible because the operations that are multiplexed take place at a considerably slower speed than the operating speed of the facility in question. Multiplexing is a key factor in effectively utilizing existing telecommunication links.

The simplest definition of multiplexing is the packing of many things to one and demultiplexing as the unpacking of the one thing into many things that were originally packed. Figure 2–18 depicts the multiplexing and demultiplexing functions in their simplest forms. There are n inputs to the multiplexer. The link is capable of carrying n communication channels. The multiplexer combines signals from the n input lines and transmits those signals

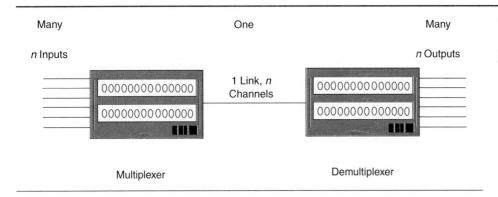

Many   One   Many

n Inputs   *n* Outputs

1 Link, *n*
Channels

Multiplexer   Demultiplexer

Figure 2–18
Multiplexing–
demultiplexing
functions.

over a higher capacity link. The demultiplexer accepts the multiplexed stream from the link and separates the signals according to each channel and delivers them to separate output lines.

Communication channels are normally grouped together in *packages* that fill the bandwidth available on different types of plants. Twelve analog voice channels may be multiplexed together, for example, to form a signal of 48 kHz bandwidth. Most telecommunications plants are built to handle this bandwidth. Other transmission media handle a far higher bandwidth than this; five 48-kHz links may be multiplexed to feed a facility of 240-kHz bandwidth. In this case, the original voice channel signal has passed through two multiplexing processes. It may pass through more than two. Ten of the 240-kHz links may meet at a point where they are again multiplexed together to travel on a facility having a bandwidth of 2400 kHz (in fact, higher because packing is not 100 percent efficient). Yet again, three of these packages may occupy an 8-MHz bandwidth. High-capacity links can therefore be created by multiplexing large numbers of channels that were, themselves, grouped together stage by stage.

The original signal may undergo many multiplexing stages and equivalent demultiplexing and thus passes through a variety of electronic conversion processes before the signal ultimately arrives at its destination, little the worse for these multiple contortions.

In telecommunication, there are three methods of multiplexing: *space-division multiplexing, frequency-division multiplexing*, and *time-division multiplexing*. Space-division multiplexing is simply the combining of physically separate signals into a bundled cable. It would not make sense to have a unique set of telephone poles or underground trenches for each telephone subscriber. Subscriber loops and trunks are therefore combined. The shared use of space is attained in space-division multiplexing. A ribbon cable is a form of space-division multiplexing. Since large volumes of space are not always available to contain cables, time-division multiplexing and frequency-division multiplexing are often combined with space-division multiplexing.

space-division
multiplexing,
frequency-division
multiplexing, and
time-division
multiplexing

Figure 2–19
The frequency range is divided into a separate path to accommodate multiple transmissions.

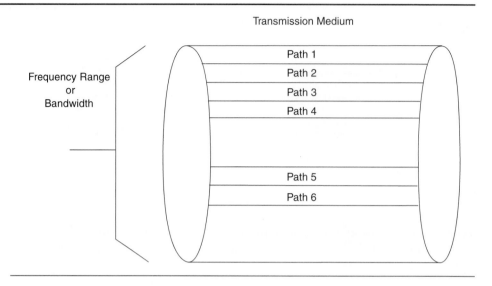

Transmission Medium

Frequency Range
or
Bandwidth

Path 1
Path 2
Path 3
Path 4

Path 5
Path 6

Frequency-division and time-division multiplexing are alternative techniques for grouping many individual signals for transmission over a single physical path. The information to be transmitted may be thought of as occupying a two-dimensional continuum of frequency and time, as illustrated in Figure 2–19. If the amount of information required from one channel is less than that which the physical facilities can carry, then the space available can be divided up either into frequency slices, or time slices, as shown in Figure 2–20.

In either case, the engineering limitations of the devices used prevent the slices from being packed tightly together. With frequency division, a guard band is needed between the frequencies used for separate channels, and with time division, a guard time is needed to separate the time slices. See Figure 2–21. If the guard bands or the guard times were made too small, the expense of the equipment would increase out of proportion to the advantage gained. In the next subsection, we present the principles of the time-division multiplexing technique.

Figure 2–20
The capacity of the transmission facility is divided into slots.

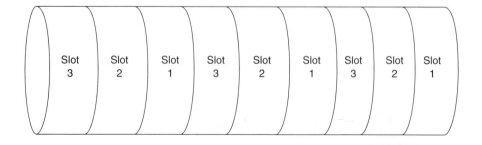

| Slot 3 | Slot 2 | Slot 1 | Slot 3 | Slot 2 | Slot 1 | Slot 3 | Slot 2 | Slot 1 |

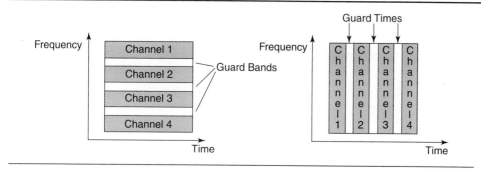

Figure 2–21
Frequency-division
multiplexing (left);
time-division multi-
plexing (right).

2.7.1 Time-Division Multiplexing

In this multiplexing technique each signal occupies the entire bandwidth of the transmission facility for a slice of time. Each device is given an opportunity to transmit for a window of time allotted to it. Then the window is given to the next device to transmit. The multiplexing circuitry scans the signals of the transmitting devices in a round-robin fashion. In time-division multiplexing, only one signal occupies the channel at any particular instant. It is thus different from frequency-division multiplexing in which all signals are sent at the same time but each occupies a different frequency band.

Time-division multiplexing operates on the Nyquist sampling theorem that was introduced earlier in this chapter. Each input signal is sampled at a rate at least twice its highest frequency component. Therefore, a sine wave input with a 1 kHz frequency must be sampled by the multiplexing device at a rate of at least 2 kHz. The useful component of the speech can be found in the 300 to 3400 frequency range of the voice signal. Thus, as stated earlier, in a practical system, the voice signal is sampled at a rate of 8 kHz.

Time-division multiplexing can be thought of as being like the action of a commutator. As shown in Figure 2–22, the arm might be used to sample the signal from three devices. Of course, the rotation speed of the arm has to be fast enough to catch the changes in the signal of each device. Each device has to be visited at a rate faster than twice the rate of change in the signals so the signals can be reconstructed accurately when they are received at the intended destination.

In terrestrial telecommunication systems, time-division multiplexing (TDM) is the technique used in digital transmission. It was developed by the telephone company to make better use of its cable plant. In the simplest operation of a TDM system, the analog voice signals from several telephone sets are digitized and combined into a single digital bit stream, which is transmitted over a TDM facility. The converse is true on the receiving end—the incoming bit stream is decomposed into several digitized voice signals. Each signal is then converted into an equivalent analog voice signal that is fed to the receiver of a telephone set.

The *T1 carrier* is one of the earliest telecommunication transmission types   T1 carrier
of equipment that used the technique of time-division multiplexing. T1 con-

Figure 2–22
Sampling of four
input devices.

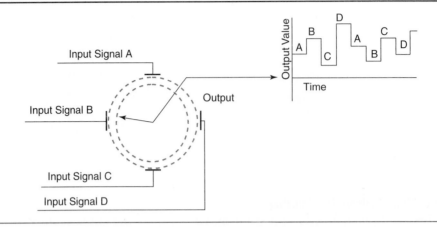

centrates 24 voice signals onto a single digital transmission link. In the next section, we will examine the principles of T1 operations.

2.7.2   T1 Transmission Basics

T1 line driver

The TDM multiplexer picks up one PCM word channel and sends it serially (one bit at a time) to another component of the system called the *T1 line driver*, which we will see later. The bit stream from TDM, then, is a series of 8-bit PCM words strung together. Each pass of the TDM strings together 24 PCM words as shown in Figure 2–23. This group of 24 PCM words is called a *frame*. Each frame is separated from the next by a bit inserted by the TDM.

superframe

Until the mid-1980s, T1 frames were further organized into groups of 12 frames called a *superframe.* Later, the superframe format was extended into groups of 24 frames. Today both superframe and *extended superframe format*

ESF

(ESF) are in use. In this section we focus on the superframe format, in the next section we will take a closer look at the extended superframe. See Figure 2–24.

Whether organized in a superframe format or extended superframe format, the primary purpose of framing the T1 bit stream remains the same. The frame structure allows us to keep track of what is going on more easily. Although digital devices are good at counting things, it is easier to count repeating patterns than to count infinite series. The frame/superframe structure provides those repeating patterns.

The framing bit in a superframe bit stream follows a specific pattern. The framing bit in odd-numbered frames alternates 1 0 1 0, and the framing bit in even-numbered frames repeats 0 0 1 1 1 0. By counting this repeating pattern of framing bits the multiplexer can always tell where it is in the bit stream. We will say more about the importance of the framing bit, but now let us take a closer look at the full T1 bit stream. See Figure 2–25.

We know that each channel will be generating a stream of PCM words. Each word is 8 bits long, and represents one sample. We know that we will be

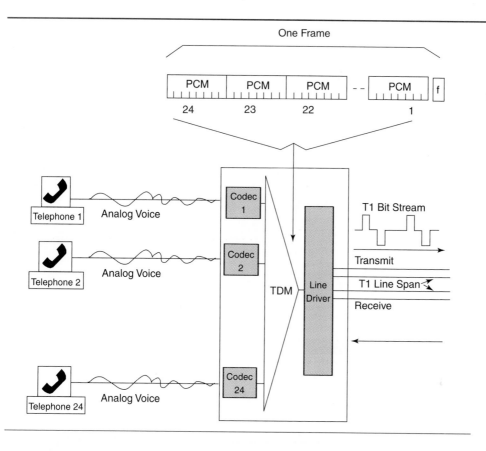

Figure 2–23
Time-division
multiplexing
over T1.

taking 8000 samples per second. For a single channel, then, how many bits per second do we generate? Based on this information, each channel will generate 64,000 bits per second (or 64 kbps) as we have already seen. Now the arithmetic may get a little tougher to do in your head. If each channel is generating 64,000 bits per second and we have 24 channels going into a T1 stream, what then is the bit rate of the T1 stream? The rate should be $24 \times 64{,}000 = 1{,}536{,}000$ bits per second. But, the actual rate is 1,544,000 bits per second (1.544 Mb/s) because of the framing bit inserted after each frame, which adds 8000 bits to our calculation. These 8000 bits are referred to as *overhead* because they are not carrying any of the user's data. This same information is illustrated in Table 2–3.

Now let us look at that 1.544 Mb/s from a slightly different perspective. From a channel/frame perspective our arithmetic looks like the following: First, the Codec samples the analog signal and generates a PCM word for each sample. Then the TDM picks up the PCM word for each channel and generates its frame. Now we may ask, How many frames per second does the TDM generate?—8000 frames per second is the right answer. At this rate each

Figure 2–24
Superframe and extended superframe of T1.

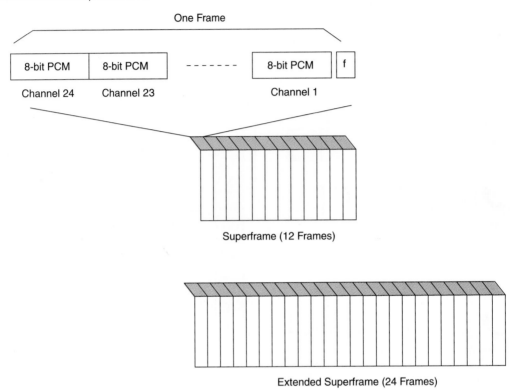

Figure 2–25
Framing bits in a superframe following a particular sequence.

f4	f3	f2	f1	f12	f11	f10	f9	f8	f7	f6	f5	f4	f3	f2	f1	f12
0	0	0	1	0	0	1	1	1	0	1	1	0	0	0	1	0

Odd-numbered Frames
Alternate 1010...

Even-numbered Frames
Repeat 000111...

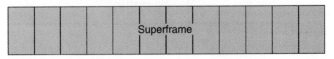

Superframe

24 channels at 8 bits/sample = 192 bits	
8000 samples/second	192
	$\times$ 8000
Subtotal	1,536,000
Overhead	+ 8000
(Framing Bits)	
Total	1,544,000

Table 2–3
Effective transmission line rate throughput = 15.444 Mbps.

channel is guaranteed to be sampled at least 8000 times per second, and thus we satisfy the Nyquist requirement. Since each frame contains 192 bits plus 1 framing bit then 193×8000 gives us the 1.544 Mb/s T1 bit stream rate.

Before we take a look at the line driver, let us introduce two new terms. As you already know, each Codec generates 64,000 bps (8×8000). This 64-kbps signal is known as the *DS-0 channel*. DS-0 stands for digital signal level 0. The next level up is the 1.544 Mb/s signal, known as the *DS-1 channel*, digital signal level 1.

DS-0 channel
DS-1 channel

T1 Line Driver   Looking again at Figure 2–23, the line driver sits between the TDM and the external interface of the system. The line driver performs two specific functions. First, it conditions the electrical characteristics of the T1 bit stream to match standard specifications, ensuring that the pulses are of uniform shape and that the voltages are correct. Second, the line driver converts the bit stream from a unipolar format to a bipolar format. The bit stream emerging from the TDM looks like what is shown in Figure 2–26(a). Notice that all the 1s are in the same direction. That is, the signal is unipolar, or all the same polarity. But the signal that leaves the line driver looks like what is in Figure 2–26(b). This is a bipolar format, with alternate mark inversion.

The bipolar *alternate mark inversion (AMI)* format alternates the polarity of each 1 (or *mark*). The first 1 is sent as positive signal; the second 1 as a negative signal; the third 1 is positive again. On a T1 line, two consecutive 1s of the same polarity are called a *bipolar violation (BPV)*, indicating an error on the line. The bipolar AMI format makes it much easier for the repeaters on the T1 line to recognize the 1s, improving the accuracy of the repeater.

AMI

BPV

Robbed Bit Signaling over T1   In early versions of T1, each PCM word carried signaling information. Seven bits of the PCM word were used to encode the analog signal. The eighth bit was used for signaling (off-hook, on-hook, ringing, flash-hook, and so on). But with only seven bits to encode samples, voice quality has to suffer. Think back to the section on PCM. If we use 7 bits, we will have only 128 quantization levels, compared to 256 levels if we used 8 bits.

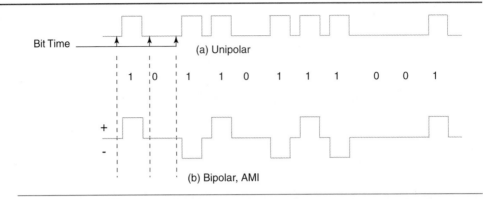

Bit Time

(a) Unipolar

1 0 1 1 0 1 1 1 0 0 1

+

-

(b) Bipolar, AMI

To maintain good voice quality, we need all eight bits of the PCM word to encode the sample. So how do we handle the signaling? Instead of sending signaling information in every PCM word, perhaps we could send it every so often? After all, if we are sampling each channel 8000 times a second, it will take 125 microseconds to generate each PCM word.

bit robbing

To handle signaling, then, we use a technique called *bit robbing*. Every sixth frame, we steal a bit from each PCM word and use it for signaling. Since the human ear cannot detect a difference in sound that occurs in 125 microseconds, this occasional bit robbing will not affect the apparent quality of the voice.

With superframe, the signaling bit in the sixth frame is called an *A bit*. In the twelfth frame, it is called a *B bit*. We will say more about the signaling bits in the section on the extended superframe. Figure 2–27 depicts the organization of a superframe when robbed-bit signaling is used. The bit we steal from each PCM word in every sixth frame is the least significant bit (LSB) of the PCM word.

Stealing the LSB from a PCM word means that the PCM word we send will be, at most, one step away from the actual sample. To see why, consider the value of a PCM word given by 10100111, which means a positive sample, segment 1, and we are at the seventh quantization step in segment 1 (step = 0111). If we steal the least significant bit, the step becomes 0110, which is step 5 within segment 1 instead.

Thus, every sixth frame, the least significant bit of each PCM word carries signaling information. The transmit side of the TDM multiplexer ignores the least significant bit as set by the codec, and changes the bit, if necessary, to indicate circuit status (on-hook, off-hook, flash-hook, ringing, etc.).

Different signaling schemes make different uses of the A and B bits. In typical E&M signaling, the A and B bits are set to the same value to indicate the voice circuit status. If both bits are high (that is, they are both 1s), then the circuit is busy. If both bits are low (that is, both are 0s), then the circuit is idle. With two bits to use, we have four signaling states.

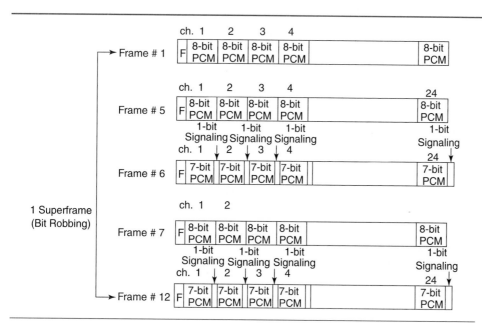

Figure 2–27
D4 superframe, the 6th and 12th frames show one bit being robbed from the sampling word (used for signaling— on-hook, off-hook, etc.).

So the T1 bit stream contains PCM words describing the samples of analog signals. It also contains framing bits to mark the beginning of each group of 24 PCM words. And, every sixth frame, it contains A or B bits for signaling and supervision.

T1 Synchronization So far we have looked at T1 primarily from the perspective of one end (the transmitter) of the time-division multiplexing circuit. Let's broaden our view to include the other end of the circuit (the receiver) and some of the equipment in between as shown in Figure 2–28.

We mentioned earlier that digital devices are very good at counting things. Since T1 operates in a synchronous environment, counting repeating patterns is a critical function. In order for the T1 to work, both ends of the circuit must be synchronized. The PCM word for channel 1 on the sending end must go to channel 1 on the receiving end. The only way for a time-division multiplexer to accomplish this task is for both ends to count accurately and at the same rate. Let us see how the framing bit can help in this process.

The framing bits provide a good way for the time-division multiplexer on each end of the circuit to verify that they are in sync. If the framing bits do not arrive at the receiving end of a time-division multiplexer at the time they are expected (every 193 bits) and in the order expected, as shown in Figure 2–25, then the multiplexer recognizes that something is wrong and asks for help. The framing bits also help the multiplexer when things are going right.

Figure 2–28
TDM transmit/
receive configuration.

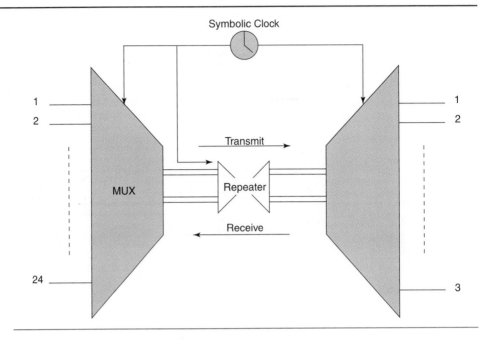

Example

Suppose that the time-division multiplexer received the framing bit pattern 1000, with 0 being received first and 1 received last. Refer to Figure 2–25, what frame will the multiplexer receive next?

The frame that the multiplexer expects to be receiving next is the sixth frame. If the pattern received was 0111 with 0 being received last, the next frame the receiving multiplexer is waiting for is frame 12. Remember that the least significant bits in the PCM word in frame 6 and frame 12 are used for signaling information.

So, by watching the framing bits, the multiplexer can tell when signaling information is about to arrive. And, as we mentioned before, framing bits (or a lack of them) can alert the multiplexer to loss of synchronization with the other end of the circuit.

Maintaining synchronization on a T1 circuit raises the issue of timing, or clocking the circuit. The easiest way to ensure synchronization is to see that both ends of the circuit are attached to the same clock as in Figure 2–28. This is usually accomplished by having one end of the circuit work in a "slave" relationship. The slave side of the circuit takes its timing signals from the "master" side. In this way the slave side can always be in sync with the master, as shown in Figure 2–29. This master/slave timing is also called *loop timing* because the slave takes timing from the T1 loop.

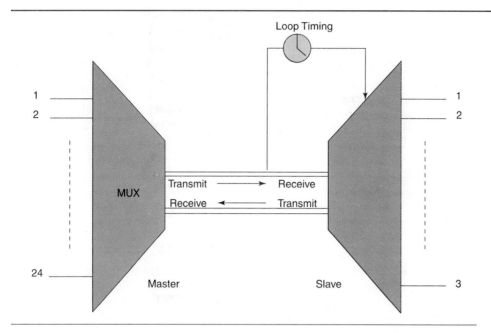

Figure 2–29
T1 circuit
synchronization.

Of course, the master–slave relationship begs the question: How is the master side of the circuit timed? There are a number of possibilities. A time-division multiplexer has its own internal clock, of course. It can also take its timing from an end office switch. Or, it can use the BSRF timing signal. BSRF stands for *basic system reference frequency*. The BSRF is an atomic clock (cesium) located in Hillsboro, Missouri. Why do you think the BSRF is located in Hillsboro? If you do not know the answer, do not feel bad. We did not understand it until someone explained it to us.

BSRF

It seems that Hillsboro is located in the geographical center of the continental United States. That minimizes the propagation delay of timing signals to any other point in the country, and that means you can time your T1 signal precisely.

The BSRF is the most accurate clocking device available for T1 circuits. As such, it is known as a stratum 1 clock, which is accurate to 0.001 parts per million (ppm). Most digital tandem switches have a stratum 2 clock, which is accurate to 0.200 ppm. Most time-division multiplexers (channel banks) use a stratum 4 clock, which is accurate to about 50 ppm. This is sufficient for most point-to-point T1 circuits. But as you begin to network T1 lines, greater accuracy is usually required.

So far, we have looked at the DS-0 level channel signal, the DS-1 bit stream, signaling with A and B bits, the frame/superframe format, and T1 timing. But we have not discussed what happens between the T1 terminal devices. In the next section we take a closer look at the T1 line itself.

T1 Repeaters   Although T1 signals are digital, they are still electrical in nature (except, of course, when we are dealing with fiber optics). If those T1

Figure 2–30
T1 circuit with
regenerative
repeaters.

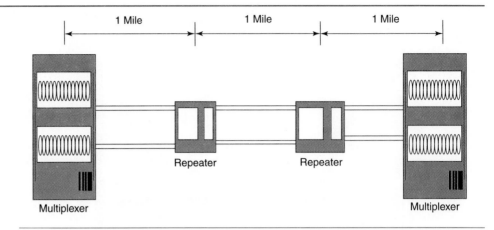

signals are being carried over copper, they are subject to the same kinds of signal degradation that affect an analog signal—attenuation, noise, distortion, and so on. Just as an analog circuit needs amplifiers to keep signal levels up, so a T1 circuit needs repeaters to maintain the integrity of the bit stream. As shown in Figure 2–30, the equipment between the T1 multiplexer terminal devices is referred to as *regenerative repeaters*. They take the T1 bit stream in, regenerate a clean, exact duplicate, and send it on to the next repeater (or T1 terminal device, if it is the last repeater in the circuit).

Over a copper facility, the maximum transmission distance without using a repeater, for a T1 signal, is approximately one mile (6000 feet is about the maximum). The spacing matches the load coil spacing on a typical analog circuit.

The repeaters have a relatively simple mission in life. They have to recognize a 1 when it arrives and send out a 1. They also have to recognize a 0 when it arrives and send out a 0. When the bit stream leaves a T1 terminal device, such as a channel bank, it looks like the signal in Figure 2–31(a). After a mile in the copper, it might look like the signal in Figure 2–31(b).

The repeater circuitry is really quite simple. It has a detector that examines the incoming signal to determine if it is a 1. If the signal exceeds a certain preset threshold value (voltage), it's a 1. If the signal does not exceed the threshold value, it is a 0. As shown in Figure 2–32(a), first the signal is "peaked up" at the middle of each timing interval (bit time). Then, if the peaks exceed the decision thresholds, they are determined to be 1s. If not, they are 0s. When the repeater has determined the 1s and 0s, it regenerates the bit stream and sends it out in sync with the original, like the signal in Figure 2–32 (b).

As mentioned, time is very important in T1. That is why the repeater must maintain the clocking of the original signal it receives. The repeaters use the incoming 1s to maintain their timing. The 1s are used because each time a 1 comes along, the repeater has a distinct mark to take its timing from. If there are too many 0s in a row, the repeater's timing circuits may get confused. The

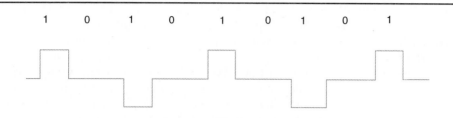

| 1 | 0 | 1 | 0 | 1 | 0 | 1 | 0 | 1 |

Figure 2–31
Degradation of a T1
bit stream signal.

(a) The T1 Bit Stream When it Leaves the T1
Terminal Device

(b) The T1 Bit Stream Signal When it Arrives at the
Repeater

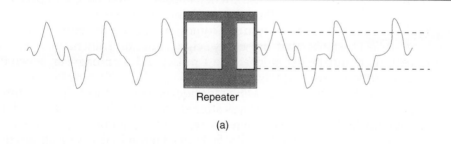

Repeater

(a)

Figure 2–32
(a) Peaking the
T1 signal, and
(b) regenerating the
bit stream.

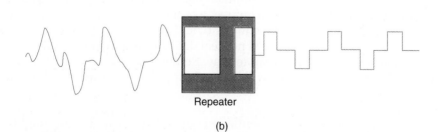

Repeater

(b)

next 1 that comes along may get sent off a little early or a little late. Then if the next repeater is taking its timing from the incoming bit stream, you can see what will happen. If there are 5 or 10 repeaters in a row, synchronization may be thrown off completely.

There are two possible solutions. We could put very accurate (i.e., expensive) clocks in each repeater. Or, we can somehow ensure that there are always enough 1s in the bit stream to keep the repeaters' clocking accurate. The trade-off is clear. Ensuring an adequate density of 1s will be far less expensive than putting accurate clocks every mile.

zero constraint
rule

The rule about 1s density is usually referred to as the *zero constraint rule.* This rule can get very complex because 1s density can be measured both as an average function and as an instantaneous function. But, in general, the rule states that you must average at least one 1 in every eight bits, and that you cannot send more than seven 0s in a row. If you do send more than seven 0s in a row, you run the risk that the repeater's clocking will get confused and synchronization will be lost. In the next section, we will take a closer look at a technique known as B8ZS (bipolar eight zero substitution), used to ensure adequate 1s density over the T1 line.

Byte Synchronization Since each sample consists of eight bits, the samples are in the form of bytes. The transmitter and receiver synchronize based on the pulses in the byte format. To maintain synchronization, we need to maintain an adequate number of 1 pulses in the bit stream. In voice communication, maintaining 1s density is not a problem, since voice generates a continuous change in amplitude voltages, which, when encoded into bytes, will produce numerous 1s. However, in data transmission, long strings of 0s occur frequently. Hence, the "1s density rule" comes into play. Put simply, this rule (used by AT&T) states that in every eight bits of information to be transmitted, there must exist at least one 1 pulse, and no more than seven 0s may be transmitted consecutively.

B8ZS

A technique known as *pulse stuffing* was developed to meet these requirements. The eighth bit in every byte was stuffed with a 1. This reduces the data transmission rate to 56 kbps since there are now only seven usable bits and one stuffed bit being transmitted (8000 × 7 = 56 kbps). Figure 2–33 demonstrates the bit stuffing technique. A technique was developed to overcome the limitation of the pulse stuffing method and still meet the 1s density requirement. This technique is known as *bipolar 8 zero substitution (B8ZS).* The T1 transmitter circuit examines the information being transmitted, a string of eight 0s is recognized immediately and replaced by the fictitious byte, "00011011." The receiving end knows this is a substitute word and not real data, because two violations to the bipolar signal are created at the same time. Remember that the bipolar rule requires that alternating voltages will be used for pulses. Figure 2–34 illustrates the B8ZS method.

T1 Bit Stream Errors We have mentioned several times that there might be some errors in the T1 bit stream. This brings us to a discussion of T1 alarms

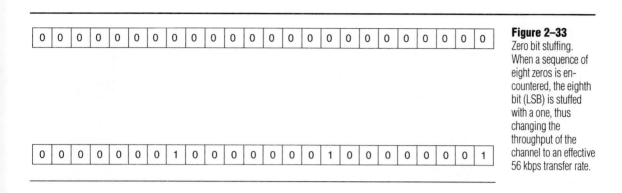

| 0 | 0 | 0 | 0 | 0 | 0 | 0 | 0 | 0 | 0 | 0 | 0 | 0 | 0 | 0 | 0 | 0 | 0 | 0 | 0 | 0 | 0 | 0 | 0 |

| 0 | 0 | 0 | 0 | 0 | 0 | 0 | 1 | 0 | 0 | 0 | 0 | 0 | 0 | 0 | 1 | 0 | 0 | 0 | 0 | 0 | 0 | 0 | 1 |

Figure 2–33
Zero bit stuffing. When a sequence of eight zeros is encountered, the eighth bit (LSB) is stuffed with a one, thus changing the throughput of the channel to an effective 56 kbps transfer rate.

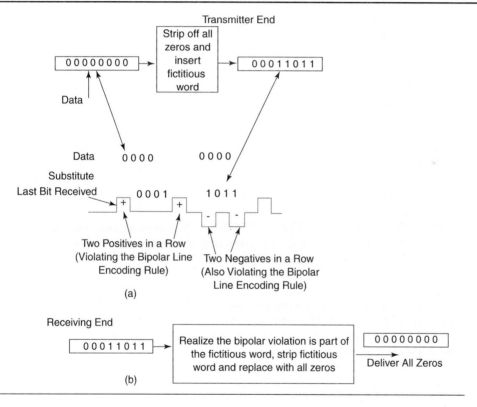

Figure 2–34
Bipolar eight zero substitution: (a) transmitter inserts a fictitious word replacing the all-zero pattern, and (b) the receiver strips off the fictitious word and replaces it with the all-zero byte intended for transmission.

and how T1 devices respond to errors when they see them. T1 bit stream errors can occur due to one or more of the following reasons:

- Synchronization can be lost.
- Repeaters might send a 1 when they really should be sending a 0.
- Hits on the line can change a bit.

Figure 2–35
T1 universal alarms.

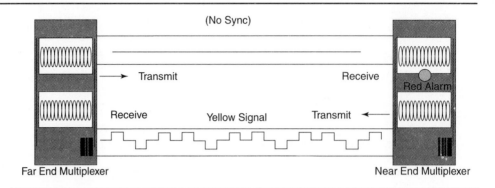

T1 terminal equipment universally recognizes at least two classes of alarms: red alarms and yellow alarms. Let us assume that one end of the T1 circuit finds that it is out of sync with the far end. The framing bits are not arriving on schedule. As shown in Figure 2–35, the near end terminal knows it is in trouble, so it presents a *red alarm*. First the near end (near end and far end are really defined by context) recognizes that it has lost sync and sets off a red alarm.

red alarm

The red alarm says, "Hey, I'm out of sync with the other end of the circuit!" But the other end (far-end multiplexer in Figure 2–35) does not know about the problem (yet). So when the near-end multiplexer goes into the red alarm state, it also sends a signal to the far-end multiplexer to tell it, "Hey, I'm in trouble." The signal it sends is called a bit 2 alarm. The near-end multiplexer forces the second bit of each PCM word in the DS-1 stream low (this is how a yellow signal is transmitted to the far end). When the far-end multiplexer sees that the second bit of every PCM word is a zero (a very unlikely event under normal circumstances), it knows that the other end of the circuit is in trouble. The far-end multiplexer then presents a *yellow alarm*. The yellow alarm says, "I'm sending stuff out OK, but the other end is not receiving it. There is a problem somewhere."

yellow alarm

The red and yellow alarms are reserved for serious problems that affect the entire T1 bit stream. The end that receives no signal, loses sync, or gets gibberish from the other end raises the red alarm. When the red and yellow alarms go off, the repair crew goes to work.

The red alarm is not displayed as soon a single framing bit is out of place. Instead, the multiplexer first tries to reestablish synchronization. If the multiplexer cannot reestablish sync within 2.5 seconds, it then presents the red alarm. The 2.5 second grace period allows the multiplexer time to recover from an occasional hit on the line, or other bit errors, without setting off unnecessary alarms.

The red and yellow alarms are generic to the T1 environment. All T1 pieces of equipment recognize and respond to red and yellow alarms. Different equipment manufacturers may provide additional types of alarms to indicate various other types of problems such as with a single channel.

AIS

There is another alarm that has become a standard. Often called the blue code, its proper name is the *alarm indication signal (AIS)*. In the early days, the

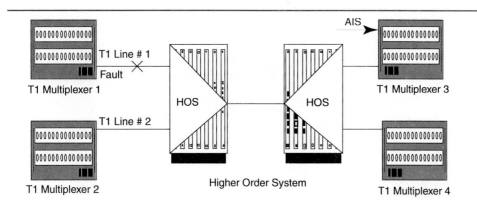

Figure 2–36
AIS signal (all ones) is used by the higher order system to tell the T1 multiplexer that there is a problem, but it is not the multiplexer's to worry about.

technology to multiplex a number of T1s together into a so-called higher order system was not available. Today, however, higher order systems (such as T3 or high-speed fiber optic systems) are common. Quite often, one T1 may "ride" one of these systems from office to office, along with a number of other T1s. Consider Figure 2–36, and suppose T1 multiplexer lines 1 and 2 are riding the higher order system to T1 multiplexers 3 and 4, respectively. If there is a fault toward T1 multiplexer 1 on T1 line 1, then T1 multiplexer 1 will raise a red alarm and the local maintenance crew will spring into action and fix it. T1 multiplexers 2, 3, and 4 will not raise any alarms because the T1 lines among them and the higher order systems are fine.

Now, suppose we lose the higher order system line? What kind of alarm should the T1 multiplexers raise? The first answer that comes to mind is, "red alarm," since they all will loose sync. That means we need to dispatch four local maintenance crews to try to locate the problem. But in this case the problem does not belong to any of the T1 links. It belongs to the higher order system. And that is what AIS is for.

AIS is a special T1 signal, consisting of all 1s, that is sent downstream from the higher order system to the lower order systems. The AIS says, "Hey, I know we are out of service, but it is my problem, not yours." The AIS avoids useless alarms and sending maintenance crews on wild goose chases. The blue code, or AIS, is but one of several important changes in the T1 network that are helping the move toward an all-digital network. In the next section, we will take a closer look at one very important change in T1 transmission: the extended superframe format.

Extended Superframe Format   In previous sections we looked at the T1 superframe (SF) format. In this section, we will explore the extended superframe (ESF) format in some detail. ESF is similar to superframe in many respects:

- Frames are still composed of 8 bits from each of the 24 DS-0 channels.
- One framing bit is placed at the beginning of the frame.
- Frames are still 193 bits in length.
        (24 channels × 8 bits/channel + 1 framing bit = 193).

Figure 2–37
Framing bit functions in both the SF and ESF framing.

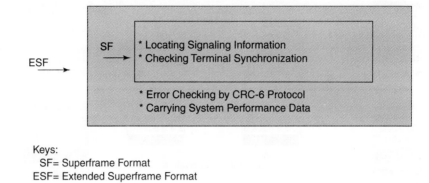

Keys:
  SF= Superframe Format
  ESF= Extended Superframe Format

F-bit

But, there are significant differences as well. The SF (or D4) format is based on a framing bit (or *F-bit*) pattern that repeats every 12 frames, but the ESF format is based on an F-bit pattern that repeats every 24 frames. As we will see in this section, the 24 F-bits of the ESF format allow much greater flexibility and accuracy in the T1 environment. Figure 2–37 illustrates the F-bit functions in the superframe format and the extended superframe format. There are two framing bit functions that are common to both superframe and extended superframe:

- Locating signaling information.
- Checking terminal sync.

But ESF adds two more functions:

- Error checking by CRC-6 protocol.
- Providing carrying system performance data.

Before we delve into the details of the extended superframe, however, let us consider a few points about F-bits. With the extended superframe 24-frame format, the ESF terminal generates 8000 bits per second. This is the T1 standard rate regardless of whether you use superframe or extended superframe format. In both SF and ESF, therefore, the F-bit provides an 8-Kbit/s channel for passing information. This bandwidth can be broken down to support the transmission of different types of information over the channel.

So far, we have only described the F-bit uses in terms of the number of F-bits allocated to a specific function. For example, in the superframe format, the six odd F-bits are used to handle terminal equipment sync and the six even F-bits to locate signaling information. Keep in mind that there are 12 frames in superframe format, thus giving us 12 F-bits.

With the introduction of ESF, we refer to F-bit usage both in terms of data rate and the equivalent number of F-bits used in each format to support a function. For example, if in reference to SF, we talk about using the full 8 kbps data rate, which is equivalent to using all 12 F-bits in each superframe. So if six F-bits are used for terminal synchronization in the superframe format, that would be equivalent to 4-kbps data rate.

Putting it another way, suppose that the data rate for terminal sync in ESF is 2 kbps. With ESF's 24-frame format, how many F-bits in each extended superframe would be used to support this function? The answer is six F-bits would be used, since 2 kbps is one quarter of the total framing bit bandwidth. One quarter of 24 F-bits in each ESF is 6 F-bits. In both superframe and extended superframe formats, whatever portion of the F-bits gets allocated to a specific function, such as terminal sync, the same portion of the F-bit data rate is also used. As we said earlier, both superframe and extended superframe transmit F-bits at 8-Kbit/s. Thus, the same total number of framing bits are always going over the F-bit channel.

Remember, there are only 12 F-bits in each superframe and their functions are already fixed (six for handling terminal sync and six for locating signaling frames). Each ESF, on the other hand, has 24 F-bits. Thus, extended superframe format can provide more functions and greater flexibility. With more F-bits, ESF should be able to support more functions, especially if there is flexibility in how the F-bits are allocated. In the rest of this section, we will explore how the extended superframe format uses its 24 F-bits.

F-bit Usage in ESF   The first two functions of the F-bits in ESF are locating signaling information and checking terminal sync.

Let us see how the extended superframe format handles these two functions compared to the superframe format. As we saw in the above sections, T1 signaling is carried in the least significant bit in each PCM word of every sixth frame. In superframe that means the sixth and twelfth frames. In ESF that means the sixth, twelfth, eighteenth, and twenty-fourth frames, as shown in Figure 2–38. This is often referred to as *in-band signaling*, because the signaling information is carried in the same bandwidth as the voice and data messages. In the superframe format, the least significant bit in the sixth frame is called the *A bit;* in the twelfth frame, it is the *B bit.* In ESF, the signaling bits are called the *A, B, C,* and *D bits.* In both cases, these signaling bit streams are often called the *A, B, C* and/or *D signaling channels,* or *signaling highways.*

in-band signaling

---

192 Bits Between Each Representing Samples from 24 Customers

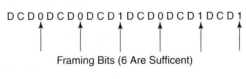

Extended Superframe Framing Word (24 Bits from 24 Successive Frames)

D C D 0 D C D 0 D C D 1 D C D 0 D C D 1 D C D 1

Framing Bits (6 Are Sufficent)

C = CRC - 6 Bit C
D = Data Link Bits (for Channel Maintenance)

Figure 2–38
Use of framing bits in extended super-frame format.

Before we go on, let us consider the additional signaling highways C and D provided by ESF. Back in the days before ESF, when telephone engineers tried to plan for the future needs of the telephone system, they anticipated an almost certain growth in the size and complexity of the public T1 network. To manage the myriad parts of this expanding T1 network, the planners believed that more signaling channels would be useful. They could have created additional signaling highways in the superframe by using the least significant bits of some of the other frames (1–5) and (7–11). If they did, however, the impact would be a degraded voice quality. So instead, they decided to extend the superframe from 12 to 24 frames. This way, ESF could preserve the standard of robbing a bit every sixth frame and allow for two more signaling highways (C and D).

In practice, however, the need for four-channel in-band signaling has not materialized. In fact, most DS-0 circuits today need only one signaling channel. Even though one signaling highway provides only two signaling states (1 = idle or 0 = busy), that is enough to convey the busy/idle status of the channel, as well as dialing information. In fact, many recently designed networks use no in-band signaling channels at all. In these situations, signaling information is carried entirely out of band via separate facilities.

Two-channel and four-channel signaling are available, however, to satisfy any special needs within the public switched network. In addition, they may be used to implement proprietary signaling schemes in private networks as well. Now let us see how these signaling bits are located in the superframe and extended superframe formats.

In the superframe format, we use a repeating pattern of the six F-bits (0 1 1 1 0 0) to locate the signaling frames. These F-bits are usually referred to as *fs bits.* This pattern helps the terminal equipment locate the sixth and the twelfth frames where the actual signaling information is carried. The receiving terminal recognizes the fs pattern as if it were two groups of 3 bits, 0 1 1 and 1 0 0. The receiving terminal recognizes that the third bit of the first half of the fs pattern (0 1 1) is also the F-bit that leads off the sixth frame. Similarly, the receiving terminal recognizes the third bit of the second half of the fs pattern (1 0 0) as the first bit of the twelfth frame.

In ESF, signaling information can still be carried every sixth frame. However, the ESF format does not require a dedicated pattern of signaling framing bits just to locate every sixth frame. Instead, ESF uses 6 F-bits known as the *framing pattern sequence (FPS)* to maintain terminal synchronization and correctly identify frame boundaries. With accurate framing, ESF equipment can locate the signaling information in every sixth frame. Let us see how this works.

As you may recall, superframe uses every other framing bit for terminal synchronization. Six F-bits out of the twelve form an alternating pattern (1 0 1 0 1 0). These terminal sync F-bits are usually referred to as ft bits. Figure 2–39 illustrates the superframe format with both the fs and *ft bits.*

After 30 years of experience with T1 transmission, and several generations of more accurate equipment, the telephone company realized that T1 equipment did not need to receive a terminal framing bit (ft) every two frames

FPS

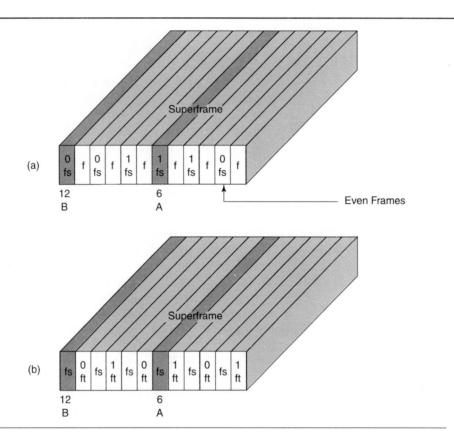

Figure 2–39
Use of the F-bits in the T1 superframe format: (a) the fs pattern, and (b) the ft pattern.

to keep terminal sync—half as often would be enough. ESF takes advantage of this fact by using an F-bit every fourth frame for terminal sync. In ESF, instead of being designated ft, these six F-bits are called the *framing pattern sequence* (FPS). The FPS framing bits are always the fourth, eighth, sixteenth, twentieth and twenty-fourth F-bits, and always have the values shown in Figure 2–40.

When the T1 bit stream is received, the ESF terminal equipment processes the bits one at a time. In each ESF, the first bit processed is a framing bit. Whenever the receiving terminal gets a framing bit, it will increment a framing bit counter by 1 (the counter is reset to 0 at the beginning of each extended superframe). Therefore, after the first framing bit is received, there are 192 bits representing PCM data from 24 channels the terminal equipment must process before another F-bit arrives. At this point, the terminal equipment encounters another F-bit. Then it increments the framing bit counter by 1, so it equals 2, and processes another 192 bits.

The process continues until the framing bit counter equals 4, and we have just received the F-bit at the beginning of frame 4. As we see in Figure 2.40, this framing bit must equal 0. If that is what the terminal equipment just received, then the bit stream is still being received in sync. So, when the framing bit

Figure 2–40
Frame pattern
sequence in the
T1 extended super-
frame format (ESF).

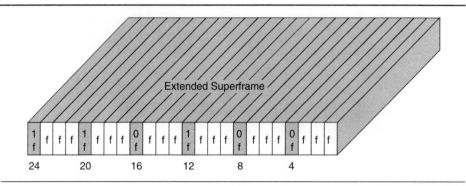

Extended Superframe

| 1 f | f | f | f | 1 f | f | f | f | 0 f | f | f | f | 1 f | f | f | f | 0 f | f | f | f | 0 f | f | f | f |

24   20   16   12   8   4

counter hits 4, 8, 12, 16, 20, and 24, the terminal equipment compares the F-bit it just received against the corresponding framing pattern sequence (FPS); as long as they match, it knows it is still in sync.

Then, let us ask this question: From what you have seen so far, how do you think the ESF terminal equipment locates the signaling frames that occur every sixth frame? Well, when the framing bit counter is set to 6, 12, 18, or 24, the terminal equipment knows that it is about to process one of the frames containing signaling information. To receive signaling information, the ESF terminal equipment processes the next 24 PCM words (192 data bits) and gets the signaling data from the LSB of each PCM word. So while superframe uses all 12 of its F-bits for signaling and terminal synchronization, ESF needs only the 6 F-bits of the framing pattern sequence. With FPS to ensure accurate framing, ESF can simply count framing bits to locate the signaling frames.

Now, let us ask another question: If ESF uses 6 F-bits (or 2 kbps) for terminal sync and does not need any additional F-bits to locate signaling frames, how many do we have left for other uses? Eighteen F-bits are left for other uses. These additional 18 F-bits are the key to ESF's improvement in accuracy and flexibility over superframe format.

CRC-6 Importance and Usage   One key use for these additional F-bits is to provide in-service error checking. ESF uses a method known as CRC-6. CRC stands for *cyclic redundancy check,* and it uses 6 F-bits per extended superframe to convey a 2-kbps CRC channel. Figure 2–41 shows the assignment of the CRC-6 bits. For many T1 users, CRC-6 is the most significant aspect of the extended superframe format. To understand why this new error checking is so important, let us look at few typical T1 transmission errors.

T1 lines, like most digital transmission facilities (except, of course, fiber optics), are prone to hits on 1 or more bits. Suppose a hit occurs on 1 bit of the byte shown in Figure 2–42. A bipolar violation has occurred, but that does not mean that the data is corrupted. What it says is that the 1 is still a 1 but its polarity has changed. Therefore, we have a problem on the line, which may or may not result in a data error. Some line hits will not be indicated by bipolar violations. If two consecutive 1s are changed to 0, the result may not cause a

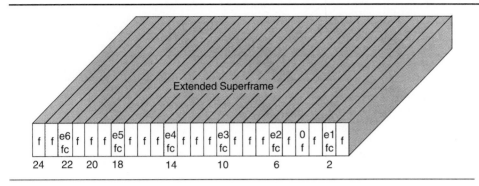

Figure 2–41
CRC-6 bit assignment in the T1 extended superframe format (ESF)

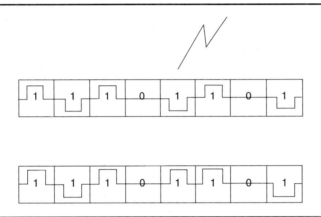

Figure 2–42
A hit on a T1 line that causes a bipolar violation in the fifth bit.

bipolar violation, even though the line hit has introduced an error into the data as shown in Figure 2–43. The hits could also become complementary, again causing errors in the data, but not causing a bipolar violation.

Therefore, bipolar violations do not provide a good indication of errors introduced into the content of the T1 transmission. A bipolar violation indicates a format error in the alternate mark inversion line code created by the line driver. Errors in the digital content of the DS-0 channels, often referred to as *logic errors*, do not correlate precisely with bipolar violations. In the superframe format, the only way to determine the quality of the T1 line, while it is carrying traffic, is to monitor the line for bipolar violations. Using a variety of algorithms, the bipolar violation count gives an approximation of the error rate.

The only accurate measurement of the line performance, however, was (and is) the *bit error rate test (BERT)*. To perform a BERT, however, the technician has to take the line out of service.

BERT

Obviously, we want to avoid taking lines out of service, except when absolutely necessary, and that is where ESF, with CRC-6, comes in. CRC-6, or cyclic redundancy check, is a mathematical algorithm by which a series of computations are made using the values of all the bits in one extended superframe.

Figure 2–43
A hit that does not
cause a bipolar
violation but causes
a logical error.

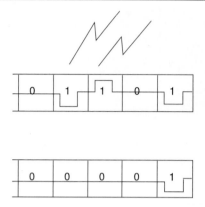

The result of the calculations is a number called a CRC code. It is expressed as a 6-bit word and is carried in the 6 CRC F-bits, as shown in Figure 2–44(a).

In ESF there is a wrinkle about CRC operation that is important to keep in mind. Since T1 is a real-time system and since we do not want to delay the transmission of the frame in an extended superframe until we finish calculating the CRC-6 for the superframe being transmitted, we send the CRC for the current extended superframe in the F-bits of the *next* superframe. So, the CRC of ESF(N) is sent in the F-bits of ESF(N+1), as shown in Figure 2–44(b). At the sending side, as the ESF(N) is assembled, a copy is made. That way, the CRC of ESF(N) can be calculated while ESF(N+1) is assembled. Then the CRC code for N is written in the CRC F-bits of ESF(N+1). When ESF(N) arrives, the receiver calculates the CRC and stores it as the *local CRC code* for ESF(N). Then, when ESF(N+1) arrives, the receiver reads the CRC F-bits and stores them as the *received CRC code* for ESF(N). The receiver then compares its local CRC code with the received CRC code to see if they match. If they match, then ESF(N) was received without error. If they do not match, then at least one error occurred during transmission.

When a single ESF is received in error, this is known as a *CRC-6 error event.* Most ESF terminal equipment will count CRC-6 error events in 15-minute intervals, over a 24-hour period, or since the CRC-6 counters were last reset to 0, if that was more recent. In addition to the *CRC-6 error events,* most ESF terminal equipment will count the number of seconds during which CRC-6 errors occurred. These may be subdivided into categories according to the number of CRC-6 error events detected (one, more than one, or many) during each second of transmission.

Not infrequently, line hits cause errors to occur in large groups called *error bursts.* Error bursts will be reflected in large numbers of CRC-6 errors. Knowing the number of CRC-6 errors and precisely when they occurred can suggest how and when a particular line segment should be tested. CRC-6 error detection is a way to monitor the health of a segment while it is still in use. Often, a T1 segment will deteriorate gradually over time long before it

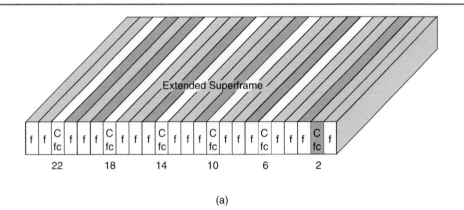

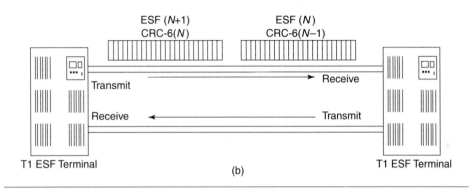

Figure 2–44
(a) CRC-6 F-bits assignment, and (b) transmission of CRC-6 of extended superframe *N* is put into the F-bits of extended superframe *N*+1.

fails to perform adequately. If many CRC errors are detected, the line segment where the errors occurred still must be taken out of service for a bit error rate test (BERT). Usually the results of a BERT are needed when reporting T1 performance problems to a T1 service provider.

CRC-6 checks the performance of individual line segments, such as the S-B segment shown in Figure 2–45. Where S is a digital switch node connecting T1 terminals. If you want an end-to-end performance check between two points while the line is in service, you must have available the CRC-6 data for all segments (both the A-S and the S-B segments) of that line.

Let us take a closer look to see how this works. In looking at any segment of the A-S-B circuit, we need to remember that each segment is a 4-wire connection with two paths. Each side transmits over one path and receives over the other. When a site, such as site A, collects CRC-6 error event data, it collects the data for its receive side. That is because it detects CRC-6 errors when it receives ESFs, not when it sends them. Site S is receiving the ESFs sent from site A, checking them for CRC-6 errors, and storing the CRC-6 error event data on those ESFs locally. So for the A-S segment, each end keeps local CRC-6 error event data for its receive side. The same is true for the S-B segment.

Figure 2–45
CRC-6 on T1 line
segments.

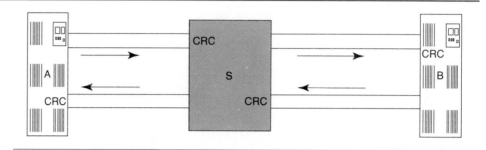

To get a complete picture on a segment, one end must poll the other and ask that it send its CRC-6 error event data. For example, site S might poll site A to get a copy of site A's CRC-6 data. Then site S would have complete performance information on the A-S segment. Site S can also poll site B for its CRC-6 data, and thus it will have a complete performance picture of the entire T1 circuit over the A-S-B segments.

In this simple example, having site S collect all the CRC-6 data for the two segments is fine. But as you can see, the switch at site S is connected to the larger network. In a complex network with lots of nodes and interconnecting T1 segments, the CRC-6 data is typically collected at a network management center (NMC). In one case, the switch at node S collects the CRC-6 data for its two T1 segments, as we saw, then sends that data to the NMC.

Another option for collecting CRC-6 data involves connecting a line monitor (LM) to both the transmit and receive sides of the T1 line segment. The line monitor can poll the devices at each end of the line segment for their CRC-6 data and then send it to the NMC. There are probably lots of other ways to get CRC-6 data on T1 line segments to the NMC, but you get the idea. With this information, the network management center can monitor in-service performance of T1 lines throughout the network.

Now whenever a site is polled for its CRC-6 error event data, it will send the data out over the T1 line via the remaining, undedicated F-bits. So far, we have dedicated six F-bits to the frame pattern sequence (FPS) and used six more to carry CRC-6 codes. This leaves 12 more F-bits that we can use for other purposes. Those 12 F-bits are the major key to the flexibility of the extended superframe format. These 12 F-bits in the ESF are known by many names. You may see them referred to as the *D channel*, the *data channel*, the *embedded operations channel (EOC)*, the *maintenance data link (MDL)*, the *frame data link (FDL)*, or simply the *ESF data link*. In this book, we will call them the ESF data link. Since the ESF data link uses 12 F-bits for each extended superframe, which is half the total F-bits, its data rate is 4 kbps.

ESF Data Link   The 4-kbps ESF data link was originally designed to allow public network carriers to gather performance statistics from T1 terminal equipment, as we saw earlier. The terminal equipment would store up CRC-6 error event counts and other performance indicators. Then periodically, the network-based monitoring equipment would poll the terminal equipment with a message in the ESF data link. The T1 terminal equipment would then return its performance report, also in the ESF data link.

In addition to collecting performance statistics, the ESF data link also provides an alarm function. As we saw earlier, in the superframe format, a superframe terminal processing a red alarm sends out a bit-2 alarm (forcing the second bit in every 8-bit byte low). The bit-2 signal causes the receiving terminal to present a yellow alarm. This is still the primary method of sending alarms in the public telephone network.

But in extended superframe format, the terminal presenting a red alarm uses the ESF data link to alert the other end. It sends a repeating 16-bit codeword consisting of eight 1s followed by eight 0s in the ESF data link F-bits. This alternating pattern in the ESF data link causes the receiving terminal to present a yellow alarm. Sending the alternating eight 1s and eight 0s will take at least 2 extended superframes, since there are only 12 F-bits used for the ESF data link in each extended superframe. The first 12 bits of the yellow alarm signal will go in the first ESF and the last 4 bits will go in the second ESF. But keep in mind that a yellow alarm signal must be sent continuously for at least 1 second. So it will actually use the ESF data link F-bits in each ESF during that 1 second (a total of 333 1/3 ESFs).

In addition to the yellow alarm signal and the CRC-6 information, the ESF data link is used to convey other important information. A terminal can send a variety of loopback messages over the ESF data link, which are signals used to initiate or terminate line testing. These codewords are carried over what is called *bit-oriented signaling*.

ESF Packets   There is another method of communication over the ESF data link known as *message-oriented signals (MOS)*. In contrast to bit-oriented signaling, the message-oriented signals carry performance monitoring information in a series of bits called a *packet*. Within each packet, the bits are organized into control and information fields. The information fields include performance data on:

- CRC-6 error events.
- Bipolar violations.
- Errors in the framing pattern sequence (FPS).
- "Slip events," or errors related to timing.

An ESF terminal designed according to the ANSI standards will collect all this performance information for each contiguous second interval. Once a second, the terminal sends out (or broadcasts) a packet containing four seconds worth of performance information. Each packet contains performance information for the most recent second of transmission, as well as for the previous three seconds.

Let us close this section by looking once again at the various types of F-bit functions. The 6 F-bits beginning frames 4, 8, 12, 16, 20, and 24 are assigned to carry the values of the framing pattern sequence (FPS). In ESF the FPS requires 2 kbps of the overall 8 kbps of F-bits available on a T1 line. The extended superframe format uses the FPS F-bits to maintain terminal synchronization and to correctly identify frame boundaries. With accurate framing, ESF equipment can count frames to locate the signaling information in every sixth frame.

The 6 F-bits beginning frames 2, 6, 10, 18, and 22, are assigned to the CRC-6 algorithm. These F-bits provide a 2 kbps CRC channel. ESF uses the CRC F-bits in each extended superframe to carry error-checking information (CRC-6 code) on the extended superframe sent before it. This allows the ESF equipment to monitor the health of a T1 line while it is still in service. The remaining 12 F-bits (The F-bits beginning the odd-numbered frames) provide a 4-kbps data link that we called the *ESF data link.* As we saw earlier, the data link can be used to transmit yellow alarm signals and collect performance statistics (CRC-6 error event data). But it can also handle other functions depending on the options made available by the equipment suppliers and the network designers. Table 2–4 summarizes the use of the F-bits in the extended superframe format.

Table 2–4
The functions of the F-bits in the extended superframe format.

Frame Number	Bit Number	Framing-Bit Use			Bit Number in Each Timeslot		Signaling Designation
		FAS	DL	CRC	For Data	For Signaling	
1	1	–	M	–	1–8	–	
2	194	–	–	e1	1–8	–	
3	387	–	M	–	1–8	–	
4	580	0	–	–	1–8	–	
5	773	–	M	–	1–8	–	
6	966	–	–	e2	1–7	8	A
7	1159	–	M	–	1–8	–	
8	1352	0	–	–	1–8	–	
9	1545	–	M	–	1–8	–	
10	1738	–	–	e3	1–8	–	
11	1931	–	M	–	1–8	–	
12	2124	1	–	–	1–7	8	B
13	2317	–	M	–	1–8	–	
14	2510	–	–	e4	1–8	–	
15	2703	–	M	–	1–8	–	
16	2896	0	–	–	1–8	–	
17	3089	–	M	–	1–8	–	
18	3282	–	–	e5	1–7	8	C
19	3475	–	M	–	1–8	–	
20	3668	1	–	–	1–8	–	
21	3861	–	M	–	1–8	–	
22	4054	–	–	e6	1–8	–	
23	4247	–	M	–	1–8	–	
24	4440	1	–	–	1–7	8	D

Key: FAS = frame alignment signal (..001011...)
DL = 4-kbps data link (message bits M)
CRC = cyclical redundancy check, CRC-6 block check field (check bits e1, ...e6)

Signaling channel designation is only applicable in channel-associated signaling. Also known as *in-band signaling.*

2.8 T3 FUNDAMENTALS

Now that T1 is behind us, you are ready to consider more complex concepts. Most organizations and users skip from a T1 directly to T3 without any of the intermediate steps of T1C/T2, fractional T3 and others. There are reasons for this quantum leap. Once the user has an understanding of the capacity and flexibility of digital transmission services, the need for more information becomes feverish. So the next logical step is immediately up to T3. As the technology becomes more efficient, prices also come down (as in the PC market).

A *T3 carrier* is all the things discussed earlier in this chapter, and even more. The structure of the digital hierarchy could be summarized by simply saying that the T3 provides 44.736 Mbps of digital service, or the multiplexed capacity of 28 T1s. But that is all too easy—it has to be more. Should you attempt to define T3, you may include the following:

T3 carrier

- *Digital service:* The T3 digital stream is 100 percent digital, just as T1 is.
- *Duplex operation:* Just as T1 is duplex, so too is T3. But, because of the larger T3 capacity, this duplex capacity is more versatile.
- *44.736 Mbps digital bit stream:* This is the multiplexed stream of 28 T1s into a single digital carrier, plus a 1.5 Mbps overhead for frame identification and more.
- *Pulse code modulation:* The PCM scheme for T1 and T3 is the same, using a standard of 64 kbps at the DS-0 level.
- *Time-division multiplexing:* The T1 TDM format uses a 648-nanosecond timeslot, of which 50 percent of the time is used for the pulse width (known as 50 percent duty cycle). In the T3 world, the digital time slot is only 22.5 nanoseconds in duration, and the pulse width (length of the electrical pulse) is 11.25 nanoseconds. Standards for transmission still use approximately the same time slots, and sample rates of 8000 samples per second are also still used.
- *Originally a carrier-only service:* Originally both T1 and T3 were designed for carriers. Carriers need higher bandwidth services, but the customers were perceived to require much less bandwidth. This remained the case until the late 80s for T3 services. Since T3 is new for customers, its full deployment and use are limited to specific areas around the country. Not all central offices or points of presence are equipped to deliver T3 services to an end user. It may require running a transmission facility 15 to 20 miles to a user's site, adding to the cost.
- *Based on high-speed media:* The T1 was based practically on any media, but was considered a 4-wire circuit over twisted-pair wires. T3, on the other hand, requires greater bandwidth than can be delivered over traditional twisted pairs. Therefore, the primary means of delivering this capability to the user are digital microwave, coaxial cables, and fiber optics. This creates some limitations for the providers, since they may require a new facility to deliver new service
- *Framed format:* Just as the T1 uses a framed format, so too does T3. However, since the capacity is so much larger, the frame is different. T3 uses a 4760-bit

frame, much larger than the 193-bit T1 frame. The T3 frame is known as the M13 (pronounced M one-three) frame, which comes from the use of the M13 multiplexer.

• *Bipolar 3 zero substitution (B3ZS) line code:* Timing and synchronization of the line is controlled through pulses in the data stream. T3 uses B3ZS line code, which differs from the B8ZS of the T1 (for clear channel).

• *Multiple protocols:* Just as all the previous digital transmission standards evolved, the T3 has gone through several phases. The T3 protocols deal with the M13 framed format transmitted in an asynchronous mode (without a synchronizing clock). However, problems exist with this form of transmission, so newer rules have been specified for implementation:

    a. C-bit parity.
    b. Synchronous transmission (SYNTRAN).
    c. Synchronous optical networks (SONET).

We can come up with more terms in defining T3, but that is enough before we reach our saturation point. Let us now see the driving forces behind T3. First, the cost per bit of transmission is declining because of the deployment of high-capacity networks (fiber-optic systems). Second, performance and monitoring capabilities are emerging that make T3 attractive. Third, because of users' insatiable appetites to consume every spare bit of bandwidth, they are outstripping the supply of transmission capacities. Finally, newer services through advances in computers require high bandwidth transport over the conventional dial-up, slow-speed networks.

2.8.1 T3 Bipolar 3 Zero Substitution

B3ZS

T3 uses what is known as *B3ZS* or *bipolar three zero substitution* line code. Recall that timing and synchronization of the line and equipment is derived from the pulses on the line. To accommodate this process, the pulse stuffing technique was used to maintain a minimum number of pulses.

The transmitter knows that the receiver must draw its timing from the pulses. Therefore, if three consecutive 0s were to be transmitted down the line, all timing could be lost. As the transmitter watches the bit stream that will be delivered to the line, it will strip sequential 0s, as in T1, and create a fictitious word with enough pulses to drive the timing.

Figure 2–46 shows the B3ZS line code and substitute wording. In this scenario, there are two possible opportunities to substitute the fictitious word depending on the convention used. First, the convention states that the number of pulses between violations should be an odd number and that as the three 0s are removed, a violation will be entered into the data stream as either B0V or 00V, where:

• If the number of pulses to be transmitted since the last violation is even, then a substitute word (101) will be inserted. This equates to B0V.

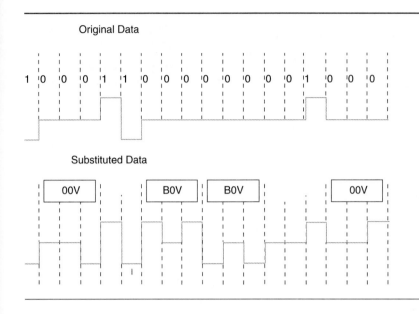

Original Data

1 '0 '0 '0 '1 '1 '0 '0 '0 '0 '0 '0 '0 '0 '1 '0 '0 '0

Substituted Data

00V . B0V B0V . 00V

Figure 2–46
Using B3ZS, three
consecutive 0s will
be replaced. An odd
number of pulses
(1s) is required be-
tween violations.

- If the number of pulses transmitted since the last violation is odd, then the substitute word (001) will be inserted. The pulse will create a violation; that is, it will be transmitted at the same voltage as the last pulse transmitted. This equates to 00V.

The need for B3ZS line code should be obvious, but the timing is even more critical to maintain synchronization under high-speed communications. The convention to handle an odd number of pulses between violations requires some (not a lot) remembering on the part of the transmitter to recall the last digital stream sequence.

2.8.2 M13 Frame

The framed format for T3 uses the M13 frame. The M13 multiplexing process is performed in two steps, as shown in Figure 2–47. The sequence is:

- Four T1/DS-1 signals are multiplexed together using pulse-stuffing synchronization to form an internal T2/DS-2 signal.
- Seven T2/DS-2 signals are multiplexed together using fixed pulse-stuffing synchronization to form a T3/DS-3 signal.

The DS-3 signal is divided into frames of 4760 (bits). Each frame is divided into seven subframes. See Figure 2–48 for the M13 frame. Once the frame is established, it is further divided into 8 separate columns, each representing 85 timeslots (time-division multiplexing).

Figure 2–47
The two-step
process to create
a DS-3.

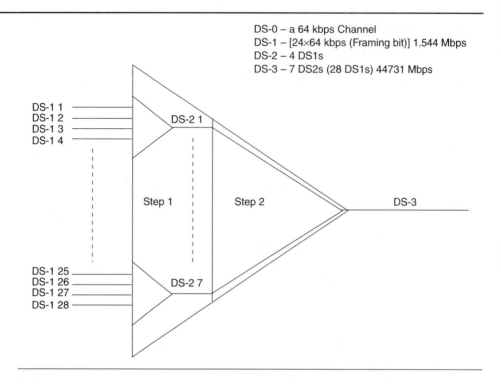

DS-0 – a 64 kbps Channel
DS-1 – [24×64 kbps (Framing bit)] 1.544 Mbps
DS-2 – 4 DS1s
DS-3 – 7 DS2s (28 DS1s) 44731 Mbps

Of the 4760 timeslots, 4704 are for user information and 56 are for over-head. The overhead is divided as follows (see Figure 2–48 for the overhead):

- Twenty-eight bits are reserved for control (called F-bits).
- Two bits are used for low-speed signals between ends (called X-bits).
- Two bits are used for parity checks on the user information. The trans-mitter counts the user data stream and inserts a 2-bit parity sequence (P-bits).
- Three bits are used for multiframe alignment in DS3. This is the same as a framing bit sequence. This helps in the location of the subframes inside the DS-3 signal (M-bits = 3).
- C-bits are used depending on the protocol used, as determined by the transmitter and the receiver. There are 21 C-bits available in the frame.

M13 synchronous protocol uses the C-bits as stuffing bits to come up to a common speed. The M13 frame uses stuffing bits in 7 of the 18 repeating frames to get common timing and clocking for the 7 DS-2s that are multi-plexed together into a DS-3. This system uses a dedicated stuffing slot and is indicated by the C-bits. Figure 2–48 is a summary of the M13 frame.

This means the two-step process, the fixed stuffing slots, and the frame format are somewhat ambiguous. If you are inserting 28 DS-1 streams into the

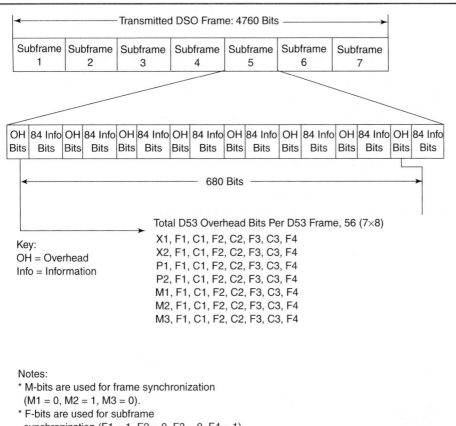

Figure 2–48
M13 frame format.

Key:
OH = Overhead
Info = Information

Total D53 Overhead Bits Per D53 Frame, 56 (7×8)
X1, F1, C1, F2, C2, F3, C3, F4
X2, F1, C1, F2, C2, F3, C3, F4
P1, F1, C1, F2, C2, F3, C3, F4
P2, F1, C1, F2, C2, F3, C3, F4
M1, F1, C1, F2, C2, F3, C3, F4
M2, F1, C1, F2, C2, F3, C3, F4
M3, F1, C1, F2, C2, F3, C3, F4

Notes:
* M-bits are used for frame synchronization
  (M1 = 0, M2 = 1, M3 = 0).
* F-bits are used for subframe
  synchronization (F1 = 1, F2 = 0, F3 = 0, F4 = 1).

DS-3 using the two-step process, the multiplexing uses a bit-oriented rather than a byte-oriented schedule. Therefore, the ability to track or trace a DS-0 or DS-1 is impossible since the bit stream varies depending on the order of input using bit multiplexing. There is also very little signaling and maintenance capability using the M13 asynchronous frame structure.

2.8.3 SYNTRAN

SYNTRAN, another set of protocols for DS-3, was developed to overcome some of the problems in M13 asynchronous framing. This protocol redefines the contents of the M13 frame format to eliminate the two-step process used to multiplex the signal (two steps include translating the four DS-1s to a DS-2 and seven DS-2s into a DS-3) and instead uses an efficient one-step process.

SYNTRAN, short for synchronous transmission, offers basic functions that were not readily available in the asynchronous protocols. Using this newer technique, bundles of DS-1s and DS-0 (a DS-0 is a single 64-kbps channel) can be isolated and tracked, since they are transported as an integral bundle. In the asynchronous mode, which uses bit multiplexing and interleaving, you could not trace a DS-0 through the network.

Further, SYNTRAN builds on a superframe format, which is similar to the evolution we saw on T1—the ability to perform maintenance and diagnostics on the T3. The superframe format is somewhat complex because the industry's standards bodies responsible for setting rules were looking to provide backward compatibility and a migration path toward the future.

2.8.4 DS-3 Superframe

Using a superframe format, SYNTRAN compresses the M13 4760-bit frame into 699 frames per superframe. The 699 frames are subdivided into groups of three frames, called triads, to allow identification of the subframe inside the superframe, as well as identification of individual DS-0s and the DS-1 in the T3 data stream. Fixed time slices of 8-bit bytes are used to create a byte synchronous format as opposed to a bit-oriented protocol.

Using the SYNTRAN protocol, the C-bits are redefined to support greater diagnostics, nondisruptive monitoring, and isolation of subrated multiplexed signals inside the DS-3. These subrated services are known as tributary signals, such as the DS-0 and DS-2 within the bundle.

Regardless of the type of service (at the DS-0 or DS-1 level), the ability exists within the DS-3 signal under SYNTRAN to provide rate adaptation. The overhead allows isolation of DS-0s and DS-1 with:

- Robbed bit signaling.
- Common channel signaling.
- D4 superframe.
- ESF.

Although these are important enhancements, SYNTRAN is still limited since it has evolved around the older cable/radio plant in the telephone company arena. The future is most likely to evolve around additional features of SONET and other capabilities to be discussed next.

2.8.5   SONET/SDH

To enhance services and overcome the limitations of existing technologies and to provide for a common connectivity interface among the local exchanges and long-distance carrier, work began in 1985 on a standard for digital transmission over optical networks called *synchronous optical network (SONET)*. BellCore, the Regional Bell Operating Companies research arm, led the effort with the *Consultative Committee on International Telephone and Telegraph (CCITT)* joining. This resulted in parallel recommendations called *synchronous digital hierarchy (SDH)* that differed from SONET in minor ways.

SONET design had four major goals:

1. Provide the ability of different carriers to interconnect.
2. Provide a way to multiplex multiple digital channels together. At the time SONET was devised, T3 operating at 44.736 Mbps was in use and T4 was

SONET

CCITT

defined, SONET's mission was to continue the multiplexing structure to gigabits per second and beyond.

3. Provide support for operations, administration, and maintenance (OAM) since previous systems did not do a good job of that.

4. Provide a means to unify the U.S., European, and Japanese digital systems, all of which were based on 64-kbps PCM channels, but all of which combined them in different (and incompatible) ways.

SONET was to be based on the traditional TDM systems, with the entire bandwidth of the fiber devoted to one channel containing time slots for the various subchannels. Thus, SONET is a synchronous system. It is controlled by a master clock with an accuracy of one part in one billion. Bits on a SONET line are sent at external precise intervals, controlled by the master clock.

When cell switching was later proposed to be the basis for broadband ISDN, it permitted irregular cell arrivals, and hence was labeled as *asynchronous transfer mode (ATM)* to contrast it with the synchronous operation of SONET.   ATM

A SONET system consists of switches, multiplexers, and repeaters, all connected by fiber. Figure 2–49(a) shows a SONET multiplexer system and Figure 2–49(b) shows a path between a destination and a source. A fiber going directly from any device to any other device, with nothing in between, is called a *section*. A run between two multiplexers (possibly with one or more repeaters in the

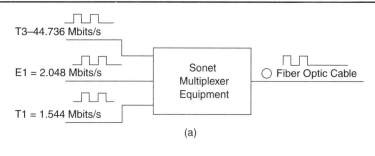

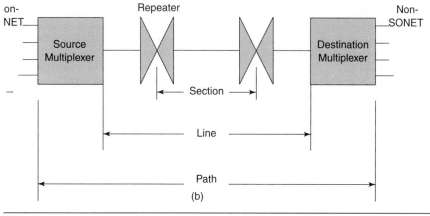

Figure 2–49
(a) SONET allows data streams of different formats to be combined into a single high-speed fiber optical data stream, and (b) a SONET path.

middle) is called a *line*. Finally, the connection between the source and destination (possibly with one or more multiplexers) is called a *path*.

2.8.6   SONET Multiplex Rates

A newer and more flexible structure of transmission rates is defined through SONET. Currently these rates range from 51.84 Mbps up to 2.048 Gbps, yet the theoretical limit is far greater. This surpasses the DS3 rate of 44.736 Mbps by a large margin. At the 2.048 Gbps rate, the industry is excited. But the specifications of transport will be exponentially improved to approximately 13.92 Gbps as SONET equipment and protocol mature. Table 2–5 is a summary of the capacities as defined today under the SONET rules. The basic building blocks for digital transmission are 64-kbps (DS-0) channels that are multiplexed together to form higher rates. Under ANSI standards, a modular signal (multiplexed services) for synchronous transport is the synchronous transport signal/level 1 known as *STS-1*. The STS-1 operates at 51.84 Mbps. The optical equivalent to the STS-1 is labeled the *optical carrier/level 1 (OC-1) signal*. Much higher rates are obtained by multiplexing the lower-level signal rates together. These higher rates are also labeled *OC-n* where *n* is the multiplexing rate. The available *n* rates presently are 1, 3, 9, 12, 18, 24, 36, and 48.

These basic signals can be subdivided into portions to carry lower-speed services at the DS-1, DS-2, and E1 standard rates. SONET is designed to be both forward and backward compatible with the transmission speeds in use today and has added overhead for control, signaling, and diagnostics.

To the casual or lower-end user, this process is transparent and users will employ T1 or T3 as normal. The network suppliers provide SONET on their backbone systems to provide high-quality services over the original copper-based systems.

2.8.7   SONET Framed Format

As with all digital transmissions, the signal and overhead are combined into a framed format. The STS-1 basic SONET frame is a block of 810 bytes put out

Table 2–5
SONET mutiplex rates.

Label	Mbps	Equivalent		STS Level
		DS1	DS3	
OC-1	51.84	28	1	1
3	155.52	84	3	3
9	466.56	252	9	
12	622.08	336	12	
18	933.12	504	18	Not
24	1244.16	672	24	defined
36	1866.24	1008	36	
48	2488.32	1344	48	
255	13920.0	6120	255	

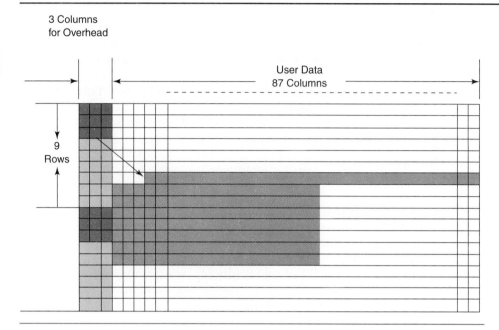

Figure 2–50
Two back-to-back SONET1 frames.

every 125 μsec. Since SONET is synchronous, frames are sent whether or not any useful data are to be transmitted. Having 8000 frames per second exactly matches the sampling rate of the PCM channels used in all digital telephone systems.

The 810-byte SONET frames are best described as a rectangle of bytes, 90 columns wide by 9 rows high. Thus $8 \times 810 = 6480$ bits are transmitted 8000 times per second, for a gross data rate of 51.84 Mbps. This is the basic SONET rate, as already stated, called the *synchronous transport signal/level 1 (STS-1)*. All SONET trunks are a multiple of STS-1.

The frame contains the overhead layers. Transport overhead consists of the first 3 bytes in each of the 9 rows, or 27 bytes of information. The transport overhead serves two functions: the 27 bytes are divided into 9 bytes for the section overhead and 18 bytes for the line overhead. The section overhead is generated and checked at the start and end of each section, whereas the line overhead is generated and checked at the start and end of each line.

The remaining 87 columns hold $87 \times 9 \times 8 \times 8000 = 50.112$ Mbps of user data. However, the user data, called the *synchronous payload envelope (SPE)* do not always begin in row 1, column 4. The SPE can begin anywhere within the frame. A pointer to the first byte is contained in the first row of the line overhead. The first column of the SPE is the path overhead (i.e., header for the end-to-end path sublayer protocol). Figure 2–50 demonstrates two SONET1 frames and column usage.

2.9   SUMMARY

Digital transmission evolved significantly in the past two decades. Analog networks gave ways to the more sophisticated digital networks, which consist of bandwidths of tens and hundreds of Mbps.

In order to transmit analog voice signals over the digital networks, we need to digitize the voice signals to convert them to forms suitable for such networks. The digitization techniques consists of three steps:

- Sampling.
- Quantizing.
- Coding.

A popular device that performs these three functions is known as the *codec,* short for coding and decoding. It is available as a single chip from many of the semiconductor communications manufacturers.

The digital facilities in the public and private networks continue to advance. T3 and optic fiber facilities are rapidly replacing many of the slower T1 spans. SONET, synchronous optical networks, is a standard that defines the formats and operations of the digital transmission over the fiber optic spans.

Maintenance is an important aspect to be considered when operating digital facilities. Remote devices must communicate the statuses of their health over the digital facilities so that operators and computers can devise a sequence of actions, either to inform someone or take corrective action to remedy the situation.

REVIEW QUESTIONS AND PROBLEMS

1. What are the advantages of digital transmission?

2. Give the steps required to digitize analog voice signals.

3. Give the PCM code word for a digitized voice signal such that the sample is in segment 2, step 4, on the negative side of the logarithmic curve of an A-law companding curve.

4. What are the advantages of using nonlinear companding techniques in quantizing voice samples?

5. State the Shannon and Nyquist theorems.

6. Calculate the capacity of a channel that has an *S/N* ratio of 100, using the Shannon theorem.

7. Describe in detail the functions and operations of a codec device.

8. Design a card that can be controlled by a PC to digitize voice signals, using a codec device and I/O hardware logic. The objective here is to capture the digitized voice samples into files on the PC and then be able to play them back on the codec card.

9. Describe the frame formats of a D4 T1.

10. Describe the frame formats of an ESF T1.

11. What are the differences in using the F-bit in both D4 and ESF T1s?

12. List all the functions of the ESF data link, and the bandwidth used for each function.

13. What are the differences between a message-oriented and bit-oriented ESF data link?

14. What is a red alarm, and how is it communicated to the far end by the side experiencing it?

15. What is a yellow alarm?

16. What is a yellow signal?

17. How can alarms be monitored by the ESF data link?

18. How is the CRC calculated for an ESF T1 and how is it transmitted?

19. Describe the different levels in the SDH hierarchy.

20. Define the following terms: section, path, line, and endpoint in the structure of a fiber optic facility.

21. Describe the SONET frame format.

22. Describe the functions of the overhead in the SONET frame.

23. What are the advantages of the SYNTRAN frame format over older formats of DS-3?

3

T1/T3 Applications and Subscriber Loop Carrier Systems

3.1 INTRODUCTION

Digital transmission coupled with advances in computer and semiconductor technologies have started what has been described as the *information age revolution.* If you are not happy with the term *revolution,* then feel free to use "very fast evolution." Digital transmission found its way into services in satellite systems offering digital services at a cost to the consumer comparable to that of the cable companies. T1 services are being used today to carry video conferencing meetings among many branches of an organization. T3 has found its way into the exploding market of Internet services. Fiber optics, using *synchronous optical network (SONET)* protocol and *asynchronous transfer mode (ATM),* are becoming common phrases of the day-to-day communications not only of those involved in the telecommunications industry but also among ordinary users of telecommunication services. Within this noisy and sometimes cloudy atmosphere, some logical decisions must be made and good judgment must be used in selecting what equipment and what vendor to provide these sophisticated and necessary communication services.

SONET
ATM

This chapter can be considered an extension of Chapter 2 but with a focus on applications of T1 and T3 systems. We begin the chapter with a general discussion regarding use of T1 digital transmission to provide the telephone company with cost savings and enhanced services. The chapter proceeds with a discussion of the components and operations of a typical digital loop carrier system. Specific examples of widely used although older digital loop carrier systems using T1 are presented, including AT&T SLC96, AT&T SLC series 5, and Fujitsu's digital loop carrier systems. More advanced digital carrier systems are being implemented using T3 and SONET. Discussion of one example from these systems is presented, specifically the Litespan 2000 system manufactured by Digital Systems Corporation. Research and development efforts are taking place to digitize what is called *the last mile* in the customer's loop. We present some technologies being considered to achieve that goal.

By now you are probably ready to take a break from the bits and bytes description of how T1 and T3 work. In the following sections, we present the criteria you will need to consider when deciding on a T1/T3 vendor. We also present the application of those digital transmission facilities.

3.2   T1 EQUIPMENT AND APPLICATIONS

In the case of T1, selecting the equipment to use can be complex. The equipment may have growth limitations; financial considerations must be factored into the decision-making process; and ongoing use and maintenance issues need to be seriously considered. The market for T1 equipment is always in a state of flux. With new entrants in the marketplace and older, less stable companies barely hanging on, the end user is in a dilemma over which option to choose. All too often, the selection criteria are based on a single issue, which could be one of the following (though many more may come into play):

- *Price.* Users tend to look at the initial cost to buy the equipment. This is particularly true when multiple pieces are involved. However, the initial cost is only a fraction of the overall system price. Long-term maintenance and the cost of the add-on upgrades should not be ignored.
- *Function.* Initially, a T1 infrastructure may be used to serve a single function, such as voice on wide area telecommunication systems (WATS) lines. The carrier may offer an equipment solution to serve a single function. But what if the dynamics of the organization change? An example might be the installation of a D4 channel bank (a system that multiplexes voice channels over T1 facilities) to consolidate WATS lines. Over time the need for superrate or subrate multiplexing capability may surface, but the fixed-function equipment may not support this application.
- *Emotion.* Unfortunately, the issue of whose product to buy may become more of an emotional rather than a business decision. This situation could cause problems. If the user becomes enamored with the bells and whistles of a hardware vendor, or if the seller and buyer become very close friends, the decision becomes more clouded. This process must be kept on a criteria selection basis to serve the organization's long-term needs and pricing arrangement.

- *End-of-life cycle.* Often, as a new product is about to be introduced, the older products are pushed heavily. This is not always bad, since some attractive financial incentives may apply. But how long will this product be supported? Will it be supported, upgraded, and maintained for the foreseeable life of the product? If not, then a second look must be taken at that choice.
- *Technological changes.* As new features are developed (such as ISDN, broadband ISDN, and frame relay) and introduced, the hardware must grow with the technology. If not, limitations may exist for future uses of the system and the network.
- *Standards.* Standards are constantly changing. The committees responsible for setting the rules are concerned with backward compatibility (newer systems must still work with older ones) and forward compatibility (current systems must work with future systems). Will the system you are buying be forward compatible or will massive and costly changes be required.

Many pieces of equipment provide T1 functions for one purpose or another. Based on the analysis of your applications you may end up choosing one or several pieces of that equipment. The T1 equipment listed below is not a complete list, but should give you an indication of T1-type equipment:

- Channel banks.
- Multiplexers.
- Network processors.
- Digital access cross-connect systems (DACs).
- Electronic switching systems.
- LAN connectivity bridges.
- WAN connectivity routers and gateways.

In the following section, we will take a closer look at some of this equipment.

3.2.1   Channel Banks

One of the first devices used by customers (users) to interface to a T1 circuit was the *channel bank.* This device was originally designed around customers' using voice-only services. The channel bank combined 24 signal sources (analog) into a single DS-1 (1.544 Mpbs) bit stream to transmit across either a customer (private) or a carrier (public) network. Since the channel bank was designed around voice, it was a fixed-function service, using card interfaces for analog inputs (voice or data) at 56 kbps. For data transmission, an analog modem was used to provide throughput of up to 28.8 kbps analog data. This used a full 64-kbps channel to carry 28.8 kbps of data. A gap existed between end-user demands for services and what the supplier provided.

channel bank

The evolution to a D4 (superframe) channel bank allowed voice or digital data input. Now a data service unit (DSU) can be plugged into the D4 channel bank cage to carry data at up to 56 kbps digitally. However, the D4 channel

Figure 3–1
Channel bank functional diagram.

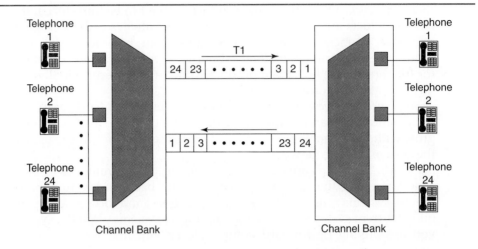

bank still had limitations on uses since it dealt with 64-kbps only services, thus making it one of the simpler devices. As such, it is also the least expensive DTE available. The cost of a channel bank (D4) averages from $2500 to $5000. Figure 3–1 shows a functional block diagram of a channel bank.

3.2.2 T1 Multiplexers

The next step above the channel bank is the MUX. The MUX provides more flexibility, in that services can be channelized (24 fixed time slots) or nonchannelized (user configurable). The MUX supports multiple interfaces via the use of a wide range of plug-in cards to provide superrate (above the basic 64-kbps channel) or subrate (less than 64 kbps) channel capabilities.

The general features of a typical multiplexer include the following:

- *Terminal interface.* Unlike the D4 channel bank where configuration takes place by setting jumpers and dip-switches, multiplexers are configured using a dumb terminal and an RS232 interface. In addition to configuring the MUX, the terminal interface is used for running diagnostics and on-the-fly changes.
- *Voice, data, and video handling.* A T1 MUX handles signals from different types of sources, including voice, data, video, and fax inputs. The reason for that is the MUX's ability to integrate serial inputs from various sources into an integrated bit stream. For example, a video call may require use of a 384-kbps stream that has interleaved bits for compressed motion video, voice, and possibly digital data.
- *Support for different standards.* With a T1 MUX, it is possible to select companding techniques, either μ-Law for North America or A-Law for Europe. It is also possible to select the type of modulation: ADCPM or standard PCM.

- *Vendor specific voice compression.* A T1 MUX may allow for high-compression voice multiplexing at 16 and 8 kbps. This allows a single 64-kbps channel to be used by up to 8 voice devices.

All T1 MUXs are based on a generic structure, including the following:

- Power panel.
- Frame.
- Common control logic to provide the control functions such as configuration, mode of operations, diagnostics and maintenance, and so on (that is why the MUX only needs a dumb terminal to operate and configure).
- Card cages to house the service cards.
- Interface cards (voice, data, aggregate cards).
- Backplane—an internal digital bus over which digital traffic flows internally until possible conversion and formatting is done in preparation for transmission over the T1 line or lines.
- Equipment interfaces—interface circuits and devices such as T1 interfaces, voice interfaces to connect telephones, data interfaces to connect computers and other equipment of that type, and possibly interfaces such as V.35 and RS336 to enable video calls.
- Management and diagnostics software.
- Special interface cards (LAN, bridges, routers, and gateways).
- Ringing generator.

Since all of the above resources are standard but may be called by different names, they will be considered basic equipment. The software that drives the features and capabilities would be the key factor that differentiates one manufacturer from another. Figure 3–2 shows a typical MUX layout using the architectural framework previously outlined. T1 multiplexers range in price from $4000 to $50,000 or more, depending on the mix of cards, the network management features, and interfaces to other services. With a spread of this magnitude, the user must be specific in defining the requirements, or the prices and variations could lead to confusion. Once again, depending on the user's needs, price should be only one factor.

3.2.3 The Channel Service Unit

The *channel service unit (CSU)*, although not a major expense, is the primary interface between the customer equipment and the digital network. The CSU works at the basic level of electrical and mechanical interface to a medium (i.e., line). It can be either a simple interface, or depending on the standards and protocol supported, it can be complex. This piece of equipment is required by the FCC and provides the following functions:

CSU

- Electrical isolation from the line.
- Signal generation compatible with the line.
- Assures that the 1s density is met. Generates (bipolar eight zero substitution—B8ZS) if necessary.

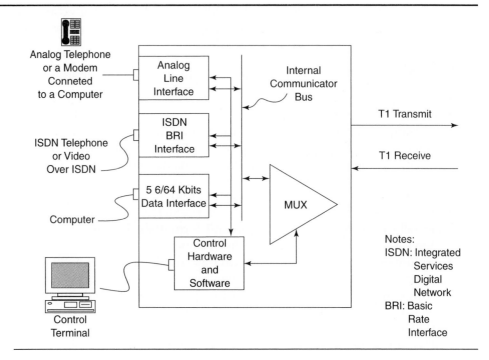

Figure 3–2
Architecture of a T1 multiplexer.

- Bipolar violation correction.
- Response for carrier (interoffice exchange carrier) loopback requests.
- Provides an interface for local loopback (toward the customer's side).
- Collection of ESF data, if equipped.
- Test pattern generation for error checking.
- Framed 1s pattern sent if equipment failure occurs.

Newer T1 equipment may integrate the channel service unit functions so that a separate CSU will not be needed, thus making such features another aspect in selecting T1 devices.

3.2.4 The Digital Service Unit

DSU
DTE

The primary function of a *digital service unit (DSU)* in the digital network is to convert a unipolar signal from the customer's *data terminal equipment (DTE)* into a bipolar signal for the network and vice versa. Figure 3–3 depicts the input and output of digital signals through the DSU. The DSU ensures 1s density, bipolar signaling, and extracting of data timing. The channel and digital service units (CSU and DSU) have evolved to the point of being combined. Since stand-alone devices are expensive and limited, a simple progression combines these functions as chip sets in a single box. The combination allows the transmission of high-speed data communications at rates from 56 kbps to the T1 rate of 1.544 Mbps. To take it further, much of the newer T1 equipment,

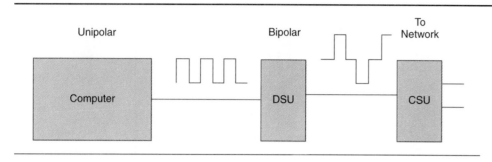

Figure 3–3
The DSU sits between the CSU and the customer's data equipment.

such as multiplexers and channel banks, integrates the functions of the CSU and DSU.

3.2.5 The Private Branch Exchange

The digital *private branch exchange (PBX)*, which will be discussed in Chapter 4   PBX
in greater detail, provides a direct termination of a T1 into a digital trunk interface. In the typical PBX environment, that digital trunk interface is a shelf inside the system that acts as a channel bank. Many of these systems can also provide a built-in channel service unit.

As an integral part of the PBX, this T1 interface is designed primarily around voice circuits. However, later PBX generations use this interface to provide services such as video and data.

The PBX offers some features that are not accessible to the multiplexer or network, including the following:

- Station message detail recording.
- Queuing a channel on a T1.
- Least cost routing.
- Redialing a busy number.
- Hunting.
- Call transfer.
- Conferencing.
- Call forwarding.

These features can be extended to both voice and data calls serviced by the switch. Figure 3–4 illustrates a typical digital PBX system.

3.2.6 Digital Access Crossconnect Switch

The *digital access crossconnect switch (DACs)* is a system that, by using a time-division multiplexing scheme, switches or crossconnects digital bit streams. The   DACS
number of inputs will equal the number of outputs allowing for totally nonblocking connectivity (i.e., all lines connected to the system may operate simultaneously without overload). These devices were developed to simplify the telephone company's administration and testing procedures, dynamic recon-

Figure 3–4
The digital PBX.

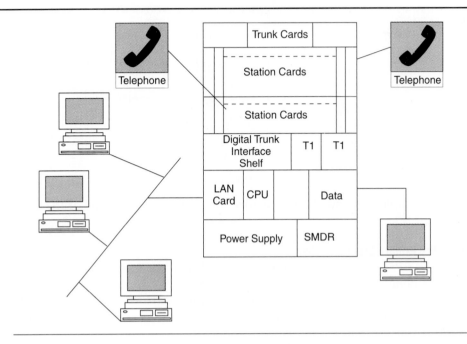

figuration of ports on the system, and also network grooming (e.g., rearranging channels within a digital span).

Although the DACS was originally designed for use in the carrier networks, larger corporations have also implemented these systems. The system allows linking traffic across multiple digital links and drop-and-insert capability (drop some channels off and insert new bit streams into channels). These systems can support very large amounts of bandwidth, such as T1 to T3, or T1 to T3 to T1 for up to 960 T3 ports.

Smaller versions of these systems are called micro DACS, housed in a PC cage, which can support up to 16 T1 interfaces. These smaller versions are considerably less expensive. Figure 3–5 depicts a typical DACS architecture.

3.3 T3 EQUIPMENT AND APPLICATIONS

Once you have decided to consider T3 technology, then you will need terminating equipment. As with T1, the availability of products and services in the T3 area is complex and varied.

Normally, if you are a beginner, you may use the standard equipment based on a vendor's recommendation, with the intent of changing the system as the application dictates. This may sound reasonable, but since these systems are usually expensive, current and future needs must be considered. The migration path for the equipment must be clearly and logically thought out.

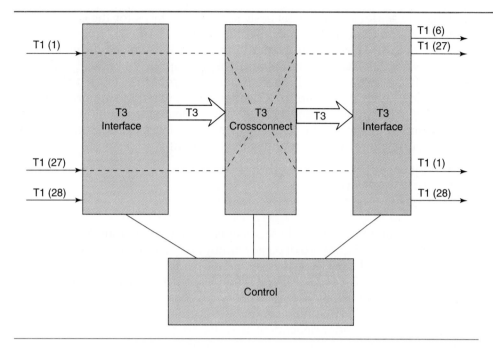

Figure 3–5
The digital access crossconnect system.

The old principle of making mass changes by buying completely new systems (called box changes) is no longer appropriate.

Services that require high bandwidth trigger the need to consider T3. Typical services that may trigger such considerations are:

- *Video.* Compressed high-quality 1.544 Mbps video conferencing.
- *Data.* Multiple 64-kbps channels for lower-speed communications to run multiple 28.8 kbps terminal connections across a 64-kbps channel.
- *Voice.* 64-kbps digital voice communications for tieline (trunk) services.

These are fairly standard services that might be transmitted across a T3 circuit. You by now realize that the use of a T3 requires very high communications interaction between locations. T3 supports 28 T1s, so the bandwidth needs of your application must be high to justify this service.

For T3 networks, the equipment suppliers offer the ability to connect through the following devices:

- Multiplexers.
- Network processors (servers).
- DACS.
- Electronic switching systems (ESS).

3.3.1 T3 Multiplexers
Multiplexers are designed primarily around the point-to-point service between two customer locations, a customer location and a carrier, or two carrier

locations. These devices began as basic service interfaces for the carriers and ultimately found their way to customer premises.

The T3 multiplexer can support a T1 and T3 environment within a single box. It appears that the wave of the future is the support of both T1 and T3 in a single platform. This may involve different approaches toward the same common goal, such as back-to-back configuration, external support at the T1 level, or internal support at the T1 level. All of these systems ultimately support a common output at the T3 level. The multiplexer is a standard system that may put the information in one of the following:

- A bit encoding or interleaved format.
- A byte interleaved format.
- A packet or frame interleaved format.

Regardless of the format used, these systems are designed around multiple inputs and interfaces. Figure 3–6 depicts a multiplexer within the domain of multiple DS-1, DS-0, and DS-3 capacities (recall in Chapter 2, DS-0 = 64 kbps, DS-1 = 1.544 Mbps, and DS-3 = 44.736 Mbps). When using a T3 service, the need for redundancy becomes critical, since many services are riding over a single digital span. Note that specific redundant components are reflected in this representation. However, it is vendor dependent as well as dependent upon whether this capacity exists. It is therefore up to you as a telecom manager, *craftsperson*, or technologist to determine what services require redundant paths.

craftsperson

Figure 3–6
The T3 multiplexer.

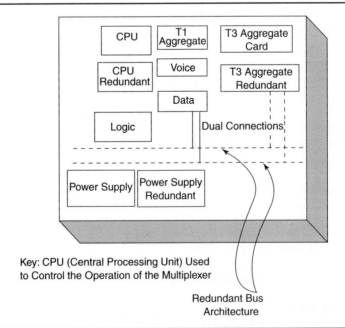

Key: CPU (Central Processing Unit) Used
to Control the Operation of the Multiplexer

Redundant Bus
Architecture

The DS-3 aggregate is the high end of the multiplexer signal and is the 44.736 Mbps digital stream, or 28 circuits operating at the T1 level. The T3 MUX can accept inputs on data cards, voice cards, video cards, and T1 aggregate cards. Keep in mind that the multiplexer's primary function is to bundle these inputs into one stream transmitted across the T3 span to another location.

3.3.2 Digital Access Crossconnect Switch (DACS) in the T3 Environment

As stated earlier, the DACS is a fully nonblocking, time-division multiplexing system that crossconnects the digital bit streams. Nonblocking means that there are as many input bit streams as there are output bit streams. In other words, the number of inputs is equal to the number of outputs. Therefore, an output path (or a channel) must exist for every channel of input. These devices were developed for use in a central office or point of presence (POP) to gain access to every digital bit stream (at the DS-0 to the DS-3 level). In the past, a physical pair of wires was run from frame to frame (a crossconnection point) to connect channels. Within the DACS this is all done electronically, meaning a channel on a T1 (called a *digroup*) can be multiplexed onto a channel from a different T1, or from one T3 to a different T3. This is switching at its best.

For example, channel 17 from the first T1 (digroup 1) needs to be crossconnected to channel 4 on T1 channel 6. Through the DACS, as the digital bit stream enters the input side, the system separates this channel and switches it through a matrix to the output side of the DACS onto the appropriate channel. Figure 3–7 shows a functional view of this process, which is also done at the DS-1 or DS-3 level.

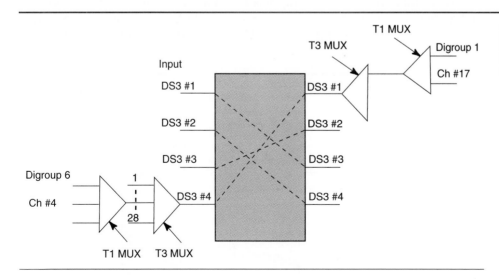

Figure 3–7
Functional diagram of DACS in a T3 environment.

In general, the DACS is designed to assist the carriers with administration and maintenance of user networks as stated previously. Other services on the new generation DACS may include:

- Digital services for the customer or carrier.
- Network administration.
- Port-by-port, channel-by-channel reconfiguration.
- Changes to user networks on demand.
- Network recovery and switching.
- Load balancing.

In many cases the carriers use the DACS to perform grooming. With grooming, a network technician can rearrange a network configuration to combine similar types of services in a single location. This might happen when multiple T3s are run from a central office to a point of presence. For example, a local exchange technician may put together all channels using extended superframe format onto a single DS-3 to the intermediate exchange point of presence. Point of presence is the location at which the local exchange office connects to a long-distance carrier.

Further, many diverse routes may exist between two points of presence in distant cities, for example between New York and Boston. A customer may be leasing several channels (DS-0s) between the two cities. As the network is installed, each of these DS-0s may be spread across several DS-3s. The DS-3s are running between the two points of presence for the convenience of the intermediate exchange. However, a new customer may come along and request a full DS-1 between the same locations. A technician may groom the network by shifting multiple channels around so that a full DS-1 can be provided contiguously for the new customer (in a transparent fashion to other customers).

This situation is outlined graphically in Figure 3–8 and 3–9 as the original and groomed configuration networks, respectively. Drop-and-insert capabilities are now easily achieved through the DACS at the carriers serving customer

Figure 3–8
Original network before grooming.

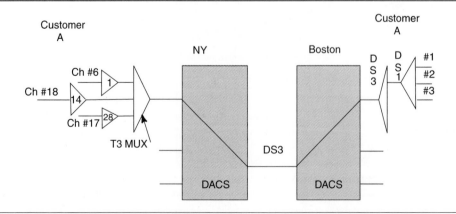

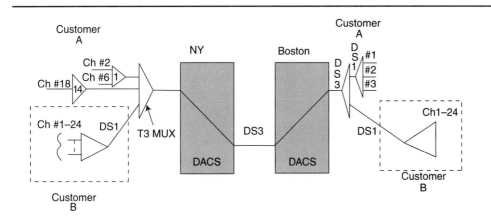

Figure 3–9
Reconfigured
(groomed) network.

needs. Larger corporations are now using DACS at the premises for full DS-3 bandwidth control at the end-user level.

As the availability of T3 services expands, the applications will follow. Users no longer feel that they must be tied to a band-limited service. They will deploy higher bandwidth services, which can be financially justified, and have extra capacity for on-demand service. One question, as a telecom specialist, you must ask is, Do I need to lease the bandwidth full time, or could I use it on a part-time basis, as needed?

The future of T3 service includes the integration of low- and high-speed communications services onto a single transport service. Some of these services include:

- Voice bundling.
- Data bundling.
- Video.
- High-speed special interface.
- FDDI connectivity.
- Frame relay.
- SMDS.
- Asynchronous transfer mode (ATM) service.
- Broadband ISDN.

Using a T3 to carry multiple voice circuits is straightforward, since the digital broadband services were originally developed to bundle voice circuits on a carrier. The telephone companies and intermediate exchanges deployed high-utilization trunks to support their need for capacity between the central office and other points of presence. Therefore, a natural extension of the digital network is to bring a large quantity of voice circuits to customer premises.

For example, the cost of T3 services has been dropping dramatically at the same time the bandwidth within the carrier networks has been increased. At one time, a T3 required a break-even point of approximately 17 to 20 T1 circuits to justify it. The drop in prices has moved the break-even point to between five and nine T1s, depending on the distance involved.

Access to a central office or POP depends on the mileage to the serving office from the customer's premises. These mileage figures still vary, but as we stated, the break-even point is between five and nine T1 circuits. A customer with needs above nine T1 circuits should seriously consider T3 service.

Clearly, when using digital service (T1 or T3), the distinction among voice, data, and video vanishes. The digital stream consists of 1s and 0s regardless of the input. Regardless of how the bundling of services is handled, it is possible to drive high-speed communication services across a T3. If a voice scenario, which also fits a data scenario, could provide a break-even point of five to nine T1 services between two company locations, then everything else gets a free ride. Whether transmitting voice or data, the T3 access will still save money.

Video conferencing services with quality as good as network television use a huge amount of bandwidth. Typically, one-way service uses 45 Mbps of transmission capacity. This of course is different from the services a corporate user might consider. Generally, video compression techniques are employed to use a 1.544 Mbps transmission capacity for high-quality video conferencing. Newer compression techniques are driving the bandwidth needs down to 768, 384, or 128 kbps. Even so, the cost of video services can quickly mount, forcing a financial justification process. As needed in the voice and data bundling examples above, video falls into the same category. The same break-even comparison can be applied.

3.4 DIGITAL LOOP CARRIER SYSTEMS

DLC

A *digital loop carrier (DLC)* system is a transport system in the local loop that uses digital multiplexing technology to place 24 analog channels together onto a single DS-1 digital signal. These systems are not new to telecommunications. The techniques they use are based on research done in the 1950s and 1960s. DLC systems first saw widespread use in the local loop in the late 1970s.

One of the most popular DLC systems used in the old Bell System was AT&T's SLC96. Because of its popularity and widespread use, its operating format and requirements were released to all manufacturers after the break up of AT&T in the form of Bellcore document TR-08 (TR stands for Technical Requirement). This document was issued in 1987 and did not propose any new options. It merely documented the existing SLC96 system. Bellcore is short for Bell Communication Research, the research arm for the regional operating companies known as RBOCs. The seven regional operating companies funded all research activities performed by Bellcore. This TR document, describing the SLC96 design and operations, allowed other switches and DLC system manufacturers to build compatible equipment.

The compatible equipment fell into two groups. The first group was switch manufacturers who could build interfaces to their digital switches that would talk to SLC96 DLC systems already installed in the field. The second category was manufacturers of DLC equipment. They could now build DLC systems that were compatible with the SLC96.

Manufacturers of equipment in both groups have seen widespread success. Although the technology used in the SLC96 system is not at the cutting edge, there is a base of thousands of SLC96 and SLC96 clones, which are still in use today to bundle voice services onto digital bit streams riding over DS-1s. The sole users of SLC96 in the U.S. are the regional operating companies.

The idea behind the SLC96 is to reduce the number of cables running from the central office to customer locations through the use of digital transmission. For example, if the telephone company wanted to provide voice-grade telephone service to 24 customers who live in one apartment building, 24 loop cables would be required for each customer location. Let us see how we can reduce the number of cables we must run to customer premises in this scenario. First, at the central office location, we bundle (multiplex) the 24 voice-grade circuits onto one digital bit stream using, say, the T1 D4 format. Then we run a single 4-wire cable from the central office to the curb next to the apartment building where the customers live. At the curb, we separate and assign each voice channel to each of the customers. Lastly, we run loop cables for a short distance from the curb to each individual customer location.

The idea here is that customers will still use the plain old telephone set known as *POTS*. Customers will not know that their loop cable extends only to the curb of the apartment. In other words, it is completely transparent to customers, fees are still the same, the cost per call is still the same, and troubleshooting is still the same as far as customers are concerned.

To provide bundling at the central office location, an electronic box will be needed to combine the analog circuits into one digital stream. This box is called the *central office terminal (COT)*, which is part of the DLC. As far as the   COT central office equipment is concerned, there is a loop cable for each customer. These loop cables, however, run a very short distance to the COT, which is located next to the central office equipment.

At the curb, another box debundles (demultiplexes) the voice channels from the T1 digital bit stream and puts the analog voice signal onto the customer loop from the curb to the apartment. This box is called the *remote terminal (RT)*, which is also part of the DLC system.

In general, a DLC system consists of two main components: the COT and the RT. The COT resides at the central office location and the RT may reside in a strong metal box on a curb in a city street or in a small building in a rural location. The COT and RT perform both time-division multiplexing and demultiplexing. Figure 3–10 depicts a generic diagram of a digital loop carrier system. From the figure, it can be seen that the COT and the central office equipment are usually placed in the same building. The DS-1 line may run for a few miles from the central office location to the curb next to the apartment building. From the RT, there are loop cables for each individual customer.

A question, however, must be raised. How do you troubleshoot the customer loop in this digital configuration? The answer to this question involves an elaborate testing scheme that we will discuss later in this chapter.

Figure 3–10
Generic digital loop
carrier system.

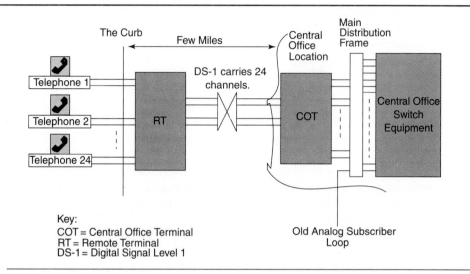

Key:
COT = Central Office Terminal
RT = Remote Terminal
DS-1 = Digital Signal Level 1

3.4.1 SLC96 Channel Bank

SLC96, known as the grandfather of the digital loop carrier systems, is basically a channel bank that converts and multiplexes 96 voice signals into digital bit streams. As described in the previous section, it consists of two main units: the central office terminal (COT) and the remote terminal (RT). Five T1 lines connect the COT and the RT; four main T1 lines are used to carry the bits for the 96 channels (each T1 carrier 24, $4 \times 24 = 96$), and a spare T1 line is used when any one of the main lines fails. The secondary T1 line provides for protection against loss of customer services. This method of redundant service is called *protection switching*.

The SLC96 COT is colocated with the central office equipment. On one side, the COT is connected to the central office equipment via a 2-wire connection for each voice loop (96 pairs of wires). The RT is located near the customers' premises. On one side, it connects to the COT via the five T1 lines, and on the other, it connects to the customers' telephones.

Each of the SLC96 units (COT and RT) consists of four shelves, designated from bottom to top as shelves A, B, C, and D. Each shelf holds certain electronic cards (plug-ins). There are two types of plug-ins: *channel units,* and *common* plug-ins. Each channel unit card serves two customers. Each shelf holds 12 channel units and 4 common plug-ins. Depending on the shelf, the plug-in may vary. Figure 3–11 depicts a diagram of an SLC96 channel bank. The common plug-ins are used for system operations. From the figure, you can see there many acronyms that deserve some elaboration of their functions. Now let us see, briefly, the function of the common plug-ins used in the SLC96 digital carrier system:

- The *alarm control unit (ACU)* controls the operation of the system and provides alarming information via LEDs.

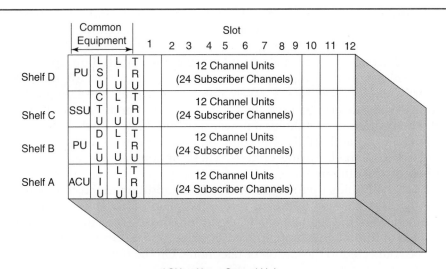

Figure 3–11
SLC96 COT or RT bank.

Key:

ACU = Alarm Control Unit
CTU = Channel Test Unit
DLU = Data Link Unit
LIU = Line Interface Unit
LSU = Line Switching Unit
PU = Power Unit
TRU = Transmit and Receive Unit
SSU = Special Services Unit

- The *channel test unit (CTU)* provides a means of troubleshooting channels.
- The *data link unit (DLU)* provides the data link for the digital loop carrier system.
- The *line interface unit (LIU)* provides interfacing to the T1 line. There are five line interface units in the system; one in each shelf and an additional one used to replaced any of the failed lines automatically. The LIU converts a unipolar signal into a bipolar signal, the format used for transmission over T1 lines. The LIU is also capable of providing T1 line loopback for testing and troubleshooting.
- The *line switching unit (LSU)* controls the automatic protection switching. By continuously monitoring the four main T1 lines, if one fails, the LSU then switches the bit stream of the failed line onto the spare line interface unit used for protection switching.
- The *power unit (PU)* supplies the various components of the system with the appropriate level of voltages.
- The *transmit and receive unit (TRU)* takes 24 pulse amplitude modulation (PAM) signals from the channels and multiplexes them into a DS-1 signal. The TRU inserts the data link onto the line.
- The *special services unit (SSU)* provides timing channel cards that are used for special services.

Four digroup shelves of the SLC96 are shown in Figure 3–11. The A digroup shelf contains an alarm control unit, a line interface unit that serves the T1 line connected to this digroup, a line interface unit that serves as a spare for the four digroups, and a transmit and receive unit. The B digroup shelf contains a power unit, a data link unit, a line interface unit for that digroup, and—similar to all other shelves—a transmit and receive unit. The C digroup shelf holds the special services unit, the channel test unit and the C line interface unit. The D digroup shelf holds a power unit and a line switching unit, in addition to an LIU and a TRU. The common plug-ins can be configured for different applications by using options. These options are set via switches or jumpers on the plug-in.

Each shelf holds up to 12 channel units in addition to the common plug-ins. A popular channel unit card serves the plain old telephone set (POTS). The functions of the channel unit plugged into the COT are different from the functions of the channel unit plugged into the RT. The channel unit at the RT must be connected to the customer telephone. On the other hand, the channel at the COT connects to the central office equipment.

A T1 digital carrier system such as the SLC96 must provide the digital transport transparently to both the central office equipment and the customer's telephone set. The central office equipment needs to think that it is connected directly to the customer telephone, and the telephone set must think that it is connected directly to the central office. In other words, as far as the central office switch equipment and the customer telephone are concerned, the SLC96 does not exist. On one side (the COT side) the system must simulate the interface of the telephone to the switch, and on the customer's side (the RT side) the system must simulate the switch interface to the telephone.

When a customer goes off-hook (closing a current loop), the channel unit at the RT side detects the loop current and inserts an off-hook signal (digitally) using the A and B signaling bits discussed in Chapter 2. The channel unit at the COT side detects the changes in the A and B bits as an off-hook signal and closes a current loop toward the central office equipment. As can be seen from this scenario, the central office equipment operates as if it were connected directly to the customer telephone.

If the central office equipment wants to apply a ringing signal to the customer's telephone, it applies the signal directly to the loop of the COT channel unit. The channel unit then translates that AC analog signal into A and B bits and sends these bits toward the channel unit at the RT side. The channel unit at the RT side interprets the status of the A and B bits as a ringing signal and closes a relay connected to a ringing signal source. That relay connects the ringing signal source at the RT to the customer's telephone and the phone rings. The same process is used for all other signaling between the telephone and the central office equipment. On one end it is converted from loop current signals to 1s and 0s, for each channel, in the designated bits. The other side of the system translates these bits into loop current signaling.

The talk path, end-to-end, within the carrier system is completely digital using the PCM techniques discussed earlier. On the analog sides of the system

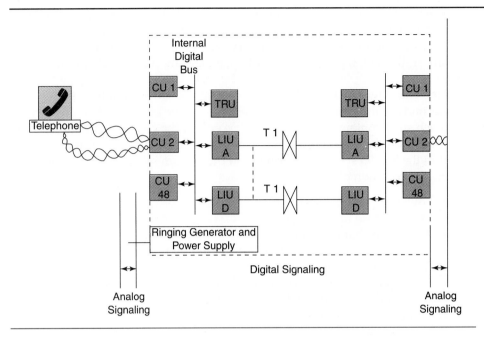

Figure 3–12
Functional diagram
of the AT&T SLC96
digital carrier
system.

the talk path is analog. Figure 3–12 illustrates this concept. It seems like a lot of work to make the system operate, but the world is generally analog (in the case of voice), and for almost one hundred years the telephone system had been analog. Many functions can be added and efficiency greatly increased through digital circuits and signals. Thus, we have a hybrid (digital and analog) system.

3.4.2   AT&T SLC Series 5 System

The AT&T subscriber loop carrier (SLC) Series 5 system was built to improve on the SLC96 system, known as the grandfather of digital loop carrier systems. Figure 3–13 illustrates the Series 5 system. Some of the advantages of the Series 5 system are:

- Twice the number of customer lines supported in the same space as the SLC 96 channel bank.
- Fewer types of channel units required to provide the same services.
- Provisioning and testing special services channel units at the remote terminal using a craft interface unit (CIU).

The AT&T SLC Series 5 DLC system is comprised of two independent 96-line systems, known as the white system and the blue system, in the same size channel bank as the SLC96. The channel bank is arranged into five shelves, with the white system occupying the top two shelves, the blue system the bottom two shelves, and common units occupying the middle shelf. Some common units are also located on the other shelves. Each independent system contains four digroups: A, B, C, and D.

The Series 5 system is available in what is called *feature packages*. A feature package is really a combination of software and hardware versions made to work together to provide particular services. Examples of these feature packages are A, B, C, D, I, and 303. We only need feature package B to emulate the SLC96.

Like the AT&T SLC96, the Series 5 plug-ins are generally grouped into channel and common units. The channel units can be either dual or single channels. Each shelf contains two digroups (except for the center shelf) and each digroup has slots for 12 plug-ins. The common plug-ins are also similar to those found in the SLC96 system. Beginning on the bottom shelf of the blue or lower system, the common plug-ins are:

- Power converter unit (PCU).
- Channel fuse unit (CFU).
- A digroup channel units.
- A/B transmit receive unit (TRU), shared by two digroups.
- B digroup channel units.
- Bank control unit (BCU), brain of the system.
- Alarm display unit (ADU).
- Channel test unit (CTU).

The second shelf of the blue system contains identical plug-ins to that of the first shelf, except that the digroups are C and D. The BCU, ADU, and CTU are double height and extend upward to that shelf, and the CFU is replaced with either an office timing unit (OTU) at the COT or a fan control unit (FCU) at the RT.

The center shelf holds only the following common units:

- Power converter unit.
- Channel fuse unit.
- Line fuse unit.
- Line interface units (LIUs) for A, B, C, D and protection (spare) DS-1s of the blue system. (The spare DS-1 is used to carry the traffic of a failed A, B, C, or D DS-1.)
- Line switching unit (LSU) for the blue system.
- Digital test unit (DTU) for each system.
- Line interface units (A, B, C, D, and P) and a line switch unit for the white or upper system.

The top two shelves house common and channel plug-ins for the white system and are identical to the bottom two shelves for the blue system except that there are no channel test units for the upper system. In other words, there is only one channel test unit card per SLC Series 5 system. At the COT, an alarm interface unit is placed to interface the systems' alarms with a remote alarm reporting system. Like the CTU, both systems in the channel bank share the alarm interface unit.

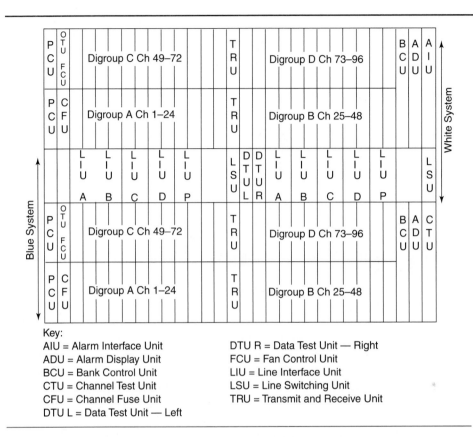

Figure 3–13
AT&T SLC Series 5 layout.

Key:
AIU = Alarm Interface Unit
ADU = Alarm Display Unit
BCU = Bank Control Unit
CTU = Channel Test Unit
CFU = Channel Fuse Unit
DTU L = Data Test Unit — Left

DTU R = Data Test Unit — Right
FCU = Fan Control Unit
LIU = Line Interface Unit
LSU = Line Switching Unit
TRU = Transmit and Receive Unit

3.5 NEWER DIGITAL LOOP CARRIER SYSTEM (TR303)

Unlike TR0208, Bellcore document TR-NWT-000303 is a forward-looking document. It details what the next generation of DLC systems should look like. In fact, in some literature, TR303 systems are sometimes referred to as the *next generation digital loop carrier systems*. TR303 (illustrated in Figure 3–14) offers compatibility between switch interfaces and the remote equipment, like TR08; however, it is less restrictive, giving manufacturers a small set of requirements and a large set of options. The advantages of the TR303-based digital loop carrier systems are:

- Use of extended superframe DS-1 format for additional overhead. As discussed in Chapter 2, the extended superframe format uses a 24-frame pattern, giving 12 additional framing bits for synchronization and communication among T1 terminal equipment. In addition, by bit robbing from the sixth, twelfth, eighteenth, and twenty-fourth frames, additional channel signaling information can be sent. The format still uses a standard DS-1 frame, and the DS-1 signal is still at a rate of 1.544 Mb/sec.

Figure 3–14
The basic configura-
tion in the TR303
digital carrier system.

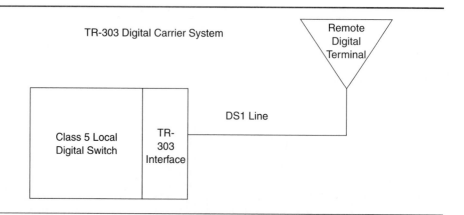

- Use of cyclic redundancy check (CRC) for protection switching instead of bipolar violation
- Fewer channel units are needed to provide the same level of services as SLC96 and SLC Series 5.
- Efficient means of providing ISDN services
- Larger system size—up to 28 DS-1s and up to 2048 DS-0s per system.
- Introduction of static and dynamic time slot interchange functions.
- Out-of-band signaling for clear channel services.
- Microprocessor-based with full remote capabilities for alarming, provisioning, and inventory.
- Optional SONET-based interface.

3.5.1 R-TEC SONET DISC*S

The Reliance-TEC (R-TEC) DISC*S digital loop carrier system was one of the first systems that was based on the guidelines of TR303. It was initially a copper-feed system, operating in a TR08 format, but it was designed for future incorporation of TR303 capabilities as the guidelines were being refined. The SONET DISC*S consists of three shelves and a fuse and alarm panel. The shelves include a common shelf, where all the system plug-ins are located, a DS1 extension shelf, which is integrated with the common shelf, and a channel shelf that houses the channel units. The SONET DISC*S is similar to the copper version in that it reuses the channel shelves and channel units and also reuses some of the common plug-ins. This helps in upgrading older systems to the SONET version.

The SONET DISC*S consists of a TR303-based digital loop carrier system mated with a SONET OC-3 multiplexer. Recall that the rate for SONET OC-3 is 155 Mbps. Although the multiplexer provides three DS-3s or 84 DS-1s, only 28 of these DS-1s can be used in the digital loop carrier system itself (this is a TR303 limitation). However, the remaining DS-1s can be extended either optically through the SONET multiplexer, or electrically by using the DS-1 extension.

The common shelf of the SONET DISC*S houses all the common plug-ins for the entire system. One common shelf can support the operation of up to seven channel shelves at this time. This means the maximum capacity of a SONET DISC*S system in its current configuration is 672 DS0 channels (7 shelves × 96 DS0 channels per shelf).

Unlike the TR08-based digital loop carrier systems, the SONET DISC*S has a one-to-one redundancy in most of the common plug-ins. The common units are:

- *Power converter unit (PCU)* to supply the system components with appropriate power levels.
- *Multiplexer/demultiplexer unit (MDU)* interfaces between the STS-1 signal of the SONET multiplexer and both the DS-1 shelf and the digital loop carrier portion of the system. It multiplexes 84 T1s into the STS-1.
- *Optical line unit (OLU)*, main and spare, provides the optical transmit and receive portion of the SONET system.
- *Transport processor unit (TPU)*, main and spare, controls the functions of the transport (SONET) side of the system.
- *Data link unit (DLU)*, main and spare, provides the TR303 data link for the digital loop carrier system.
- *Central processor unit (CPU)*, main and spare, controls the overall operation of the digital loop carrier portion of the system.
- *Signaling processor unit (SPU)*, main and spare, processes signaling and supervision information of the channel units of the digital loop carrier system.
- *Transmit/receive unit (TRU)*, main and spare, as in the AT&T SLC 5 system, combines the PCM signals from the channel units together into a DS-1 signal.
- *Maintenance unit (MU)*, no spare, provides all alarm and testing functions for the system, as well as local access via a craft interface device.

A DS-1 extension shelf to the common shelf allows the copper extension of up to 84 DS-1s from the optical SONET equipment. It does this by using a universal DS-1 unit (UDU), which derives four DS-1 circuits per plug-in. Universal in this context means that the DS-1 format supported by this unit is called for in the telecommunications standards. The shelf holds three low-speed groups of plug-ins, which derives 28 DS-1s per group. Thus, all 84 DS-1s of the OC-3 system could be extended. There is also a protection or electrical spare per group and a DS-1 protection switch unit per group.

The last shelf in the SONET DISC*S system is the channel shelf. Each channel shelf contains four digroups (A, B, C, and D) of 24 channels each. Thus, each channel shelf derives 96 channels. Up to seven channel shelves can be controlled by a single common shelf in the current configuration of this system. This means the maximum capacity is 672 DS-0 channels.

3.5.2 DSC Communications Litespan 2000
Another new generation of the digital loop carrier systems based on the TR303 guidelines is the DSC Communications Litespan 2000. Unlike the R-TEC

DISC*S system, the Litespan 2000 was initially designed to be optically fed via an integrated SONET multiplexer. The Litespan 2000 consists of two shelves and a fuse and alarm panel. The shelves include a common shelf for the system common units and a channel shelf, which houses channel units. Similar to the SONET DISC*S, the Litespan 2000 consists of a TR303-based digital loop carrier system mated with a SONET OC-3 multiplexer. Also similar to other TR303-based systems, only 28 of the 84 available DS-1s can be used for the digital loop carrier portion of the system. However, the remaining DS-1s can be extended either electrically or optically to other locations.

3.6   LINE TROUBLESHOOTING IN DIGITAL LOOP CARRIER CONFIGURATIONS

It became evident early on that dispatching maintenance personnel to trace and follow the customer's loop (2-wire) from the central office to the customer's premises was neither practical nor economical. Therefore, the operating companies devised ways to perform the bulk of the troubleshooting from the central office location. Dispatching maintenance personnel was done only after determining the location of the problem in the loop. The trouble can be on the customer's premises, somewhere in the loop, or in the office. Remote testing systems were developed to automatically test the line for particular conditions such as on-hook, off-hook, talkpath, short, and open lines.

These systems had sophisticated electronics to perform measurements on the line and report the failure condition and the location of the failure in the loop to the maintenance personnel via computer monitors. In addition to testing the lines when trouble was reported, this remote testing equipment were used to perform routine tests on the lines in the late night hours (when telephone calls are least likely to be made since it is not possible to use the telephone during testing).

When the digital loop carrier systems came along, it was necessary to make them compatible with the same systems used for testing the plain old 2-wire loops. This requirement placed a level of complexity on the design of the system. In this section, we will use the AT&T SLC 5 digital loop carrier system as an example to see how remote testing is performed.

To test the plain old telephone lines on the AT&T SLC 5, a system that was initially designed to work with the 2-wire copper loop and the SLC 96 is used. This system is called the *pair gain test controller (PGTC)*. This system interfaces with the central office terminal of the digital carrier system in the same manner of interfacing to the plain old 2-wire loop. Since the POTS channel units housed in the central office terminal, COT, side of the digital carrier system appear to the switch side like the plain 2-wire loop, then all the things the PGTC did with the plain loop can also be done for lines on the digital carrier system. The same conditions that the line was tested for in the plain 2-wire loop are applied to lines over the digital loop carrier system. Now let us see how this works.

PGTC

In the digital loop carrier system configuration, the complete loop to the customer's premises includes three sections:

1. From the central office to COT of the carrier system.
2. The digital loop carrier system itself, including the central office terminal (COT) and the remote terminal (RT) and the DS-1 lines connecting both terminals.
3. The connection between the customer telephone and the remote terminal.

Therefore, the primary goal of the PGTC is to isolate the trouble to the particular section in the loop. Furthermore, the PGTC must point to the location within the isolated section. Figure 3–15 shows the three sections of the loop and the PGTC connectivity to the digital carrier system. The PGTC, when commanded to perform the testing of the line, automatically runs a number of tests, including the following:

- COT channel unit test.
- COT–CO loop test.
- RT channel unit test.
- RT–telephone loop test.
- Permanent off-hook test.
- Permanent on-hook test.
- Channel loss test.
- Impedance measurements to determine the location of the fault within a particular section of the loop.

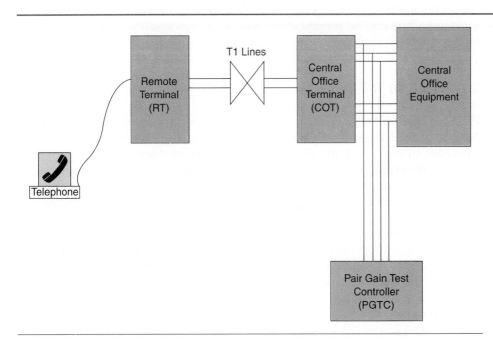

Figure 3–15
Connection of the PGTC to the digital loop carrier system.

On the COT channel unit side, analog signals are applied in a particular sequence. The COT hardware converts these conditions into a digital signal (bit-levels) in the signaling bits and transmits this pattern to the RT channel unit over the DS-0 assigned for that channel unit. The RT in turn performs maintenance tasks based on these digital bits. The end result is a reading or measurements seen by the PGTC. From these readings, the test is marked "pass" or "fail."

As a result of these tests, if for example a COT channel unit is found to be bad (failed), then it can be easily replaced. The maintenance person simply walks to the COT and makes the replacement since the COT and the central office equipment are colocated. However, if the fault was found in the RT channel unit, then a maintenance person has to be dispatched to the RT location to make the replacement. Furthermore, if the trouble is located in the customer premises (the PGTC is capable of concluding that from the series of tests it runs), then the dispatch is made to the customer's location. The advantage here is that, before the dispatch is made, the trouble is already known and the maintenance person merely makes a replacement or simple repair, if necessary.

XTC

For special services channel units, such as digital data communication units, used for 56-kbps digital data transfer between the customer premises and the central office, a more sophisticated system is used. This system is called the *extended test controller (XTC)*. This system performs all the tests that are performed by the PGTC as well as additional tests specifically designed for special services. The interface to the digital loop carrier system is similar to the PGTC interface with additional features designed for enhanced services. The XTC can perform bit error rate tests on the data service units. One of the enhanced features of the XTC is its ability to monitor the digital channel, meaning a line can be tapped digitally and a conversation can be listened to without being interrupted.

3.7   FIBER TO THE HOME

We have seen in the previous sections the effort made by the telephone company to digitize the subscriber's loop by using the digital loop carrier systems. However, in these systems we saw that digitization stops before the loop gets into the subscriber's home. The loop from the remote terminal to the home is the plain old copper loop. Issues arise with this copper loop between the remote terminal and the home when the subscriber demands higher bandwidth than that provided by the copper loop.

In this section, we consider three systems designed to address the "last mile" problem: Assuming a high-speed backbone network is available, how can residential or business customers be connected to it in a cost-effective manner? The optical subscriber loop systems connect customer premises to a central office with optical fiber. The proposed systems allocate a channel with a fixed bandwidth to each customer location. Typically, the bandwidth of the channel is 64 kbps, 192 kbps (the ISDN rate), or 1.54 Mbps (the T1 rate). The

AT&T SLC Series 5 was upgraded to use fiber to replace the copper loop from the RT to the subscriber's home. The British Telecom TPON system multiplexes and demultiplexes the signals optically. Other proposals include a wavelength-division multiplexing system called the *passive photonic loop,* which we will discuss later.

The telephone company has an economic justification for using fibers in the subscriber loop when fibers are less expensive than copper, or when new services require a larger bandwidth than that provided by copper. The cost of installing a fiber loop is comparable to that of a copper loop. Furthermore, optical loops are less expensive in the long run because they have a longer lifespan. As a result, local operating companies often deploy optical loops for new installations. Some are replacing old copper loops with fibers when maintenance is needed or anticipated. The use of fibers does entail additional cost for optical-to-electrical conversion. Also, as we will see later, because it is difficult to amplify optical signals, only a small *pair gain*—the number of subscriber lines per feeder line—can be achieved with an all-optical distribution system.

Some people believe that the installation of fiber will be justified by the anticipated new services that require high-speed access. Examples of these services are ISDN, TV distribution, home shopping, and library access. Working at home (telecommuting) and distance learning may contribute greatly to the demand for high-bandwidth services. Although the technology is now available for wide implementation of bringing fibers to the home, the business justification is still absent. This should put a damper on how fast fiber to the home will become a reality for the average consumer.

3.7.1   Fiber to the Home over AT&T SLC 5

In Section 3.4 we discussed the AT&T SLC 5 system. The system we discussed did not extend the digital signal all the way to the customer's home. Instead, TDM was used to convert the digital signal between the COT and the RT. At the RT, the system converted the digital signal into an analog voice signal suitable for transport over the copper loop between the RT and the subscriber's home. In the late 80s and early 90s, the SLC 5 was upgraded to convert the electrical signal at the RT into an optical signal that is transported over a fiber loop between the RT and the subscriber's home. The RT converts the optical signal from the subscriber's home into a digital signal that is multiplexed and then sent into the central office. (See Figure 3–16.) By contrast, the subscriber signals in the British Telecom system are multiplexed in the optical domain, saving equipment needed for electrical-to-optical conversion.

The AT&T SLC 5 system allocates a DS-1 (1.544 Mbps) channel to each customer location, which runs all the way from the RT to the customer's home. As a result, additional lines and ISDN services are easily provided to the customers.

3.7.2   The British Telecom System

Figure 3–17 depicts the British Telecom telephony on a passive optical network (TPON) system, which uses optical multiplexing to reduce the cost per customer. The TPON topology is a double star that attaches 128 (16 × 8) customers

Figure 3–16
Fiber to the home using AT&T Series 5 subscriber loop system.

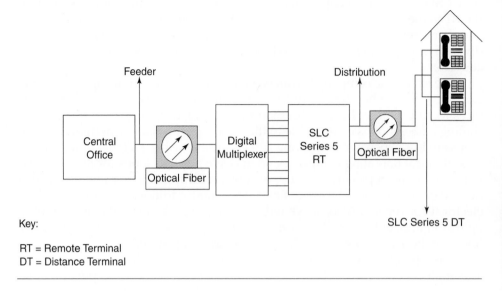

Key:

RT = Remote Terminal
DT = Distance Terminal

Figure 3–17
The British Telecom TPON system.

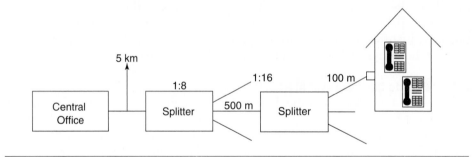

on each fiber from the central office. The two splitters in the figure are optical (contrast with Figure 3–16 where multiplexing and demultiplexing are done electronically, requiring electrical-to-optical conversion).

3.7.3 Passive Photonic Loop

The passive photonic loop (PPL) uses wave-division multiplexing (WDM) instead of the time-division multiplexing (TDM) discussed in Chapter 2. It allocates one pair of wavelengths of light to each subscriber. Thus, n subscribers need $2n$ different wavelengths. A subscriber transmits using one wavelength and receives using the other wavelength. A block diagram of the PPL system is shown in Figure 3–18. In the figure, subscriber 1 is allocated λn, $\lambda n + 1$, subscriber 2 is allocated wavelength $\lambda 2$, $\lambda n + 2$, and so on. The subscriber can modulate light of one wavelength to send information to others, and others

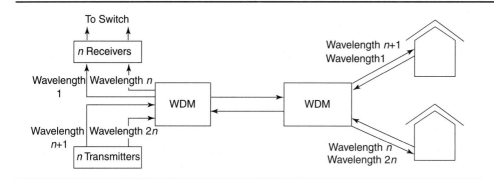

Figure 3–18
The PPL architecture allocates one pair of wavelengths of light to each subscriber.

can modulate light of a second wavelength to send information to the subscriber. There is a modulation bandwidth around each wavelength, which defines the maximum transmission speed. In this sense, WDM is similar to frequency-division multiplexing (FDM), which we discussed in Chapter 2. The bandwidth available to each subscriber is huge, on the order of GHz, although current technology limits utilization to a few Gbps.

The number of subscribers that PPL can accommodate is limited by the number of wavelengths that can be reliably differentiated using tunable lasers and optical filters. Current technology places an upper bound of 50 subscribers. PPL and other WDM-based networks exist today only as laboratory experiments. Many advances are needed before such networks will be commercially viable. However, WDM seems to be the only feasible modulation scheme that can effectively utilize the enormous bandwidth of optical fiber.

3.8   SUMMARY

This chapter extended Chapter 2's introduction of T1 and T3 into specific applications and systems as well as criteria for selecting specific systems. Among the systems discussed were SLC96 and its use in reducing the number of cables from customers, and the Series 5 enhancement of the SLC96. We discussed in detail the use of channel banks, including their various units, to interface between customers and the digital network.

The features and interfacing of PBXs were introduced. A substantial portion of this chapter was devoted to the digital access crossconnect switch (DACS) and its use with T3 systems. TR303, a newer digital loop carrier system, was carefully explained. Also, the SONET DISC*S, based on TR303, was introduced.

The chapter concluded with an extensive discussion of troubleshooting, the DSC Communications Litespan 2000, and fiber to the home.

REVIEW QUESTIONS
AND PROBLEMS

1. What are the criteria one must consider when purchasing T1 equipment?

2. Which criteria might you consider more important when purchasing T1 equipment: (a) ability to upgrade in the future to more advanced features, or (b) pricing?

3. Give three examples of equipment that is solely based on T1 digital transmissions.

4. Briefly describe the core components of T1 digital transmission equipment.

5. What are the differences between unipolar and bipolar digital signals? Describe why one format is preferred over the other in digital transmission.

6. List four of the features that a T1-based PBX might provide.

7. Is it possible to use a PBX as a multiplexer? Explain why.

8. T3 is rapidly becoming available: (a) What media can be used to transport T3 signals, and (b) What is the equipment that makes use of the T3 digital transmission format?

9. What are the problems one may encounter when troubleshooting a T3 digital transmission facility based on an M3 system?

10. Briefly describe the main differences between an electronic switching system and a digital access crossconnect switch (DACS).

11. Who might be the main users of a digital access cross-connect switch?

12. Show an example of how a DACS can be used in digital transmission and explain the reason it is used in the manner you describe.

13. Are there principle differences between a T3 DACS system and a T3 multiplexer system that has drop-and-insert capabilities?

14. Give three reasons why the operating companies went with digital loop carrier systems.

15. What are TR08 and TR303? Compare and contrast the two.

16. Give one example each of a DLC that uses TR08 and a DLC that uses TR303.

17. Describe the connectivity between the customer's premises and the central office using a digital loop carrier system.

18. Describe the conversions that take place from when a customer starts dialing a number until a call is disconnected in a successful telephone call over a digital loop carrier system.

19. What are the advantages of a digital loop carrier system implemented using the technical recommendations 303 versus those implemented using the technical recommendations 08?

20. What are the differences related to troubleshooting a line in a plain two-wire loop and a line in a digital carrier loop system configuration?

21. What are the advanced features of the XTC?

22. Describe how line troubleshooting is done over a digital carrier loop system using TR08 recommendations.

23. Is it more difficult to troubleshoot a line over TR303 systems than TR08 systems? Explain.

24. This problem challenges your general understanding of this chapter. Using the knowledge you gained so far:

a. Design a digital loop carrier system that uses a PC as a controller, and a number of off-the-shelf PC-compatible T1 interfaces and channel units to support 380 customers. Explain in your design how you are going to test the lines and how you are going to price the system. Is such a design using a PC feasible? Justify your answers.

b. In this project, would it make sense to use SONET? Why or why not? Keep in mind the limitations of the technologies you are working with.

4

Switching Systems

OBJECTIVES
Specific topics covered in
this chapter include the
following:

- The need for switching,
  types of switching sys-
  tems, and evolution of
  switching technology.

- The functions of
  switching systems.

- The concept and tech-
  nology of circuit-
  switching systems.

- The applications of circuit-switching systems
  for exchange switching
  and customer
  switching.

- The concept of packet
  switching compared
  with circuit switching
  and message
  switching.

- Packet switching
  technology.

- The concept of com-
  puter telephony
  integration.

- A call center setup and
  its industry-related
  applications.

4.1 INTRODUCTION

Networks use switching to achieve connectivity among users while
sharing communication links. Using switches, networks can establish
higher-capacity communication paths among users with fewer links and
at a lower per-user cost. In this way networks can take advantage of the
economics of scale in communication links.

When we view a switch from the outside as a black box, it appears as
a device with several input ports and output ports that terminate incoming
and outgoing links. Focusing on digital switching systems, an incoming
link carries multiplexed bit streams of several users. The switch guides
each of those bit streams to the appropriate output port. Networks differ in
the methods used to switch and multiplex signals. For example, the telephone network uses circuit switching and time-division multiplexing. On
the other hand, data networks use packet switching and statistical multiplexing. ATM uses fast-packet switching and statistical multiplexing.

In this chapter, we explain the operation and design of switches, con-
centrating on two types of switches: circuit switches and packet switches.
We start with the concept behind switching and the need for it. We then
study both circuit switching and packet switching. To illustrate the concept
of circuit switching in telephony, we study the architecture of a particular
implementation of a circuit switching system: the Definity private branch

exchange (PBX) switch by Lucent Technologies. We then proceed with the store-and-forward switching techniques, in particular datagram and virtual circuit connection methods. We close the chapter with a study of a computer telephony switching application in a business customer setting.

4.2 SWITCHING CONCEPTS

As we stated in Chapter 1, the earliest telephones were not connected by central office switches as they are today. Initially, every telephone was connected to every other telephone. Obviously, this could work only for very small networks. Over time, the concept of distributed (rather than totally centralized) switching systems evolved, which provides efficient and cost-effective access of every telephone to every other telephone in a network. Finally, the switches were interconnected to provide the nationwide telephone network envisioned by Alexander Graham Bell. Today, software control and distributed processing have given switches the power to support numerous customer services and provide a hint of what will come in the following years.

The need for switching is the economic trade-off among transmission, switching, and terminal equipment to provide telecommunication services. If we consider communication between two users without any switching, we need separate telephones and transmission links between any two users, as illustrated in Figure 4–1. For example, a two-subscriber system would require one transmission link and one telephone per subscriber, for a total of two telephones. A three-subscriber system requires three transmission links and two telephones per subscriber, for a total of six telephones.

Centralized switching is aimed at reducing the number of interconnecting links, as is conceptually illustrated in Figure 4–1. By introducing a centralized common switch point, the number of interconnecting links is reduced from $N(N-1) \div 2$ to N links between subscribers and the common switching point, as we discussed in Chapter 1. The same concept can be used to develop a network for switching systems. As you can see here, the trade-off is the addition of switching equipment, but the number of transmission links is reduced for the total system.

The nature of telecommunication service is that not everyone is likely to communicate at the same time. By taking advantage of the random nature of service demand, we can design switching systems economically by defining a given level or a grade of service. The random nature of service demands includes request for services and duration of services. Many factors influence what we refer to as the *call origination process*. By call origination process we mean how long a customer waits for a dial-tone, how many attempts a customer makes before succeeding in making a telephone call, and so on. The state of the system may have a strong influence on the call origination process. For example, heavy system overloads will result in repeated attempts. Most switching systems are designed based on the assumption that call origination

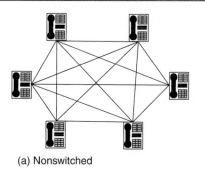

(a) Nonswitched

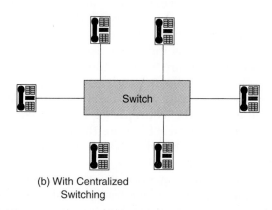

(b) With Centralized
Switching

Figure 4–1
The need for
switching.

is a random process when the population universe is viewed as a whole. A customer originates a call independently of other customers and independently of the state of the system. However, you can see that this assumption may break down under special situations.

The *duration of service* includes the dialing time, ringing time, and the conversation time. Conversation time is the largest element of holding time. Holding time is the interval measured from the moment a person picks up the telephone and dials a number until the call is disconnected. For voice communications, duration of service has general characteristics of exponential distribution. For data communications, the traffic has somewhat different characteristics. It generally:

duration of service

- Occurs in bursts.
- Has a variable holding time from short to very long.
- Is time sensitive to the information latency (the time the information. takes to travel from one point to another).

Therefore, in designing the switching system, types of services and applications are important considerations.

circuit switching, message switching, and packet switching

Generally there are three types of switching systems: *circuit switching*, *message switching*, and *packet switching*. In circuit switching, communication is provided by setting up a dedicated physical path (the circuit) between communicating customers. The connection exists between two communicating parties until they decide to terminate the connection. This system is used in the vast majority of voice-based networks today, regardless of whether they are digital or analog.

store-and-forward switching

Message switching is a *store-and-forward switching* technique in which customers exchange information by sending discrete messages. The message is transported as a whole. Message switching is designed for one-way delivery of messages and generally not for real-time interconnections. The telegram service is an example of message switching.

Packet switching is similar to message switching except that the complete message is not sent at one time. Messages are divided into segments called *packets* for transmission through the network. Each packet must have additional information, referred to as the *packet header,* to identify the packet and indicate its destination. Packets are treated individually, each forwarded along the best available path through the network at any given instant. At the destination, the individual packets are reassembled into the original complete message. Figure 4–2 illustrates the concepts of circuit switching and packet switching. Circuit and packet switching are the most commonly used techniques in telecommunication networks today. For this reason, we will emphasize these two types of switching methods.

In the case of circuit switching, once a connection (circuit) is established, the network is transparent, acting like a transport pipe. In the case of packet switching, the network must perform certain transactional processes. As illustrated in Figure 4–2b, user A's message is divided into packets X and Y. User C has a single packet Z. Packets X, Y, and Z are transmitted independently through the network. Acknowledgments are sent back for confirmation. Packets are retransmitted if an error occurs. Packets are reassembled at the receiving end and delivered to the user. In this example, user B receives a message from A and user D receives from C.

Packet switching was invented in the early 1960s to provide secure voice communications for the U.S. military since packetized voices cannot give useful information to someone tapping the transmission lines. In the late 1960s and early 1970s, the ARPANET was developed. In the 70s, 80s, and 90s, packet switching experienced a growth period due to computer technologies and the distribution of computer resources. As more information has to be transmitted at a higher speed, wideband packet switching will have many data applications.

Let us quickly refresh our memory with the call scenario described in Chapter 1. The simplest call sequence is between two telephones connected on the same switch. Walking through this simple scenario will illustrate the important aspects of centralized switching. A user originates a call by going off-hook,

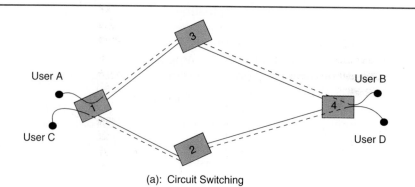

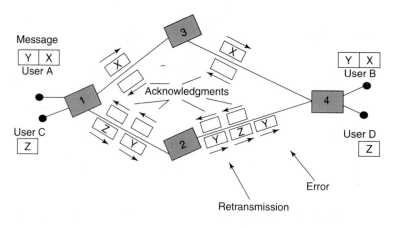

Figure 4–2
Types of switching techniques.

that is, by picking up the telephone handset. This change of line status is detected by the line circuit as a request for service, and an off-hook signaling message is sent to the switching system. The system checks the line class marks (i.e., the type of user terminal and type of signaling).

Assuming the terminal uses *dual-tone multi-frequency (DTMF)* for signaling, the system then finds a free DTMF receiver and sets up a communication path from the line to the receiver. A dial-tone is then placed on the line. When the dial-tone is heard, the user dials the digits (number) for the called party. As soon as the first digit is received, the dial-tone is removed. When sufficient digits are collected, the system examines the dialed number and routes the call to the terminating station.

If the called party is idle, the ringing signal is applied to the terminating user and the call progress signal (ring-back tone) is applied to the originating user. If the called party is not idle, a busy signal is applied to the originating

DTMF

user. The system monitors both originating and terminating users for abandonment and answer. If the called party answers, the system detects the answer, removes the ringing signal and the ring-back tone, and establishes a talking path between the originating and the terminating users.

Finally, when either the originating or terminating user hangs up (goes on-hook), the circuit connection is torn down.

From the preceding example, we see that a switching system must perform the following functions, which are taken by order of progress:

- *Attending.* When a subscriber raises his or her telephone handset to originate a call, the switch must be able to recognize this action. A similar attending function is also required from the trunk side for request of service from other exchanges.
- *Signal reception.* After the switching system responds to the request, it receives information to address the desired called station. This information is normally the numerical digits.
- *Path selection.* The switching system must select an idle link or a series of links within the system for connection purposes.
- *Route selection.* The switching system must determine the trunk group to which a path is to be established.
- *Busy testing.* This function determines whether a particular link or trunk is busy.
- *Path establishment.* This function controls and establishes the desired interconnection. This function normally requires some form of memory to remember the connection for the duration of the call.
- *Network interconnection.* This function physically provides the connection path.
- *Signal transmission.* On interoffice calls, address signals must be transmitted to the distant office.
- *Alerting.* This function informs the called station of the distant office that a call is being received. On calls to stations, this function is called *ringing;* on interoffice calls, this function transmits the attending signal.
- *Supervision.* This function detects when the connection is no longer needed and releases it.

4.3   CONCENTRATION IN CIRCUIT SWITCHING

We will now discuss the switching network. We begin by introducing the concepts of concentration, distribution, and expansion. The network in a switching system is the means by which any line or trunk may be connected to any other desired line or trunk in a telephone switching exchange. One of the basic facts of telephony is that most telephones are idle most of the time. The subscribers are not talking with all other subscribers at the same time and call origination is a random process. Therefore, most switching networks have been designed to provide full access for every customer to every other customer, but there is no need to ensure that every line and trunk can originate and terminate a call at the same time. Thus the concepts of concentration, distribution, and expansion for switching networks were developed. The different stages of concentration, distribution, and expansion make up the whole switching network.

    Concentration means that the number of inputs is greater than the number of outputs. Typically, it takes place between lines that originate calls to parts of the switching network. *Distribution* means that the number of inputs and outputs is equal. Typically, it takes place in the network between concentration and expansion. *Expansion* means that the number of inputs is smaller than the number of outputs. Typically, it takes place between parts of the network to the terminating lines. Concentration, expansion, and distribution can be arranged in a number of ways. A basic switching network example is shown in Figure 4–3. The four types of circuit-switching networks are: space division, time division, time multiplexed space division, and frequency division. We present these switching techniques in the proceeding sections.

concentration

distribution

expansion

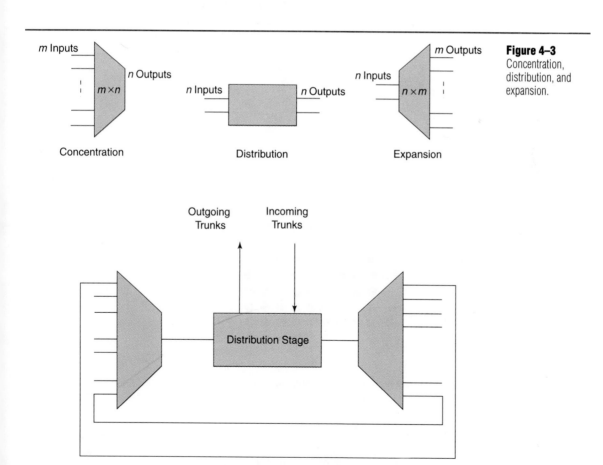

Figure 4–3
Concentration, distribution, and expansion.

4.4 CIRCUIT SWITCHING: SPACE-DIVISION SWITCHING

Space-division switching is an old technology that is being replaced by more sophisticated digital switched networks. For the sake of completeness and appreciation of the newer technology, we briefly describe this type of switching. The major characteristics of space-division networks include:

- A physical, spatial link is established through the network.
- Links are maintained for the duration of the information transfer.
- Signals transferred can be digital or analog.
- Metallic devices or semiconductor network elements are used.
- Networks may be made up of several stages.

Gross motion switches, such as step-by-step and panel switches, were used in early electromechanical systems. Fine motion switches, also known as *coordinate switches*, include crossbar switches and *ferrite* switches. Figure 4–4 il-

Figure 4–4
Space-division
switching.

Customer Lines

Output Link

A 4-point Rotary Switch

Gross Motion Switch

Inputs

Outputs

A 9-by-9 Coordinate Switch

Fine Motion Switch

Input

Control

Output

Electronic Switch

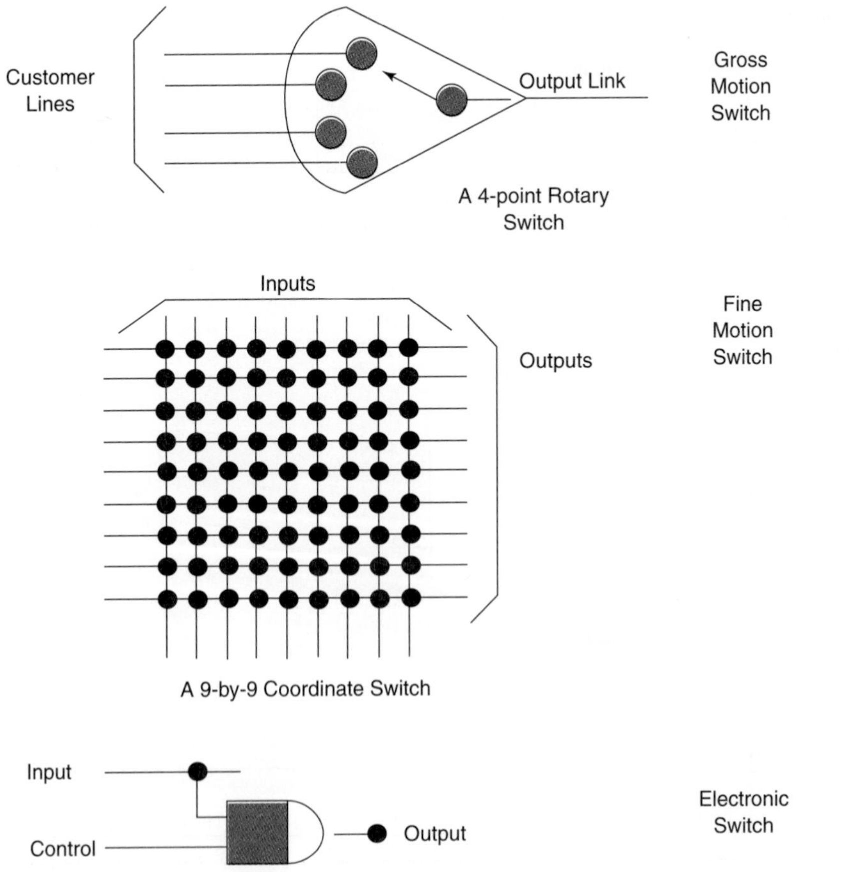

lustrates this concept. The top diagram is an example of a rotary switch typically used in step-by-step systems. Coordinate switches are typically arranged in a matrix format; therefore, switching networks are often referred to as *switching matrices.* Shown in the middle of the figure is a nine-by-nine square coordinate switch. The bottom diagram of the figure is an example of an electronic switch using the electronic AND gate. The electronic gates can be used to form a coordinate switch.

4.5 CIRCUIT SWITCHING: TIME-DIVISION MULTIPLEXING

We now discuss the concept of time-division switching networks and illustrate how this concept forms the basis for digital switching. A time-switching network is based on two major concepts. The first concept is time sharing of the transmission medium. As we saw in Chapter 2, the details of this idea of time slicing were presented when we discussed time-division multiplexing. The second concept is *time-division switching,* where the multiplexed information carried on the transmission medium must be transferred from input points to desired output points (see Figure 4–5).

time-division switching

4.5.1 Time Slot Interchanger

As shown in Figure 4–6, the *time slot interchanger (TSI)* can be thought of as having storage locations (stores or buffers) associated with specific time slots in the incoming TDM bit stream and the outgoing multiplexed bit stream for each direction of the transmission. A transfer (read-write) operation can change the order of the time slot by changing the order of the information from the input to the output stores. The transfer operation is determined by routing requirements and is directed by the call processing software, which is part of the switch architecture. The transfer instruction for a given time slot is set for the duration of a call on that time slot. A mirror transfer is performed for the other direction of transmission through the TSI. It is worth noting that an n-channel TSI is equivalent to an n-by-n space-division switch, because the n inputs have full nonblocking access to the outputs.

Figure 4–6 shows samples from four calls that are in time slots 1, 2, 3, and n of the input multiplexed bit stream being routed through a TSI. The binary information from that line is directed to location 1, 2, 3, and n of the input store. The routings for these calls have established that the calls should be moved to location $r, r + 1, 2,$ and 1, respectively, of the output store and thus, time slots $r, r + 1, 2,$ and 1 of the output stream.

In its simplest form, a time-division switching system may consist of one TSI, with n customers attached to the input side and the same n customers attached to the output side.

Time-multiplexed switching is generally used in digital switching systems. Figure 4–6 shows the concept of the time slot interchanger (TSI), and Figure 4–7 shows a time-multiplexed switch configuration during one time slot. Inputs 1, 3, and m are connected to output lines 3, 1, and m respectively. This type of switching network is time-multiplexed space-division switching,

Figure 4–5
Time-division
switching.

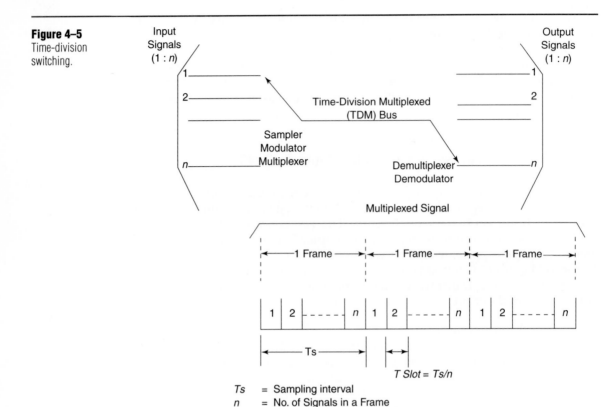

Figure 4–6
Time slot inter-
changer (TSI).

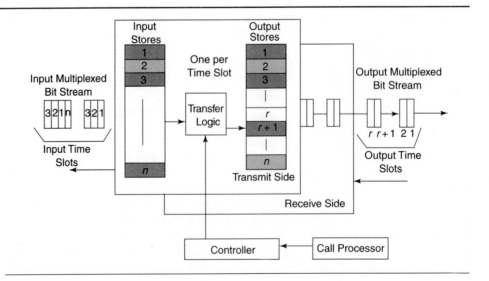

Figure 4–7
Time-multiplexed switch.

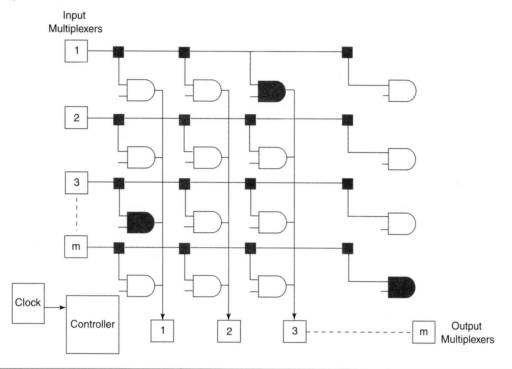

commonly referred to as the *time-multiplexed switch (TMS)*. It is a space-division network but is time-shared. The switch provides a communication path for each of the time slots of the multiplexed signal.

A digital switching network may consist of the time element only, the space element only, or a combination of time and space. The diagram in Figure 4–8 is an example of a time-space-time digital network. AT&T's 4ESS and 5ESS use this architecture. This datagram shows the connection of two communication channels, each of which is represented by a time slot at one of the TSIs. In this example, the transmission direction has been separated prior to arriving at the TSI. Samples from channel A arrive in time slot 3 on TSI A. Samples from channel B arrive in time slot 1 on TSI B. Each TSI has a mirror image (the shaded squares in the figure) that handles the reverse direction. To connect channels A and B, the control processor establishes a path between TSI A and TSI B during time slot 25 of the TMS. TSI A transfers channel A samples to TMS time slot buffer 25. During time slot 25, the TMS connects the output buffer of TSI A to the input buffer of TSI B. TSI B then transfers the samples to channel B's time slot, time slot 1. The other direction of the transmission path progresses similarly and simultaneously on other (shaded) halves of the TMS, using the same time slot of the TMS.

Figure 4–8
Time-space-time
switching.

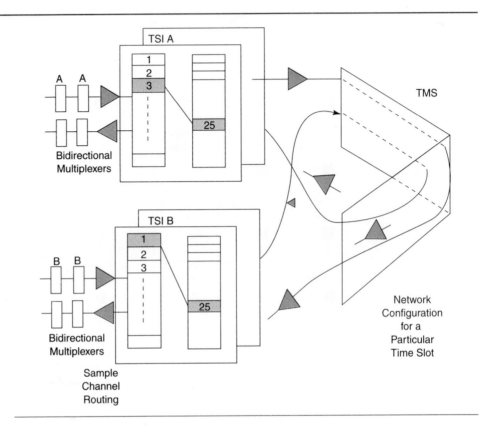

4.6 BASIC CIRCUIT-SWITCHING SYSTEMS

4.6.1 Functions

In order for a switching system to communicate with various types of lines and trunks, it must provide a variety of interfaces among inputs, outputs, and the switching system. Before we describe the various types of interfaces, we must first define the basic functions every switching system must provide. These functions may be characterized by the tasks required in the interface circuit as the following:

- *Battery feed* provides a power source to the line to allow detection of station (telephone) status and power for station signaling and transmission. If you recall, signaling includes going off-hook, on-hook, and so on. This battery feed is usually 48 volts.
- *Overvoltage protection* must be provided to prevent damage to sensitive circuits when the system is subjected to high voltage, such as lightning or power line crosses.
- *Ringing signals* must be applied when required for alerting purposes.

- *Supervision* includes functions such as monitoring the status of the line (e.g., on-hook or off-hook).
- *Codec* provides analog-to-digital and digital-to-analog conversion. This function is necessary for digital switching systems as we saw in Chapter 2.
- *Hybrid circuits* separate the incoming and outgoing signals going to or issuing from, respectively, the receiver or transmitter.
- *Testing* provides detection of faults on the lines and trunks. As discussed in Chapter 3, testing can be performed periodically or on demand. It is all done at the switching system site.

These functions are referred to as the *BORSCHT* circuit. (The mnemonic is traceable to J.E. Iwerson of AT&T Bell Labs.) These functions are common for all interface circuits. Specific functions depend on the type of interfaces, which is the subject of our next discussion.   BORSCHT

4.6.2   Station Line Interfaces

Station line interfaces are the interfaces provided to the customers to attach their equipment. They include analog and digital voice terminals, hybrid voice circuits, data lines, and *integrated services digital networks (ISDN)* lines. The analog voice terminal (telephone) interface includes the traditional telephones where the interface is often called a *tipring* interface. The digital voice terminal interfaces are used in digital switches to provide digitized voice transmission. Data line circuits are used to interface with data terminals, especially for digital switches. The ISDN *basic rate interface (BRI)* includes two bearer channels and can be used for voice, data, image transmission, inexpensive video conferencing, high-speed Internet access, and many of the multimedia networking applications geared toward high-end consumers. All of these interfaces are designed to connect the switching system to the subscriber's line and, depending on the service requirements, the appropriate interface is used to make the connection.   ISDN   BRI

4.6.3   Trunk Interfaces

Trunk interface circuits are provided to interface with various types of trunks. Trunks are the transmission links that interconnect switching systems to provide the overall hierarchical network. Trunk interfaces are classified based on the connectivity they provide and include the following: trunk interface circuits

- *Central Office trunks* may include loop-start or ground-start trunks. Loop start and ground start are methods used to alert the switch of a service request (for example, going off-hook).
- *Direct inward dialing (DID)* trunks are used to connect a central office switching system to PBX or centrex services. These trunks provide calling in one direction only, from the central office switch to the PBX or the centrex system.

- *Digital and analog tie trunks* are used to interconnect PBX systems to form a private network for large corporate customer use. A T1 facility can be used as a tie trunk to provide digital connectivity between two PBX systems.
- *Auxiliary trunks* are used for special services within a PBX system, such as paging, chiming, or music on hold. You have probably encountered paging and chiming in malls, hospitals, and large department stores.
- *Local area network interfaces* may provide connectivity between a switching system and a private local area network. This interface bridges the local area networks (LAN) and *wide area networks (WAN)*.
- *ISDN primary rate interface (PRI)*.

4.6.4 Service Circuits

Service circuits are used as common resources to support switch functions. They include:

- *Tone detector circuits* are used to collect dialed digits and to detect call progress tones and modem answer-back tones.
- *Tone clock circuits* apply call progress tones, touch-tone (DTMF), answer-back tones, and clock signals for synchronization in digital transmission.
- *Pooled modem* is used to interface digital switching systems (on a shared basis) with outside analog facilities. Modems such as these may be used to provide dial-up access to remote computer hosts (for example, a student dialing into a university computer from home).
- *Recorded announcements interface* is a circuit that holds digitized speech to play back specific messages related to the status of a call. For example, if you dial a telephone number that has been changed, you may get an announcement stating that the number has been changed and a new number may be given. Other uses of these circuits are in voice mail, automated attendant services, automated calls made by computers, and many other similar services.
- *Recording* for traffic measurements and billing interfaces is used to collect call traffic data and billing information. Call traffic data are used by the switching system to devise a particular call routing pattern based on the busy-hour call rate.

4.7 CIRCUIT-SWITCHING SOFTWARE

Software plays a very important role in a circuit-switching system. It provides the call processing, administration, and maintenance functions in addition to many of the advanced features. A switching software package generally consists of a large number of programs, routines, and databases. The supervisory programs, which schedule and supervise the execution of other programs, are generally referred to as *operating systems.* The collection or aggregate of many programs is referred to as the *generic program.*

In a basic call example, the call processing software is required to provide communication paths between customers on request. We will use the basic interoffice call example from Chapter 1 (as shown in Figure 4–9) to illustrate some of the basic functions of a circuit-switching system. In this basic call, the system involves the following seven steps:

1. Origination—detecting the request for service.
2. Digit collection.
3. Digit translation—interpreting the dialed digits and translating them into systems addresses.
4. *Routing*—establishing the connection for a talking path.
5. Alerting—applying the ringing signal toward the called party.
6. Answer.
7. Drop—disconnecting the calling and called parties.

routing

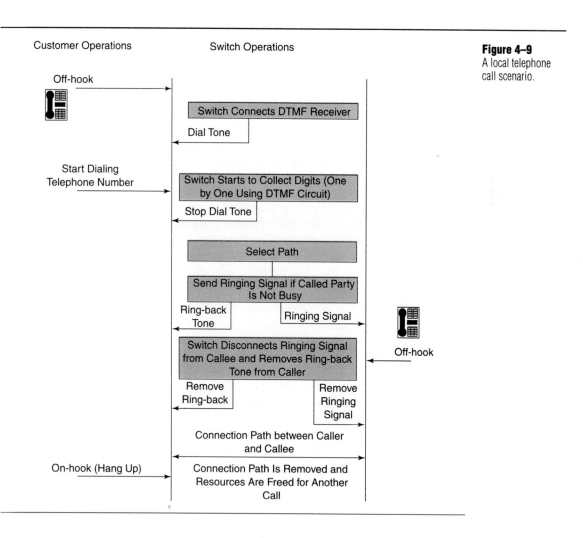

Figure 4–9
A local telephone call scenario.

Other services such as interoffice calling or custom calling features are extensions of these basic functions.

System administration and management are a software design consideration besides call processing. System administration and management is an area that is important to the operation of the switching system. This part of the software makes up a large percentage of the total software required to operate a circuit-switching system. Following are some of the features handled by this type of applications software:

- Installation—A critical component in a system.
- Testing—A complex procedure that needs to be designed carefully to reduce downtime for the telephone company, subscribers, and business customers.
- Traffic engineering—An area that contributes to efficiency and maximizes the economic benefits derived from a switching system
- Services changes and rearrangement—A continuous process that deals with subscribers' movement, system upgrades, and so on.
- Billing—What can be more important than collecting the money? Accurate billing and call records are some of the things that the operating companies take pride in.
- Feature selection—Which features to turn on and off in a switching system is another part of the software that must be dealt with during design stages.
- Database management—A very important component of the software needed for operations and services and many other functions.

Maintenance software provides reliable service, which is one of the most important features of a switching system. Modern switching systems have extensive software built in for maintenance purposes. The maintenance software generally is a collection of maintenance activities under the control of the operating system. The objective is to detect, report, and clear problems as quickly as possible and with minimum disruption of normal service. The major functions of maintenance software can be grouped into the following categories:

- Fault recognition.
- Fault recovery.
- System integrity.
- Auditing.
- System recovery.

The fault recognition and fault recovery functions normally work together. For example, if a hardware failure or a software bug is encountered, the fault recognition program must identify the nature of the problem. The fault recovery program then remedies the situation. Faults are detected by routine tests, by continuous checks, or by tests performed during the processing of calls.

System integrity software detects and responds to software faults to minimize the impact of such faults on system performance. It is responsible for detecting and handling system overload. Auditing is a technique that checks internal consistency between the states of hardware and software system resources. System recovery is used to bring the system back when it encounters a major systemwide problem. The system recovery is designed in levels, beginning with the lowest level and escalating to higher levels. The worst case, or highest level, may be a total reload of the program from permanent storage.

There are other important considerations to the maintenance software. For example, human-machine software provides communication between the person responsible for operating and managing a switching system and the switch to access administration and maintenance procedures. Also important are *remote* maintenance capabilities, which include dial-up connection to a centralized maintenance center and automatic alarming. In such a setup, the switch automatically reports its ailing states to a remote maintenance center. The software typically keeps a history or log of errors and alarms in the system.

4.8 CIRCUIT-SWITCHING APPLICATIONS

We now discuss the applications of circuit-switching systems. Switching systems are designed for specific applications. Factors taken into consideration include the type of users and traffic patterns. Generally, switching systems can be classified as:

- Local exchange switching.
- Trunk switching.
- Gateway switching.
- Customer switching.

Local exchange switching systems are designed primarily to serve the general public. Telephone lines for residential and business users are connected to the switching system and other local exchanges or trunk exchanges. *Trunk exchange switching systems* are used to connect trunks; that is, no end users are connected directly to the trunk switching systems. There are various names for trunk switching systems, such as *tandem switches, transient switches,* and *toll switches.* Since there are no station lines, the incoming and outgoing traffic pattern is quite different from that of the local exchange switches.

> local exchange switching systems, and trunk exchange switching systems

Gateway switching systems are used to route traffic between national and international calls. They provide interface functions to accommodate different signaling formats between different countries. They also provide international operators' assistance services. The gateways are also points on which billing for international calls are based. *Customer switching systems* are similar to local exchange switching except that they are designed specifically for business users. Therefore, user features and system management features, as well as costs, are primary concerns. These systems are known as *private business exchanges (PBXs).*

> gateway switching systems

> customer switching systems

4.8.1 Call Translation in a Local Exchange Setting

Translation is the process, in any software-controlled circuit-switching system, that determines how a call should be handled by examining available input information. Three-digit translation, in local exchanges, examines the first three digits registered (the local exchange code). The output information from the translation process may include any of the following:

- Outgoing trunk location on a switching machine.
- Required digit manipulation before outputting.
- Next-step information in case the chosen trunk is not available.
- Location of additional translation information when further translation is necessary.

The example shown in Figure 4–10 shows the translation process at exchange 464:

1. The first three digits (372) are examined.
2. A translation takes place to select a trunk that can complete the call (which connects to the tandem switch).

Figure 4–10
Local exchange translation of a local call.

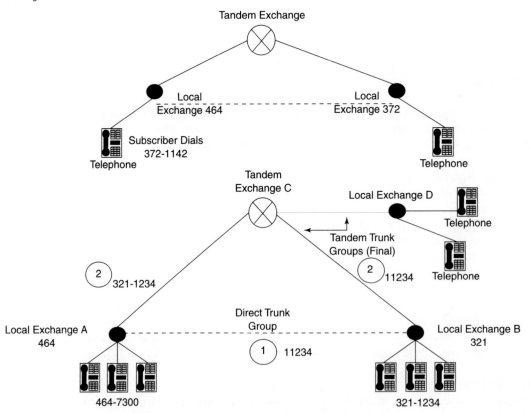

Code conversion is the process in which the received address information has digits added, deleted, or substituted to process the call over the route selected. The amount of processing required depends on the switching exchange functions. How this process is accomplished is a function of the switching machine design. In the example, a call is made from one exchange to another, but the call may be routed through a tandem exchange. The following sequence of events occurs (see lower left of Figure 14–10):

1. The customer at local exchange A (464-7300) dials 321-1234.
2. When 321 has been stored in the local exchange A, a translation indicates that the call can be completed over the direct trunk group to exchange B.
3. *Scenario 1:* A trunk is available in the group direct to exchange B.
   - A trunk in this group is seized.
   - The first two digits of the seven-digit code are deleted and five digits, 11234, are sent to local exchange B. The five digits allow groups of 10,000 customers to be identified within exchange 321.
4. *Scenario 2:* No trunks are available in the group direct to exchange B.
   - A second translation indicates that the trunk group to tandem exchange C is the next choice.
   - A trunk in this group is seized and 321-1234 is output to tandem exchange C.
   - The initial translation in exchange C examines the digits 321 and selects the trunk group to the 321 exchange at B.
   - The first two digits of the seven-digit code, 32, are deleted and five digits, 11234, are sent to local exchange B.

4.8.2 Trunk-Switching Evolution

We learned in the preceding section that completing a call may involve switching system types other than the local office exchange. One such switching system type is referred to as *trunk-switching systems.* The earliest trunk exchange was the step-by-step switching system. Although it was invented in 1889 by A. B. Strowger, it was not used extensively by the Bell system until the 1920s. This system had a stage of switching associated with each dialed digit. As the customer dialed, successive stages of switching responded to the dial pulses that represented successive dialed digits and progressively selected a path through the network. This system was referred to as *direct progressive control.* The dialed number was not reusable because dialed digits were not saved after they were used to determine the position of a switching device.

The crossbar Tandem (1930s) and N. 4A Crossbar (1943) were electromechanical devices that used the principle of common control. These devices used control equipment only to establish the path. The dialed digits were stored on a register. Logic associated with the register was then used to make the connection. Once the connection was made, the register was cleared and could be used to serve other originating calls. The 4 ESS (1976) and the 5ESS (1982) switching systems developed by AT&T are electronic switching systems using stored program control, where processors are used to execute software programs that initiate sequencing operations to establish a calling path. New services are implemented by making changes to the stored program. The 4 ESS is a large-capacity, four-wire tandem system for trunk-to-trunk interconnection, which uses a time-division/space-division digital switching network. The 5 ESS is a digital time-division switching system designed for modular growth, which can accommodate both local and trunk switching. We will see later in this chapter the architecture of the 5 ESS switch.

4.8.3 Gateway Switching Systems

The international gateway provides access between countries by connecting national and international networks with different operating parameters. The gateway must provide the conversion between the national and international signaling formats.

Due to impedance mismatches in the communication path (at junctions of different gauge cables, at intervening switch junctions, at the two-wire to four-wire junctions, etc.), echoes are generated and returned to the caller. On a long route, as in an international call, echoes occur after a long propagation delay. Delay is proportional to the length of the communication path. Echo return is quite annoying to the caller when delay exceeds 45 ms. A 45-ms delay is equivalent to a path length of about 2400 km. Echo is controlled at the gateway by using a digital echo canceler to eliminate it.

Revenue derived from international traffic is accounted for at the gateway. Automatic and manual capabilities to detect and correct faults arising from hardware failures, software defects, and personnel procedure errors are provided in the gateway as basic operations, administration, and maintenance.

4.9 CUSTOMER SWITCHING

Let us start by asking the question: What is customer switching? Customer switching describes special arrangements permitting flexibility in providing telecommunications services that can be tailored to customers' specific needs in terms of capabilities and features. It includes both voice and data communications. Customer switching can be provided either by special systems that are owned or leased by the customers (usually business customers) and placed in customers' premises, or from the local exchange. In the former case, a typical example of such equipment is referred to as the *private business exchange (PBX)* and the *key telephone systems (KTS)*. In the latter case, it is referred to as the *Centrex service*. There are advantages and disadvantages of each and a trade-off must be made when deciding on a particular system or service.

KTS

Although the specific communications needs of business customers depend on the size and nature of their business, most requirements can be divided into the following broad categories:

- Interlocation calling.
- Incoming calls.
- Outgoing calls.

- Communications systems and management.
- Office automation and management.

Many businesses need to communicate between stations on the same premises. For voice communications, this means calls between people in different offices. For data communications, this may involve communications between a terminal and a host computer or another terminal on the same premises. Incoming calls are important to business customers, because they often represent new or additional business opportunities. It is essential to have enough facilities and attendants to ensure that incoming calls are promptly answered and efficiently handled. Control of outgoing call permissions from selected stations is important in controlling telecommunications costs, as well as routing calls over the most cost-effective facilities. It is also important to make those communications as convenient and friendly as possible, which in turn will significantly improve work efficiency. Station equipment with button-operated features and switching systems with automatic processing routines meet these needs.

Communications systems management involves various needs. Since communications can be a major expense of a business, modern customer switching systems often also provide communications management features to handle administration and maintenance. Office automation, on the other hand, involves improvement of office efficiency, such as voice and text message processing capabilities and directory services.

Types of customer switching include:

- Key telephone systems.
- Private business exchanges.
- Hybrid systems.
- Centrex services.
- Special customer switching systems.
  a. Automatic call distributor.
  b. Telephone answering systems.

We will examine the features and operation of these systems in the following sections.

4.9.1 Key Telephone Systems

Key telephone systems use the concept that each business customer may subscribe several lines from the local exchange, but that individual users need flexibility to access and control particular lines from the user terminal (telephone). See Figure 4–11 for a diagram of a typical key system. For example, by pressing a button, the telephone closes a switching point and connects to a particular line. A key telephone system is a user-controlled switching system that uses keys or buttons on the station sets to access the central office (or PBX) lines or special features. The term *key telephone* is a carryover from the early days of telephony and telegraphy, when mechanical arrangements used to close a circuit were called *keys.* Similarly, since the buttons on a key telephone were designed to close electrical contact, they too were considered keys.

Figure 4–11
Key telephone
system.

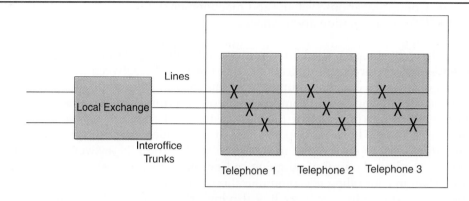

X: Switch Point

A key telephone system is basically an arrangement that permits the station user to operate the switches (or keys) to perform the following functions:

- Multi-line access.
- Hold.
- Lamp signaling.
- Common audible ringing.
- Intercom.

Multi-line access represents that most basic concept of letting one telephone user answer or originate calls on more than one line. The line to be used is selected by a switch (or key). A group of telephone users must be able to share two or more lines. The user must be able to *hold* a call in progress on one line while another call is answered or initiated on a different line. By pressing one of the buttons on the station set, the station user receives a dial-tone directly from the central office. In addition to multiline and hold, lamp signaling is used to inform the user of the status of each line, such as line pickup, hold, busy, or ringing. All lines share the audible ringing function for alerting progress. Other features, such as intercom and paging, are also implemented in key telephone systems.

4.9.2   Private Branch Exchange (PBX)

Private branch exchanges are switching systems owned (or leased) and operated on customer premises. Thus, it is remote from the local exchange and is dedicated to the private use of that customer. That is why the name is referred to as *private branch* or *private business exchange (PBX)*. Station users are connected to the PBX and the PBX is connected to the local exchange via trunks or a digital transmission facility such as T1 lines as shown in Figure 4–12. In the telecommunication network hierarchy, PBX is classified as customer premises equipment (CPE) below the level of the local exchange. A PBX provides the following basic functions:

- Station-to-station calling capability on customer premises, with abbreviated dialing plan.
- Concentration of customer stations to central office trunks.
- Attendant handling of incoming calls from central office trunks.

PBXs have evolved from simple manual systems to advanced stored-program controlled digital systems, but the basic functions remain essentially the same. Aside from the basic functions, today's PBX systems provide a host of enhanced features and capabilities. In the following subsection we will examine one of these enhanced features.

Automatic Call Distributor (ACD)   *Automatic call distributor (ACD)* is used to automatically switch large volumes of incoming calls to attendant (answering) positions. ACD applications include directory assistance, intercept operators, airline reservations, and department store catalog departments. ACD has the following characteristics: ACD

- Calls are distributed among attendants to maximize the efficiency of the attendant group.
- When all agents are busy, incoming calls are queued and answered in their order of arrival ("Please stay on the line, your call is important to us").
- Management information data are collected to administer both switching facilities and the size of the agent group.

4.9.3   Example of a PBX System:
Lucent Technologies' Definity PBX

Lucent Technologies' Definity PBX is a switching system that evolved from earlier AT&T systems, system 75 and system 85. The Definity system is based on a modular design allowing for scaling in features and pricing based on customer needs. Scaling allows the system to be used in midsize and large businesses. In a large size configuration, many Definity systems can be networked using T1 digital facilities to provide a larger system with distributed attendant services and *feature transparency.* By feature transparency we mean a user does not know which PBX in the network is providing the service. The following features of the Definity system are referred to as *core features:* feature transparency

- Call conferencing—Up to six parties can talk simultaneously.
- Call forward—A user can forward incoming calls to particular stations.
- Hold.
- Transfer.
- *Hunt groups*—A number of users can be configured in a group to answer incoming calls in a round-robin manner based on their availability. hunt groups

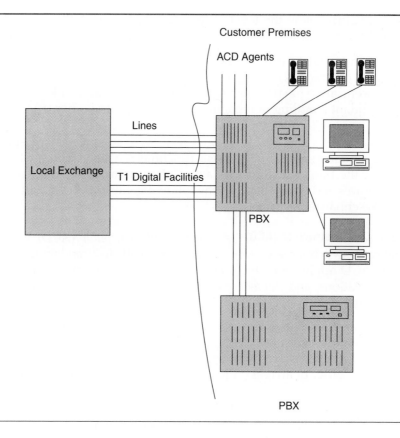

Figure 4–12
Private branch
exchange.

- Attendant services—Primary attendant and night service attendants can be configured such that day calls are handled by a primary attendant and night calls are routed to the night service attendant, which may be located in a different building on the customer premises.
- Digital, analog, and ISDN station types support—The system provides for different types of stations. The attendant console is an example of special station types having many features that the system must support.
- *System administration terminal (SAT)* interface—To allow the system operator to administer, maintain, and configure the system. We describe this interface in more detail later in this section.

SAT

Architecture of the Definity PBX System   The Definity system architecture includes two equally important components, the software and hardware. The system uses time-division switching to handle the switching functions. Figure 4–13 shows the architecture of the system. As can be seen in Figure 4–13, the system uses a 256-time slot TDM bus. Connected to the bus are two types of interfaces: the switching processor element and the angels. The switching processor element includes an embedded controller (Intel 80x86 processor), a maintenance board, memory boards for the stored program, and a flash memory card interface or a tape drive to load the system operating software and system configuration, called *system translation.*

The angel interfaces are also known as the *port boards.* They are the line interfaces with each interface supporting a particular type of line. The port boards include: analog lines, digital lines, ISDN lines, DS-1, announcements, central office trunks, analog tie trunks, direct inward dialing (DID) trunks, and tone/clock boards for tones and the collection of dialed digits. The system accepts a mix of these port boards based on the particular customer configuration. For instance, the mix outlined below is an example of a configuration that a midsize customer may require:

- 100 16-port analog boards supporting 1600 analog lines.
- 10 8-port digital boards supporting 80 digital lines.
- 2 8-port ISDN boards supporting 16 ISDN-BRI stations.
- 25 8-port central office (CO) trunk boards supporting 200 CO lines to handle incoming calling traffic.
- 10 DS-1 boards used to interconnect to another Definity system residing in a nearby customer building.
- 20 tone clock boards.

Intrasystem Communication   Communications between the angels (port boards) and the archangel (the switching processor element) are carried out over five dedicated time slots of the TDM bus. Those time slots are referred to as the

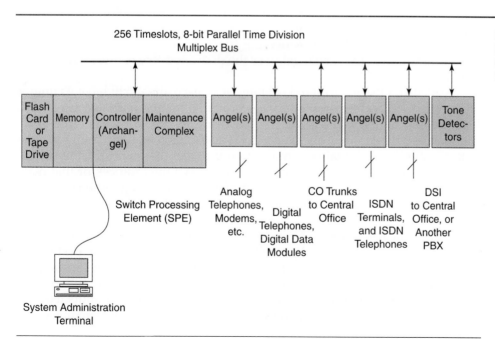

Figure 4–13
Basic architecture of a Lucent Technologies' PBX.

common channel control messaging (CCCM) interface. A set of uplink and downlink messages are defined for communication between each angel and the controller, or the archangel. The system has only one archangel and many angels. The downlink and uplink message set varies depending on the type of angel (port board) being serviced. Angels cannot talk to the archangel at any time they choose; instead, they have to wait their turns when they are polled to put their uplink messages on the TDM bus. Let us now examine an off-hook scenario on an analog port board.

The user lifts the telephone handset. The analog port board detects the off-hook condition. A microcontroller on the analog port board formats an uplink message indicating the off-hook condition, waits its turn on the TDM bus (dedicated 5 control channels), then puts in its request to the archangel. The archangel determines the availability of a talk time slot (one of the other 251 time slots) and the availability of a DTMF receiver to apply a dial-tone to the user. The archangel then sends a downlink message to the angel telling it which time slot it should listen to, so the dial-tone can be applied to the user. The angel then makes the appropriate path between the user's station and the DTMF receiver. The call then progresses following the same manner—uplink messages are exchanged between the angel and the archangel until the call is completed. This simple call scenario gives the idea of call processing operations.

Architecture of the Angel   Regardless of the service(s) an angel provides, it must have a common angel interface to be able to talk on the TDM bus. The common angel interface consists of two main components, the sanity and control interface referred to as the *SAKI,* and the common channel control interface. The SAKI provides functions related to the health of the board, reporting to the archangel on the status of the health of the board. The common channel control provides the functions necessary for the angel to communicate with the archangel over the dedicated five control time slots of the TDM bus.

Aside from these two standard interfaces, each angel will have its unique architecture depending on the services provided by that particular angel. For instance, the digital line's angel has hardware to provide the interfaces between a digital telephone and the angel. The digital telephone communicates with the angel via digital communication protocol that is proprietary to Lucent Technologies. The digital line angel also has a microprocessor that manages the operations of the board, its interfaces with the digital lines, maintenance of the board, and communication with the archangel.

Power Supplies and Ringing   As in all PBX systems, dial-tone, battery feed to the user's lines, and dialed digit collections are provided by the customer's switch and not by the central office; the Definity switch provides the signals necessary for these functions. The system has power supply and ringing generator units common to the system. The signals from these units feed the port boards with the voltage necessary for their operation, the ringing signal to ring users, stations, and the battery feed needed to operate the user's lines when applicable. The system has a battery back-up system to keep it in operation for the duration of time when a power failure occurs.

System Administration Terminal One of the good features of Lucent Technologies' Definity PBX is its system administration terminal. Through the system administration terminal, the person operating and maintaining the switch can configure new lines, install system software, translate stations and trunks, troubleshoot the system, perform tests on particular line interfaces, and so on.

The user interface that the system administration terminal uses is a form-like interface. For instance, if a new station is to be added to the system for a new user, the craftsperson enters the command *add station.* A station template form with each field on the form has a corresponding blank space for the craftsperson to enter the appropriate values. Examples of such fields include station type, station extension, level of outside access permission, button assignment on the station, and so on. In a typical PBX, hundreds of those forms are associated with a particular system administration command.

Samples of the system administration commands include:

- Add station—to configure a new station.
- Change station—to reconfigure a particular station.
- Display station—to show current station parameters.
- Test station—to perform on-demand testing of a particular station.
- Test trunk—to perform on-demand testing of a trunk group.
- Test board—to perform on-demand testing of a port board.
- Test port—to perform on-demand testing of a port on a port board.
- List configuration—to display the software version and the system release information.

Those are a few of the hundreds of commands available to the person involved with operation and administration of the system.

Troubleshooting The system administration terminal is the first-level tool available for troubleshooting and diagnosing problems with the system. When the system administration terminal cannot be used to diagnose the problem, other tools, such as protocol monitoring equipment, may be used for further troubleshooting. Laptop PCs with the proper hardware are sometimes used by field engineers to take traces of the communications over the internal TDM bus to determine the offending component in the system.

When using the system administration for troubleshooting, sets of maintenance commands are available to the user, including test board, test port, test DS1, test trunk, test link, display software errors, display hardware errors, etc. When on-demand testing is performed on a port board, the test results are tabulated and displayed on a form showing the test number and the associated status of each test. Fail is displayed for failed tests, pass for passed tests, and abort for tests that cannot be performed because the resource is being used in a call. If a test fails, an error code is displayed next to it. The maintenance person can then consult the maintenance documentation to interpret the failure and determine the necessary remedies. As with any complex system, troubleshooting requires patience and common sense. The maintenance person should always start from an end-to-end approach, going down the line with fault isolation until the failure is determined.

4.9.4   Circuit Switching: One More Word

Circuit-switching networks have a number of advantages and disadvantages. Circuit-switched connections provide a dedicated path, thus no extra delay as information flows through the network. These connections are optimal for voice or interactive/real-time data calls. Constant data transmission is possible over these dedicated connections and no extra overhead is associated with transmission after the call setup.

The dedicated nature of the circuit-switched connection means that the user is billed for resources that are allocated, even when not in active use by the customer. This system is possible because a resource used for one customer cannot be shared by another. If the user's data application requires that many short data calls be established in succession, it is possible that call setup overhead will be significant compared to the duration of the data call.

Finally, circuit-switching networks usually provide only a communication path between users. Therefore, code conversion, speed conversion, and error control must be provided by the end users (i.e., a circuit-switched network is transparent to the user data).

4.10   STORE-AND-FORWARD SWITCHING

Store-and-forward switching is an alternative technique to circuit switching. A store-and-forward network, which consists of data terminal equipment (DTE) interconnected by a communication subnetwork, is now examined. The communications subnetwork shown in Figure 4–14 consists of a set of point-to-point channels connecting switching elements called *nodes*. A point-to-point subnetwork (subnet for short) accepts input messages from users or other nodes and stores them until an output channel that takes the information toward the correct destination is available. Such a network is said to be a *store-and-forward system*.

The nodes in a store-and-forward network provide a number of functions, one of which is routing. A node must determine which user or node in the subnet is meant to receive the data, and it must switch the data to the appropriate device.

Nodes in store-and-forward networks are concentrator-type devices. That is, the nodes are statistical multiplexers with large buffers. These devices allow data from many input lines to be multiplexed together into a few output lines. This process will almost certainly cause some delay in the

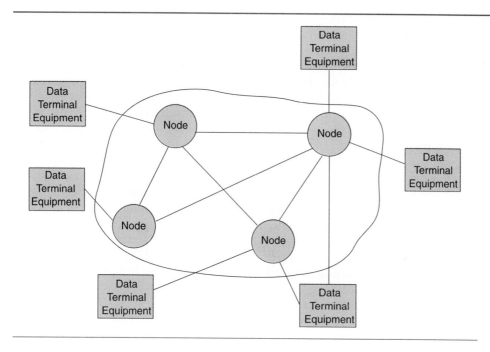

Figure 4–14
Architecture of a store-and-forward switching network.

transmission of the data. Message switching or packet switching is commonly used in data networks. Figure 4–15 is a diagram illustrating the concept of switching at the node.

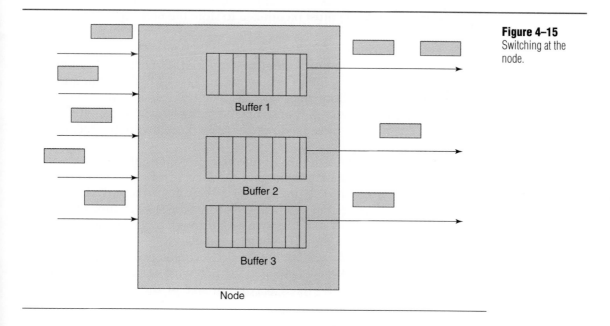

Figure 4–15
Switching at the node.

4.10.1   Message Switching

Data communications are typically composed of the exchange of entities called *messages*. Telegrams, electronic mail files, and data files are all examples of messages. In message-switching networks, a user assembles a message and passes it to the first node in the subnetwork. Each node stores the message, then passes it through the network until the message is delivered to the destination.

Message switching is a very effective communication strategy for a number of reasons. First, no delay is associated with setting up a call. Since resources are not allocated, a route does not have to be preestablished. Second, since resources are not dedicated to specific connections, there may be more active users than physical channels, due to the possibility of resource sharing. Third, the network can easily deliver one message to multiple destinations.

A message-switching network plays more of an active role in data delivery than a circuit-switched network does. Unlike a circuit-switching network, a message-switching network can provide additional functions, such as error detection and correction, protocol conversion, and flow control. Furthermore, a message-switching network can store a message for later delivery to a host that is busy or down.

There are also major disadvantages to message switching, however. First, the delay associated with the delivery of messages makes this type of network inappropriate for interactive or real-time applications. A second disadvantage is the lack of message-length definition. That is, messages may vary in length from a few octets to several gigabytes. Although message length will be constrained by the protocol implemented in the host systems, the network implementation must provide flexibility for all users—after all, we design networks to serve customers. Since a simple message-buffering scheme cannot be easily devised for entities of unknown length, slower secondary storage devices (such as tapes and disks) must be utilized to store the message.

4.10.2   Packet Switching

Packet switching is a communication strategy first implemented in the early 1960s that attempts to capitalize on the advantages of both circuit and message switching while minimizing their disadvantages. It is similar to message switching except for one important refinement: the largest transmitted unit is a packet, an entity with some maximum length. Messages are fragmented into packets that are commonly about 128 octets long.

Since packets have a fixed maximum size, it is straightforward to devise an efficient buffer allocation scheme that can take advantage of a computer's fast internal cache memory. Utilization of fast primary memory rather than slower secondary storage devices further reduces the storage time of a packet at the node. Overall delay of a message sent through a packet-switching network is less than that of a message-switching network.

Packet switching is the commonly applied technology in today's wide area networks (WANs). Packet switching does not waste resources like circuit switching does. Due to the fixed maximum size of the transmitted entity, efficient buffer allocation schemes make packet switching viable in the interactive environment, unlike message switching.

Packet switching is not well suited for constant data transmission, however. It is best suited in applications with bursty data transmission. Bursty data transmission is described as an application where 80 percent (or more) of the data is transmitted in 20 percent (or less) of the time. Put another way, 80 percent of the time no data is being transmitted; that is, the data comes in bursts.

Packet-switching networks may be classified by the types of services they provide (as seen by the user at the interface to the network) and by their implementation (the internal architecture of the network). In the next two subsections we will examine two implementations of packet switching.

Packet Switching: Datagram Implementation The architecture of a network is the network implementation, or the internal view of the network. In a datagram implementation, packets are treated independently. Thus, packets may be routed differently (even though they are being sent to the same destination), require full destination addresses, and may arrive at the destination node out of sequence. Refer to Figure 4–16, which illustrates this concept. The key features of datagram services are as follows:

- Sequential delivery of packets is not guaranteed.
- Packets may not be delivered at all or may be duplicated.
- All packets must contain the full destination address.

Packet Switching: Virtual-Circuit Implementation In a *virtual-circuit network*, data terminal equipment (DTE) places a call to another DTE by sending a special packet into the network, which indicates its desire to establish a connection with the other data terminal equipment. The network then sets up a route for subsequent packets on this virtual circuit and all such packets follow this route.

virtual-circuit network

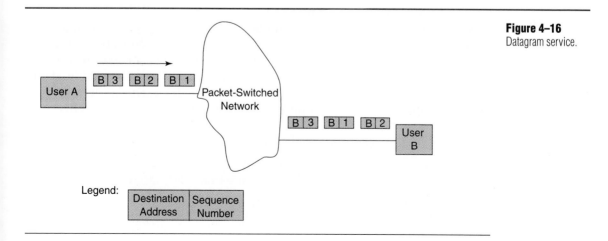

Figure 4–16
Datagram service.

Figure 4–17
Virtual-circuit
service.

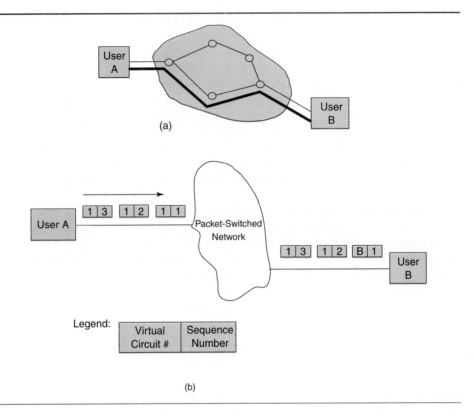

All packets belonging to this logical connection (or virtual circuit) are given a common identifier. This identifier is the virtual-circuit number located in the packet header. Packets are delivered to the recipient in the order in which they are transmitted. A virtual-circuit connection is an end-to-end connection between two users over a store-and-forward network. All packets follow the same route, need not carry a complete address, and arrive in sequence. Figure 4–17(a) and Figure 4–17(b) show a virtual-circuit connection. Thus the main feature of the virtual-circuit connection is that the user sees the packets delivered in sequence, and no decision as to routing needs to made at the nodes. Note that this can still be achieved over a datagram implementation providing that the destination node sequences the packets before delivery to the user.

Examples of packet-switched networks include:

- Virtual-circuit service on a virtual-circuit architecture (e.g., British Telecom, TYMNET system architecture from IBM, and ACCUNET packet-switching service of AT&T).
- Virtual-circuit service on a datagram architecture (e.g., Internet and U.S. Sprintnet).

- Datagram service on a datagram architecture (e.g., digital network architecture from DEC).
- Datagram service on a virtual-circuit implementation. A connection-less-mode service and a connection-oriented implementation; e.g., Internetworking (IP, datagram service over X.25 subnet).

4.11 COMPUTER TELEPHONY SWITCHING APPLICATION

Computer telephony integration (CTI) is a marriage of convenience between telephone and computer processing systems, which began in the mid-1980s as digital PBXs began to take hold. Initially, mainframe computers attached to incoming call centers provided an efficient way to link incoming customer calls with agents. Although these automatic call-routing and customer-service applications were state of the art in the 1980s, an entire suite of applications has since emerged. Computer telephony has opened an entire world of business opportunities by linking the power of the computer with that of the telephone. Computer telephony is now a multibillion-dollar industry, as commercial and utility-service providers explore new ways to integrate and deploy the telephone and the PC.

CTI

Computer telephony applications are derived from three basic ingredients: telephone access, computer/database access, and access to an agent to handle a call or an intelligent response. Computer telephony integration has even extended to unified messaging capabilities. These capabilities provide support for a single mailbox, which can use voice, fax, E-mail, or binary data as a means of leaving a message. The mailbox can, in turn, be checked from the traditional telephone, fax, desktop, or notebook PC.

Some capabilities of computer telephony systems include:

- Call setup and takedown.
- Information receiving.
- Call routing.
- Speech recognition.
- Information provision.
- Network function support.
- Message forwarding.
- Tone detection.
- Access to agents.
- Tone generation.
- Call supervision.
- Voice and machine detection.

Call-processing applications vary across industries and support the gathering and dissemination of information in sectors such as the automotive and construction industries, sales and marketing management, government, education, electronic media, entertainment, finance, health care, hospitality, transportation, real estate, retail, social services, and telephone and utility services to name a few.

4.11.1 Call Center Integration and Management

Managing call center operations has become more sophisticated as companies refine these processes as a cost-effective way of doing business. Refining the automation of call center operations has set the stage for many other computer integration innovations.

Figure 4–18 illustrates a fully integrated call center operation linking a PBX equipped with automatic call distribution (ACD) capabilities with enterprise local area networks (LANs), a mainframe database, and sophisticated servers. In this example, computer telephony integration (CTI) provides immediate caller information to service representatives. The system's computer telephony integration functionality links the PBX's automatic call distributor with several computer platforms across the entire enterprise network. (Enterprise network refers to a network that combines both local area and wide area networks into one network.) Automatic number identification and dialed number identification service originating through the public network provide information that can be retrieved from the system so that a call can be routed and information can be retrieved from the company database. This information is then presented through the service representative terminal as the call arrives at the agent's desk. Voice response units are incorporated to handle calls that may not require the full service of an agent. Voice response units can be used to disseminate information to customers and to satisfy information requirements. In this way, more routine calls can be handled quickly and efficiently. A separate

Figure 4–18
Integrated call center.

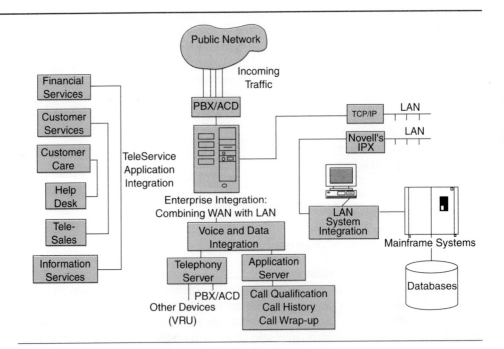

management system tracks and analyzes components of the call management center(s) to optimize and streamline operations.

A standards-based *application programminginterface (API),* together with a common object request broker architecture (discussed in Chapter 5), provide the support necessary for an organization to create specialized client/server applications to expand or modify existing applications.

API

4.11.2 Design Concept

The concept of the model in Figure 4–18 is designed to establish a custom call pathway for each call and couple the call with specific caller data to provide customized service. In addition, this system provides the facilities to qualify each incoming call for special skills-based routing. Specific information can be exchanged between applications based on computer telephony-generated call information.

Configured with the proper software and hardware resources, users can leverage their investment in systems by integrating application information across the company's network. There is an added advantage of an established security arrangement with the ability to support a fully redundant network structure.

A fully integrated environment provides the basis to establish cross-selling and integrated customer and product support. Voice response systems extend resources to ensure full support for all incoming traffic and support for fully automated wrap-up.

4.11.3 Computer Telephony Integration
Service Applications

Education   At large universities, special-purpose PBX switches support student and class registration. A PBX coupled with a call distribution system and a Unix-based computer platform provides an integrated system a university may use for student registration. Students register for a class by calling to enter their student ID along with their requests for courses and class assignments. This process supports the dynamic scheduling of classes, while applying valuable information to other administrative applications dealing with financial and physical resource management.

Financial News Agency   Private networks supported with a UNIX-based computer platform and voice response units provide support for real-time financial information, including stock quotes, portfolio values, and market trends. This service allows subscribers to track their stocks.

4.12   SUMMARY

Switching is a function that a network node must perform to efficiently establish the connections among network components. In general, switching is divided into two categories: circuit switching and store-and-forward switching.

Store-and-forward switching is further divided into two types: message switching and packet switching. In circuit switching the established links remain active for the entire duration of the calls that are using them. In packet switching messages are fragmented into smaller parts called *packets*. Each packet can take a different route on its way through the network toward its destination. The receiving node must reassemble the packets to reconstruct the original message.

Examples of circuit switching systems include central office switches and private business exchanges. Routers used with the Internet are examples of switching systems that use packet switching to establish communications. The distinction between systems that implement circuit switching and those that implement packet switching is becoming less clear. There are many examples where a switching system implements both types of switching technique.

REVIEW QUESTIONS AND PROBLEMS

1. Why do we need switching?

2. How was switching performed in the old public telephone network?

3. Which scheme of switching is used in the public telephone networks?

4. Which scheme of switching is used in data networks, for example X.25 networks?

5. What are the functions required of a central office switching system? How are those implemented?

6. List the service interfaces needed in a central office switching system.

7. What are the major differences between a central office switching system and a gateway switching system? Describe the design issues that must be addressed in a gateway switching system.

8. Is it possible to use a central office switching system in gateway switching? Justify your answer.

9. What would the advantages of digital private branch exchange be compared to an analog private branch exchange?

10. List and give a brief description of each of the private branch exchange switches.

11. Using a system-level diagram, show a configuration in which a PBX is networked into another two PBXs to provide a private circuit-switched network.

12. Consider yourself a networking and traffic engineer who is assigned the task of engineering a private branch exchange network among four sites with the following requirements:

    Site 1: 350 analog stations, 200 ISDN basic rate interface terminals (each requires 3 DS0s), and 25 digital stations

    Site 2: 1640 analog stations, 190 ISDN terminals

    Site 3: no stations, but the PBX on that site acts as a hub for the other three using T1 connectivity only

    Site 4: 800 analog stations and 140 digital stations

    Give a design for that network, and answer the following questions:

    a. How many T1 facilities will you use to interconnect the switches?

    b. What is the blocking rate you assumed for your design?

    c. Give some estimates of the cost of such a network (you can research the prices of PBXs). Note that the cost of a PBX may also depend on the features you decide to include.

13. What are the advantages and disadvantages of the Lucent Technologies PBX architecture?

14. If you were to improve the Lucent architecture, what would you recommend and how would you implement your recommendations?

15. What are:
    a. Routine/scheduled maintenance?
    b. On-demand maintenance?

16. How are routine and on-demand maintenance accomplished in the Lucent Definity switch?

17. How would the tone clocks and the tone detector boards assist in maintenance testing in the Lucent Definity switch?

18. Give two examples of computer switching applications.

19. Describe how a PBX would be used in an automatic call distribution center application.

20. What are the differences between a Centrex and a PBX switching system?

21. Under what conditions would you recommend using Centrex as compared to PBX?

5

Computer Telephony and Applications

5.1 INTRODUCTION

In most offices and in many homes, we find a telephone set and a personal computer or workstation on the same desktop. Why not exploit this situation by enhancing the telephony functions with computer intelligence and power? There are two technical approaches to achieve this goal.

First, the computer may act as an auxiliary, a complement to the ordinary touch-tone telephone for certain functions such as dialing. Incoming calls, however, are still handled by the telephone, the microprocessor and speaker of which will usually be used once the call is set up.

In contrast, the desktop computer may totally replace the telephone set and act as the dialing, voice capture, and playback device. This approach is more awkward in practice, as it implies that the computer remains up and running specifically to receive incoming calls. Note also that if a speaker is used instead of headphones, the system must be provided with an echo-canceling mechanism, implemented either in hardware or in software.

Many applications found it necessary to integrate the two approaches. For example, a telemarketing application may employ a computer to dial potential customers, but when a customer answers, the computer sets up a telephone circuit between the human agent and the answering customer. In this application, both approaches are used to automate a function that otherwise takes longer to execute. Of course, most of us do not take pleasure in answering these types of machine-initiated calls, but nevertheless, this is the state of affairs in the information age in which we live.

Computer telephony allows people to interact with a computer through their telephone. You probably come into contact with computer telephony systems several times each day. For example, you call the main number of a business and you are greeted by a recorded announcement and instructed to dial the extension of the person you want to reach. If you do not know the extension of the party, you may be further prompted to key in the person's name on the touch-tone pad of the telephone you are calling from to retrieve the telephone extension. In this scenario you are using an automated attendant application that employs the art of computer telephony. Sophisticated software and hardware are used to queue your call until the computer discerns what it should do with your call. When your phone rings while you are eating dinner and someone tries to sell you life insurance, the call was probably placed by a predictive dialing system, a computer attached to a dialer system that queries databases of potential customers then makes the calls. When the call is answered, the computer connects the called party with a human agent. You do not notice the computer's role, because when you pick up the receiver, there is a human on the other end who knows your name and life style information and habits you do not wish to reveal.

The installed base of telephony applications is steadily increasing. Auto attendants used to be limited to large or midsize firms. Today, however, many companies with fewer than a dozen employees are installing voice mail/auto attendant systems. Interactive voice response systems used to be limited to touch-tone banking at large banks. Today, movie theaters are commonly using audiotex systems that let you spell the name of a movie and hear show times. Even local golf ranges are installing interactive voice response tee-time reservation systems. Fax-on-demand systems used to be restricted to technical support questions or brochures at large corporations. Now even local realtors are building systems that will automatically fax you descriptions of properties that meet your criteria.

A typical computer telephony system performs three functions: (1) Process an automated phone conversation with a caller; (2) Monitor and log calls; and (3) Access external data or processing.

A computer telephony application communicates with callers through automated functions (functions performed by a computer instead of a manual operator) such as:

- Picking up the line.
- Placing an outbound call.
- Playing voice files.
- Prompting for digits
- Prompting for a message.
- Detecting a hang-up signal.

Computer telephony is a powerful trend in communications, and new applications are being developed and installed. To implement your own computer telephony system, you have four main choices: buy a turnkey application, use a fill-in-the-blank application generator, use a high-level development toolkit, or develop the software that controls the voice hardware directly. Turnkey applications are easy to use, but they might not always suit your needs in terms of capacity, functionality, or flexibility. Fill-in-the-blank application generators allow you to simply define your application but still limit you to the options they present. If an option you need is not available, you may be at a loss, especially after making the expenditure.

For maximum flexibility, you can control the voice hardware (usually boards that fit into a PC chassis) directly. To do this you must interface to the voice board software drivers. These drivers are usually low-level software packages that need to be installed on the PC hosting the voice boards and the application. Your software will need to control the boards via the installed drivers.

To develop computer telephony applications that accomplish tasks such as dialing out, receiving calls, recording incoming telephone messages, querying databases, and many more, a set of enabling technologies has to be integrated. Voice processing, speech synthesis, database management, intelligent networks, the Internet, and the digital networks are some of the enabling technologies that permit the integration of the processing power of computer and telephony to produce applications that benefit both the consumer and businesses.

Computer telephony applications can be classified into two types: desktop computer-assisted telephony and business applications. We enumerated many examples of business applications in the preceding paragraphs. Although the enabling technologies are essentially similar, the requirements for each type are different. Capacity, for example—in a business application the computer is required to handle multiple simultaneous calls (usually as many as possible or as the power of the computer allows). On the other hand, in a desktop computer telephony application, where the computer is used to perform enhanced telephony functions, most likely only one or two calls need to be managed.

5.2   COMPUTER-ASSISTED CIRCUIT TELEPHONY

We begin this section by posing this question: Is it true that a personal computer can replace a telephone set? The answer to this question is a qualified "yes." It is awkward to have a computer replace the telephone set because of the aforementioned reasons of having to keep the computer on all the time and the problem of echo cancellation. On the other side of the spectrum, however, telephone set designers are putting more computing processing power into the telephone set such that the distinction between the computer and the telephone is becoming blurred. In this section, we will limit our discussion to computers performing telephone functions. In the following subsections, we

Figure 5–1
Two models of
computer-assisted
telephony.

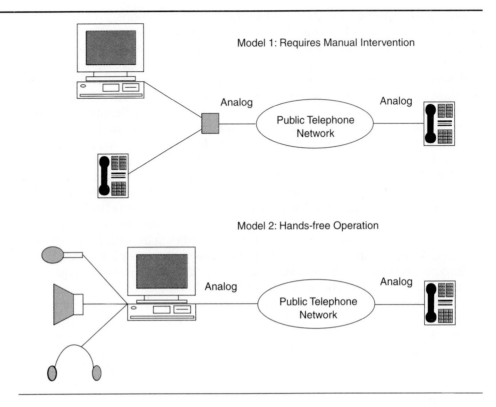

Model 1: Requires Manual Intervention

Analog

Public Telephone Network

Analog

Model 2: Hands-free Operation

Analog

Public Telephone Network

Analog

will examine the telephony functions performed by the personal computer and the hardware and software required for implementing these functions. Figure 5–1 illustrates the idea of computer-assisted telephony.

5.2.1   Directory Functions

The user may easily set up his or her own electronic directory of frequently called or calling numbers. Practically, such a directory is a list of numbers, each possibly associated with an easy-to-remember identifier and a short text. The directory is used for calling, generally by pointing and clicking on the selected entry in a list of displayed names or numbers. It may also be possible to enter a reference—a mnemonic identifier. This functionality is sometimes called *dial by name.* In the home environment, several family members may share the system; individual directories may allow privacy and customized use.

The name and associated information of the calling end is displayed if the directory look-up is successful. This feature is also abbreviated to caller ID. The individual called can decide to accept or refuse the call. This function implies that the computer is up and running when the incoming call is received.

This provision of directory functions does not necessarily require a personal computer. Advanced telephone sets provides similar functions, though on smaller screens and with smaller directories.

5.2.2 Digital Recording of Incoming Messages

Incoming messages, especially when the subscriber is absent, may be stored digitally on the magnetic disk of the personal computer for later playback. You may ask, So what's new? My answering machine does this as well. The difference lies in message access. Access to pending messages is no longer sequential. When the user asks to see the messages received, he or she is presented with a list of calls, each with calling number, call data and time, duration of the message, and the name of the caller if found in the private directory. The user can then decide which message he or she wishes to listen to first, or even which ones he or she will definitely not listen to and discard. The user can group pending messages into folders, according to their type or priority. This might be useful before as well as after playback if the message requires an action that the user cannot undertake immediately after listening. When filing a voice message into a folder for further processing, the user can generally add a short annotating text, or sometimes a voice annotation. Readers with minimal experience of networked computing will have noted that:

> The digital recording of incoming telephone messages offers potentially the same processing functionality as with electronic text-mail—presentation of a list of new messages; direct access to a specific item; and marking, discarding, and filing into folders.

5.2.3 Direct Call-Back

This function is analogous to the reply function in electronic text-mail. After playback of a received message, a simple command can generate a call-back without requiring the user to dial or to provide any reference to the number dialed.

5.2.4 Digital Recording of Outgoing Messages

Of course, any message to be sent, including replies to calls received, may be prerecorded and stored in the computer. This capability may be useful to determine whether a given call was actually made, at what time, and exactly what was said. When coupled with flexible dialing functions, the option of digital prerecording can significantly improve the efficiency of telephone communications.

Some of the facilities presented so far (recording of incoming and outgoing messages, direct access to individual voice messages) are in practice similar to those available on commercial voice-mail systems. The main difference lies in the use of individual computers to provide the service. This approach avoids the user having to interrogate a central voice-mail server, or the central server having to generate electronic text-mail to notify the recipients of incoming messages. Also, navigation in the message database may take full advantage of the graphical user interface available on personal computers or workstations.

5.2.5 Flexible Automated Dialing

Three main features are discussed in this section:

• *Programmed calls.* Imagine you are in California and need to leave a message for someone in Europe who has no answering machine. Instead of waking up at 3:00 A.M., you can record your message and program its automatic forwarding.

• *Repetitive calls.* Assume that another correspondent, who also has no answering machine, is absent when you call. You may record your message and program an automatic dialing retry, say, every hour over the next 12 hours.

• *Multiple calls (dialing lists).* This feature is considered one of the most important new functions of computer-assisted telephony—automatically distributing the same telephone message to a list of recipients instead of calling each of them individually. Imagine you are organizing a big party for Sunday, a sports event, or a barbecue for a 100 people. It is Saturday evening and the weather forecast is disastrous, so you decide to cancel! You record your cancellation message, set up a distribution list—if it does not already exist—by clicking on entries in your directory, launch the automatic list dialing sequence, and have a drink (you can enhance the process with a call retry function). After 1 hour, you return to see which destinations have not yet been reached, and go to sleep. On Sunday morning after your coffee, you simply check that everyone has been warned. Such functionality meets an extensive potential market in professional areas where warning, security, or emergency messages may have to be sent quickly to lists of recipients. Figure 5–2 illustrates this concept of operation.

5.2.6 Audio Editing

If the computer software is provided with audio editing capabilities (most modern workstations are), the user is able to create messages by using cut-and-paste functions of digitized voice sequences. For example, an incoming voice message may be aggregated to a heading comment you enter before you forward it to someone else. Or you may extract relevant sequences from several messages and compose an aggregated summary for storing or forwarding.

Unfortunately, there is a limiting factor to the use of such functions. The level of integration of editing tools with applications software supporting telephony activities is not yet always sufficient. The user will not edit voice messages regularly if he or she has to first copy them to working files, open the

Figure 5–2
Automatic message distribution.

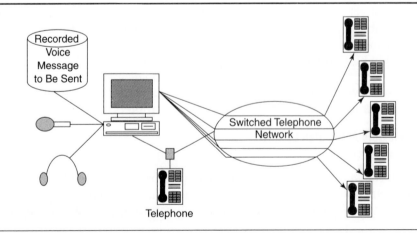

files with different tools, perform the editing, store the result, and return to the telephone support application for forwarding. In fact, this type of concern is the same that pioneering users experienced in the late 1970s for editing text-mail with early electronic mail implementations.

5.2.7 Using the Telephone to Access Personal Computers

So far, we have considered applications where the user is at home or in the office and uses his or her personal computer for improved functions in outgoing or incoming telephone calls. A computer-assisted telephone system, however, has the potential to be used the other way around. When out of the office, the user may very conveniently access his or her management environment from pay phones in gas stations, airports, hotel rooms, planes, and so on. Once communication with the user's personal computer is established, the user may listen to voice- and even text-mail, consult or make entries in the electronic calendar, send a voice message, or place a call. The important idea of such systems is:

> The entire management information and functions are made available not only locally through a visual user interface but also remotely through a telephone interface. The telephone is therefore acting as a computer terminal.

The usual computer-assisted telephone systems do not provide such facilities, which require the following features:

- Simple speech recognition.
- A touch-tone input interface.
- Full text-to-speech synthesis.
- Voice- and text-mail filtering.
- Integration of many personal management functions available on a computer with a telephone interface system.

Systems of this type have existed in experimental environments since 1993.

So far the systems we have discussed use the plain old telephone system to establish the calls. All features described are functions of the computer and are geared for the analog and ISDN telephone networks. In the next section, we will examine a different type of network; computer-assisted telephony systems will take advantage of its versatility. These networks are the packet-switched networks we regularly use in our Internet activities. In the next section, we examine how computer-assisted telephone systems may work over such networks.

5.3 PACKET TELEPHONY

This class of applications presents similarities in terms of functions with computer-assisted telephony. One main difference exists, however. The underlying network is not the telephone—conventional or ISDN—but the ordinary packet network used to interconnect personal computers for regular data communications. Note that when used outside customers' premises, such services may pose regulatory problems that will not be discussed in this text.

5.3.1   What Is It?

The idea is simple. In most organizations, the regular working tool on the desktop is the personal computer or workstation—or the terminal attached to a shared computer. This device is generally provided with a speaker. Then why not also use the personal computer or the workstation to support telephone-like conversations, at least with some correspondents? Of course, this function requires a microphone. You may think: We have seen this already in the last section. That's computer-assisted telephony! Not exactly. Here, nothing is assisted, the telephone network is not used at all, and no telephony technology is involved.[1] The computer alone supports the entire service. What is different from telephony is that the concept of two correspondents holding a voice conversation after having set up something resembling a telephone call is not used.

5.3.2   How Does It Work?

From the user's point of view, it works almost like computer-assisted telephony. The user has access to private or shared directories of frequently called numbers. The difference is that these numbers are not ordinary telephone or ISDN numbers, but *data packet network addresses.* This difference is transparent if the user calls through the directory, since he or she simply selects an entry by name. It is not transparent, though, if the user places a call outside the directory. Why should he or she do this? For example, because the number is not yet in the directory, or it will never be if it is an outside number and the directory is local only. Then, the user has to face the data network address format. As we see later, such applications essentially operate over the internet protocol service, which is itself provided with directory service to convert network names of recipients into network addresses. A network name (such as john.smith@worldcomp.uk) is not quite as user-friendly as a plain name, but this poses no problem to the user. Why? Because the user of such applications is also, by definition, a regular computer network user, and the form above is that used for the electronic text-mail the user is accustomed to seeing.

Of course, another difference is that the user always talks and listens to his or her computer in a hands-free mode, as no telephone set is involved; in contrast, only what we called model 2 in Figure 5–1 allows hands-free usage in computer-assisted telephony. The functions of caller discovery, incoming and outgoing message recording and editing, call-back and flexible dialing, and more generally voice-mail features are similar to those found in computer-assisted telephony.

---

[1]Except for speech encoding, which often follows standards designed for telephony, such as G.711 or G.722 from ITU-TS.

The user sees a final difference indirectly as it is a technical one; therefore, we shall not dwell on it at this stage. There is a standard for telephony, but there is none for packet voice conversation. This is not only a question of numbering, as this is being fixed by the marketplace, which is in the process of adopting the Internet and the E164[2] formats. Conventions are also needed for encoding and possibly the compression of the voice signal, as well as the way in which a call is placed, accepted, or refused. Fixing that is in no way an insurmountable task. In the meantime, packet voice conversation cannot be envisioned as a service as universal as conventional telephony.

5.3.3 Which Packet Networks Are Involved?

In theory, any underlying technology supporting packet data communication is involved—connectionless shared media such as ethernet, token ring, connection-oriented X.25, frame relay, ATM—and any packet format may support packet telephony. By packet format we mean level 3 format—Internet IP, ISO, and others discussed in Chapters 7, 9, and 10. Initially, use was mainly restricted to local area networks.

You may think: Why mainly LANs? Is it a question of available bit rate? Partially yes, because only 64 Kbps is necessary—even as low as 6 Kbps with compression techniques—which any LAN can accommodate. There is also no problem for many wide-area packet networks that operate at T1 speed (1.544 Mbps) to support such a call. Of course, since one single dual-party communication consumes from 0.5 to 4 percent of the available T1 bit rate, not many such calls can be established simultaneously. The main reason, however, lies in the lack of performance guarantees of most wide-area packet networks. Let us now switch gears and get back to the idea of computer telephony integration.

5.4 COMPUTER TELEPHONY INTEGRATION CONCEPTS

CTI, computer telephony integration, is a term that is widely used by telecom managers and telecommunications engineers. CTI is a concept in which one can assemble a computer telephony solution by integrating various telephony and computer resources not necessarily provided by a single vendor. For example, a system development group can specify a design for a small-size PBX in which the following telephony resources are obtained from outside vendors:

CTI

- T1 interface boards, possibly provided by Mitel Corporation.
- Analog line interfaces and Codec boards, possibly obtained from Natural Microsystems.
- DSP boards for tones and computer dialing functions from Dialogic Corporation.
- A Pentium PC.
- UNIX operating system and software resources provided by SCO Corporation.

---

[2]Numbering scheme used for the ISDN and ATM-based services.

The system developer's job in this case is to develop the system's software, say, in the C language, to provide the glue that makes these resources work and *interoperate* to provide the features desired in our PBX system. We will explore this example further as a case study, but first we need to study the CTI functions and abstraction. We will do that in the next sections.

5.4.1   CTI Abstraction

An abstraction is a myth that we create in order to make something manageable. In the case of CTI, the abstraction of telephony functionality makes it possible to design software that will work with more than one telephone system and support more than one user's paradigm. In the past (before the development of a robust abstraction of telephony functionality), software developers, installers, and customers who wanted to build CTI systems had no choice but to be aware of the internal designs and proprietary terminology, concepts, rules, and behavior of every telephone system.

The development of a universal abstraction allows any implementation, regardless of its size, to be described in the same terms. One way to think of this concept is that the telephone system presents a *façade* that appears to all observers to behave precisely as the universal abstraction dictates. Behind the façade, however, the telephone system is doing whatever is appropriate to translate between its own internal representations and those of the façade. To the observer of the façade, there is no difference between this telephone system, shown in Figure 5–3, and one with the same functionality that might be built by another vendor but with a different internal representation. This concept is illustrated in Figure 5–3. This translation between the universal abstraction (the façade) and the actual implementation is the role of the CTI (computer telephony integration) interface.

5.4.2   Manual versus CTI Interfaces

A good way to think of a CTI interface is as an alternative to the standard observation and control interface provided by the telephone set. Observation may include seeing lamps lit or hearing the telephone ring. The control function may involve pressing a sequence of numbers on a pad to originate a telephone call.

Figure 5–3
Telephony abstraction can be viewed as a façade. Same façade but different implementations.

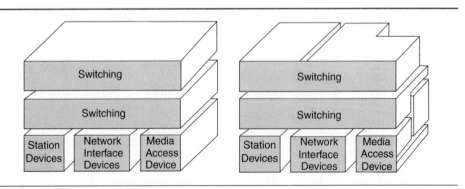

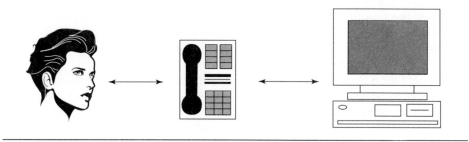

Figure 5–4
Multiple interfaces to telephone functionality.

In the example illustrated in Figure 5–4, a CTI interface is being used by a computer to control the activity associated with a particular telephone. Like a human sitting next to the telephone, the computer can see all of the lamps that are lit, the buttons that may be pressed, the text on the display, and so on. Similar to a person using the telephone, the computer can place calls, answer calls, and press buttons. Limited only by the capabilities of the CTI interface in question, the computer can, in fact, do anything the human can do with the telephone—and possibly more. It is as if the computer can reach out an invisible electronic arm and do the same things that humans can do to the telephone set. In this example, both the computer and the human have free access to the telephone set and can manipulate it independently.

This example illustrates an important point about the way that CTI interfaces generally work. To the extent that the CTI interface allows observation of a particular device, a computer may be one of many observers (either humans or other computers) of that device. To the extent that the CTI interface allows controlling of the device, that control is not exclusive. The computer cannot prevent a human from pressing a particular button, lifting the handset at a given instance, or performing some other action. Both the computer and a human are effectively peers in the telephone system. At first glance this concept might seem simple, but it does pose a few technical challenges of which you should be aware.

In application software development, the concept of multiple simultaneous control interfaces is not a new one. A single application running on a computer might be controlled through its graphical user interface with inputs coming from a mouse and a keyboard, through an independent speech recognition interface, and possibly through a scripting interface. The application must be prepared to combine all of these requests into a single stream and deal with each in its turn. This function is called *application factoring*. It separates from the core application code (which simply takes the next command from the outside world, processes it, and updates all the interfaces appropriately) all the code that creates and manages each of the different interfaces (different façades, if you will) and corresponding input paths.

Multiple interdependent or conflicting inputs may be presented to a telephone system in a near simultaneous fashion. To return to the first example, the computer might observe a call being presented in the alerting (ringing) state and might react by requesting that it be deflected elsewhere. Meanwhile, however, the human sitting next to the phone (or some other computer) might already have answered it in the time it took the computer to make its decision. This situation does not occur frequently in practice, but it is another important aspect of the CTI abstraction.

Observation and control are independent. Any request to manipulate a resource should be based on the last observed state of that resource, but the request may or may not be successful because other activity may have taken place as the request was being issued. Therefore, the results of a request should never be assumed; there is no substitute for observing what actually takes place.

The portion of the telephony feature set to which a CTI interface has access varies among telephone systems implementations. In telephone systems designed specifically to support CTI interfaces, the portion of functionality that can be accessed typically is much more than what can be accessed through the system's telephone sets. In systems where CTI functionality was not part of the initial design, the portion of the implemented telephony features that can be accessed through a CTI interface varies dramatically, from all to a small subset. In some cases, a system will offer both a proprietary CTI interface and a standard-based CTI interface, where the latter has less functionality than the former for historical and time-to-market reasons.

5.5 THE CTI INTERFACES

A CTI *interface* is a telephony resource that creates a door through which other telephony resources can be observed by CTI components (software and hardware) and through which these CTI components can request that features be set and services be carried out. CTI interfaces may also issue requests to the computer-side CTI component to perform certain tasks.

5.5.1 CTI Messages

A CTI interface operates by generating, sending, receiving, and interpreting messages containing status information and requests for services to be performed. This system is illustrated in Figure 5–5. Messages are used in either direction both to provide information and to issue requests. The structure, content, and rules governing the flow of messages back and forth through a CTI interface are defined by a CTI protocol.

We have already seen that abstraction plays a key role in making CTI possible. In fact, with a few constructs (i.e., devices, components, calls, and connections) and a reasonably small vocabulary of types, states, and attribute values, we have been able to describe and model the vast majority of telephony functionality with little effort.

Our abstraction of telephony resources, features, and services can now be expressed in concrete terms through parameters that are placed into messages. This concept is illustrated in Figure 5–6.

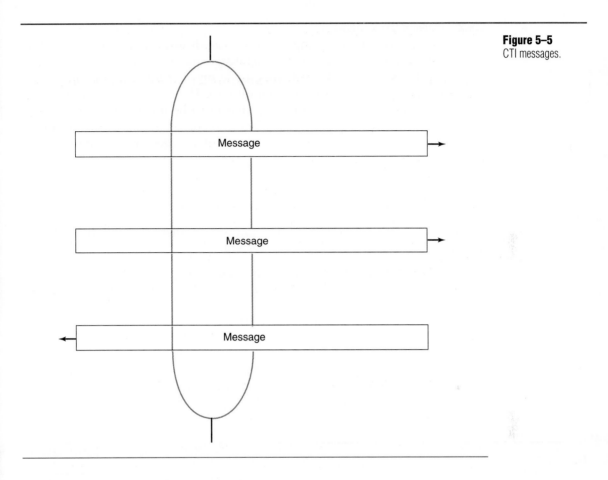

Figure 5–5
CTI messages.

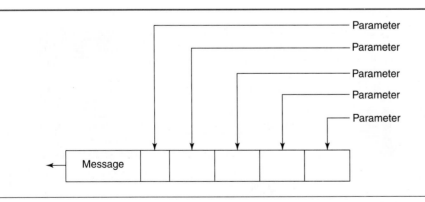

Figure 5–6
Parameters in a CTI message.

5.5.2 Modular CTI Systems

A general goal for implementing a CTI interface is to allow any combination of hardware and software components to be assembled into a CTI system of any size. Even the smallest CTI system is made up of many components, which means that the system itself contains multiple CTI interfaces. A CTI interface is needed between each CTI component that must be integrated with another. Figure 5–7 builds on the example we viewed earlier. This example is among the simplest of all CTI systems, but it still involves three distinct components:

- The telephone.
- The computer.
- The CTI software running on the computer.

Because there are three distinct CTI components, there are two different CTI interfaces at work in this simple CTI system:

- Between the telephone and the computer: This CTI interface uses protocol.
- Between the CTI software and the computer: This CTI interface uses a programmatic interface.

Once a CTI system is assembled, it becomes difficult to say where the telephone system begins and the computer system ends. All the components that make up the system are working together in a cohesive fashion to form what can be viewed on one hand as a more sophisticated telephone system, or on the other hand as a more sophisticated computer system.

Figure 5–7
Computer telephony interface in a computer telephony system.

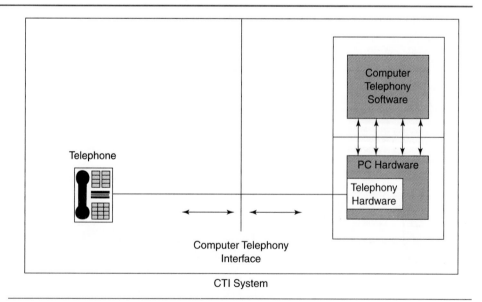

5.5.3   CTI Service Requests and Events

Four kinds of messages pass through a computer telephony integration interface:

- Events.
- Service requests.
- Positive acknowledgment.
- Negative acknowledgment.

Every CTI message is defined in terms of its kind, which message among those of its kind it is, and its parameters. For example, the CTI message corresponding to a request for the *set lamp mode* service can be shown graphically as in Figure 5–8.

Each parameter in the list of parameters appropriate for a given message may be optional, mandatory, or conditionally mandatory. Most parameters are identifiers that reference a particular resource (or resource attribute), or they are variables representing a state, status, or setting values. What is the meaning of all of that? A parameter may identify a touch-tone dialer (a resource) or a status of a telephone line: busy, idle, or ringing.

Computer Telephony Integration Events   *Computer telephony integration event messages,* or events for short, are messages sent from computer telephony modules to the computer telephony application software to indicate transitions of states and changes in the status setting of an attribute in the computer telephony module. Events are primary mechanisms used by the computer telephony application software to observe activity within the computer telephony resource modules. For example, if the connection state of a particular connection transitions from *alerting (ringing)* to *connected*, the computer telephony integration message *established* would be sent to the application software to indicate that the connection in question had transitioned to the *connected* state. The established event, with a partial parameter list, is shown in Figure 5–9.

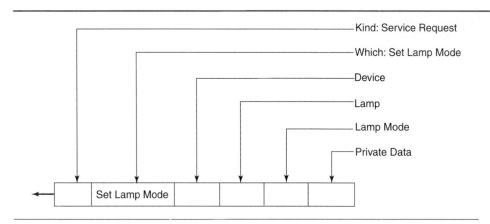

Kind: Service Request

Which: Set Lamp Mode

Device

Lamp

Lamp Mode

Private Data

Figure 5–8
Set lamp mode service request message.

Set Lamp Mode

Figure 5–9
Established event
message.

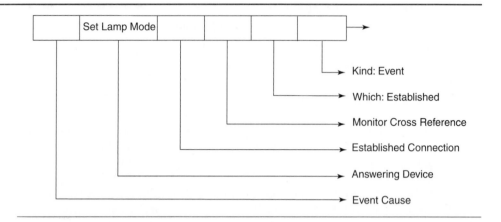

In order to receive event messages that are relevant to a particular call or device in the computer telephony application, the application must request them by *starting a monitor* on the item in question. The monitor cross-reference identifier parameter in the event identifies what previously established monitor caused this event to be sent. The definition of event messages includes:

- The meaning of the event and, in the case of events that reflect state changes, its context in terms of implications for other related connections.
- The possible causes to which the event can be attributed. (The cause code of a given event is an essential parameter in the event messages, and in many instances it represents a very important clarification of the meaning of an event.)

Event messages are defined for every type of item that has a state, status, or setting that can change. Events are also defined to indicate that new information has been received, such as an update to the related information associated with a call or the detection of a DTMF (dual-tone multifrequency, used in dialing and other telephony activity) tone. Table 5–1 includes more examples of events.

The examples of computer telephony events are typical of the messages required to detect the statuses of a telephone line (connected to a computer). A sequence of service requests and event status messages are usually *exchanged* among modules in computer telephony software, which orchestrates the function of the entire system to provide those services expected by a user of such a system. The software is continuously monitoring the states of connections, the statuses of lines, and the health of the hardware. Therefore, when a request arrives (such as an incoming call), the telephony software is capable of knowing which part of the system hardware and resources should be required to handle such an event. An off-hook event message may indicate the arrival of a telephone call. When the application software receives this indication, it instructs

Table 5–1
Event message
examples.

Event Messages	Computer Telephony Resource	Event Indicates
Call cleared	Call	Call no longer exists
Call information	Call	Update call-associated information
Bridged	Connection	Transition to *queued* state
Connection cleared	Connection	Transition to *null* state
Delivered	Connection	Transition to *alerting* state
Digits dialed	Connection	Transition to *initial* state
Established	Connection	Transition to *connected* state
Failed	Connection	Transition to *fail* state
Held	Connection	Transition to *hold* state
Offered	Connection	Transition to *alerting* state
Originated	Connection	Transition to *connected* state
Queued	Connection	Transition to *queued* state
Retrieved	Connection	Transition to *connected* state (after retrieve)
Service initiated	Connection	Transition to *initiated* state
DTMF digits detected	Connection	DTMF digits detected
Telephony tones detected	Connection	Telephony tones detected
Button pressed	Button component	Button was pressed
Display update	Display component	Update display contents
Hookswitch	Hookswitch component	Change in hookswitch state
Lamp mode	Lamp component	Change in lamp mode
Microphone mute	Microphone component	Microphone mute attribute updated
Ringer status	Ringer component	Change in ringer attributes
Speaker volume	Speaker component	Speaker volume attribute update
Agent logged on	Agent	Transition to agent logged on status
Agent not ready	Agent	Transition to agent not ready status
Do not disturb	Logical element	Do not disturb setting changed
Forwarding	Logical element	Forwarding setting changed
Out of service	Device configuration	Device is out of service

other parts of the system to take the proper action to respond to this event. In this case applying a dial-tone to the calling party will be the proper action. To do that the software may have to initiate a service request message to the hardware in response of the event received. Service request is the topic of the next section.

Service Requests   Service requests are messages sent from the computer telephony application software to the telephony resources (hardware/software) in order for certain services to be performed. Service request messages may also be exchanged among various software components of a large computer telephony application that may encompass numerous resources and modules. There are numerous categories of service requests:

- Service requests associated with the telephony features and services:
  a. Call control services.
  b. Call associated services.
  c. Logical device services.
- Service requests that manipulate or check the status of physical elements components:
  a. Pushing buttons.
  b. Getting and setting button information.
  c. Getting lamp information.
  d. Getting and setting lamp mode.
  e. Getting and setting display contents.
  f. Getting and setting message waiting indicator.
  g. Getting auditory apparatus information.
  h. Getting and setting hookswitch status.
  i. Getting and setting microphone gain and mode.
  j. Getting and setting speaker volume and mute.
  k. Getting and setting ringer status.
- Service requests that are specific to the computer telephony integration interface:
  a. Capabilities exchange.
  b. System status services.
  c. Monitoring.
  d. Snapshot services.
  e. Routing.
  f. Media access.
  g. Vendor specific extensions.

Service Request Messages   The definition of service request messages includes:

- The service that is invoked or the feature that is set.
- Required initial states and possible final status for any connections on which the service acts.
- Mandatory, optionally, and conditionally mandatory parameters of the service requests.
- Possible outcomes of the service.
- The sequence of events that should be expected if the service completes successfully.
- The completion criteria used to determine if a service completed successfully.
- The possible reasons for an unsuccessful service request.

When a service request message is issued, the telephone resource modules must respond to the computer telephony application software with either a positive or negative acknowledgment. Independent of these acknowledgments, if the service results in any action that affects the state of one or more connections, or the status or setting of some resource or attribute, the appropriate event messages must be generated. Thus, the application is always informed of the state of the system components at all times.

The event sequence defined as part of the service completion criteria for each service is *normalized event flow.* It is very important because it allows the software to verify that a service has taken place. The computer telephony software is not the only source of commands that manipulate items within a telephony application. If another interface, such as a telephone set, is used to issue commands, the computer telephony application software sees the command as an event sequence.

Negative Acknowledgments   If a service request is unsuccessful for some reason, the resource attempting to carry out the service indicates the lack of success by sending a negative acknowledgment message. This negative acknowledgement message contains an error value that provides an explanation of what the problem was. An example of a negatively acknowledged service request is illustrated in Figure 5–10.

Positive Acknowledgment   A positive acknowledgment indicates that a service request is being or has been acted upon. There are two basic types of positive acknowledgments, depending on the nature of the service request:

• One type of service request is a direct request of the CTI interface to return information about something in the telephone resources and hardware. In this case, the positive acknowledgment not only indicates that the request was completed successfully, but it also includes the requested information.

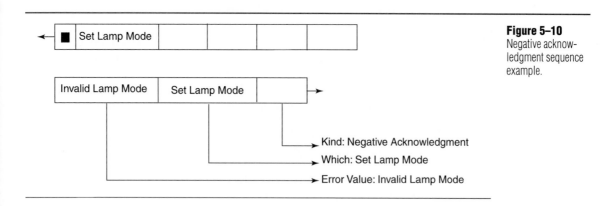

Figure 5–10
Negative acknowledgment sequence example.

- If the service request involves some manipulation of resources to take place, the positive acknowledgment indicates that the CTI interface has passed the request to be carried out. If the service in question is carried out in an atomic fashion, the positive acknowledgment also indicates notification that the service was completed successfully.

Atomic and Multistep Services   The implementation of services in the computer telephony application software may be atomic or multistep. The nature of the implementation determines the correct interpretation of positive and negative acknowledgement messages. For example, if the switching module in a computer telephony application implements atomic services, the service requests are treated as follows:

1. Validation: Are the parameters valid? Are all application connections in the correct state? Are all necessary resources available? If the service request is invalid for any reason, the switching domain sends back a negative acknowledgment message to the higher-up software to indicate that the request is invalid.

2. Execution: The switching domain then attempts to carry out the service requested.

3. Acknowledgment of success or failure: If the requested service is completed the switching domain sends a positive acknowledgment message to indicate it has succeeded. Otherwise it sends a negative acknowledgment indicating that the service did not succeed and why it did not. The format of a negative acknowledgment is shown in Figure 5–10.

If the atomic service does not succeed, no resources in the switching module are affected in any way and, therefore, no events will be generated. If the service is successful and it affects statuses, settings, and so forth, then appropriate events are generated. For example, if the forward call service is implemented as an atomic service, the complete flow might take place as shown in Figure 5–11.

5.6   CTI SOFTWARE COMPONENT HIERARCHY

Almost every imaginable CTI hardware component is based on computer technology and, thus, is managed by software that is responsible for sending, receiving, interpreting, generating, and handling CTI messages. In assembling CTI systems and building individual CTI components, however, we are concerned only with software environments designed for installation or addition of software developed by third parties. Within such an environment, CTI software components are arranged in a functional hierarchy—a CTI software framework. In that framework we are concerned primarily with:

- Operating systems and/or network operating system software.
- Products from telephony software developers.

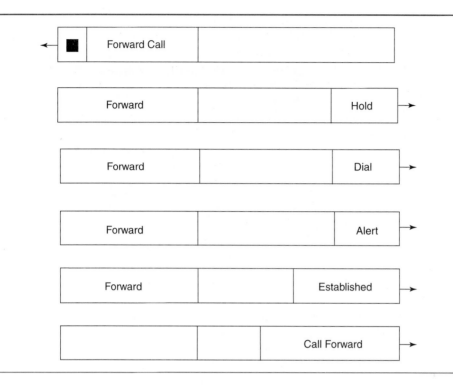

Figure 5–11
Atomic implementation of call forward service.

5.6.1 Modularity

A CTI consists of a collection of individual CTI components in which each communicates with its neighbor(s) using CTI messages that travel across an intercomponent boundary. Modularity is as important for software components as it is for hardware components. System integrators, customers, and individuals want to be able to plug and play with their software, just as they can plug and play with their hardware. This requirement is fundamental to the concept of the CTI software framework.

For CTI hardware components, the intercomponents' boundaries are in the form of CTI protocols that travel over communication paths between hardware components (as we will see in the next section in our case study). For software components, they are in the form of programmatic interfaces through which CTI messages are communicated. In general, computer telephony hardware is available in the forms of telephony cards and associated device-driver software responsible for the operation of these cards. On top of the device drivers, reside the computer telephony application-programming interfaces, providing the telephony application software with access to the resources in a simplified and orderly manner. Figure 5–12 illustrates a four-tier computer telephony integration scheme in which modules from different vendors are assembled into a single system.

Figure 5–12
Four-tier computer
telephony system.

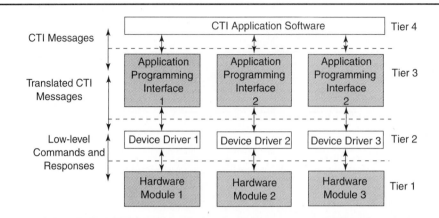

5.7 CASE STUDY

Let us now put what we know about computer telephony integration to use. In this section we will design a small-scale private exchange switch (PBX) using a PC platform and a number of third-party telephony resources. We will start by specifying the requirements for our system:

- The system must support at least 48 users.
- The system must use digital switching.
- The system interconnects to the central office via two T1 interfaces.
- The T1 circuits will synchronize their clocks with the central office.
- The system must support a number of core features, including:
  a. Conferencing for up to 6 users.
  b. Call forwarding.
  c. Call transfer.
  d. Hold.
- The system will provide maintenance access via the application software to test the various objects of the systems.
- The system will support the ability to administer (configure) a user telephone from system configuration software.
- The system will provide the ringing signals required to complete a telephone call.
- The system will be restricted to the use of touch-tone dialing (using DTMF circuitry).
- The system should be able to expand to a larger number of users than what is currently specified.

Examination of the feature specifications suggests that some features will be implemented in software and others will be in the hardware. The hardware is responsible for providing the telephony resources that will be managed by the software to implement the overall system. We will impose on ourselves a requirement consistent with the goals of computer telephony integration—no hardware will be designed in-house. This requirement suggests that we will have to look for outside vendors to provide hardware resources and as much as possible of the software when applicable. This brings into question whether we should design the architecture of our system first, and then look for what is available from third parties or do the reverse—find out what is out there, then design the system around available hardware to satisfy the requirements within the economical goals of the project. We think the latter choice is preferred in this case; we need to compare what is available, decide on the third-party vendors, and then architect the system around what is available. Let us see how we can accomplish this task.

5.7.1 Hardware Telephony Resources

One of our requirements is to connect to the central office via T1 interfaces. Several vendors supply reasonably priced T1 cards that work within the PC architecture. We decided on a dual T1 card from Mitel Corporation, which supports two T1 interfaces. This card is software configurable, meaning we can select the features of the T1 operation via software commands. Access to the A, B, C, and D signaling bits (used to convey the on-hook, off-hook, ringing, etc., status of the individual lines) is software-based, thus the status of each telephone line can be monitored and its state can be altered by the software. Additional features of this T1 card will be provided later in this section.

Our system will need to monitor the lines and provide dial tones when they go off-hook, provide busy tones when intended destinations are busy with other calls, and collect digits when users dial a number. This telephony resource will be provided by a digital signal processing (DSP) card that implements the tones and tone detection for the purpose of DTMF dialing. This PC card resource supports up to 24 channels. That means we support 48 telephone lines, but only 24 of them may go off-hook at the same time. This scenario, of course, is unlikely, and if you recall from our discussion in Chapter 4, the premise of switching is that not everyone requests service at the same time. Service requests (off-hook) are interleaved with relation to time, so our DSP resource will be able to handle all 48 lines.

So far we have decided on two major components of the hardware required for our system, now we need to choose our codecs. Recall that the codec converts the analog telephone voice signal to a digital bit stream that can be carried over the T1 facility. In addition, we will need a ringing generator to provide the ringing signal necessary to alert the called lines. NMS, a small company in Boston, Massachusetts, makes a card that supports 24 lines; thus, we will need two of those codec lines. This company also provides an internal (PC module) ringing generator that we will use in our design.

Figure 5–13 illustrates the hardware configuration of our system. It shows one T1 card that supports two T1 interfaces, one DSP card for DTMF implementation, two analog line interface cards supporting the analog-to-digital conversion for 48 lines, and a ringing generator. Another element that is shown in

Figure 5–13
A high-level hard-
ware configuration of
PBX resources.

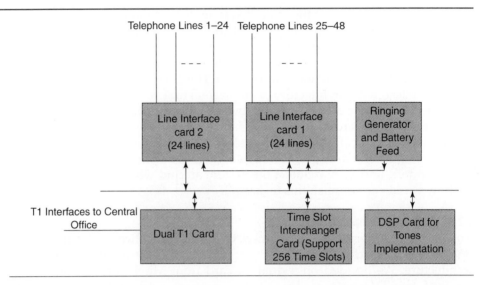

Figure 5–13 is critically important to the function of digital switching in our system—the time slot interchanger. This device is also a PC card that is controlled by the software to interconnect the digital paths among the local telephone lines on our switch or among the telephone lines on our switch and external lines via the central office to complete the telephone calls. Also shown in Figure 5–13 are double-arrow lines that interconnect these resources to a communication path. These lines are part of the hardware interfaces of each of the telephony resources we need. In the next section we examine these interfaces in more detail.

5.7.2   Module Interfaces

Two distinct interfaces are available for each of the modules in our system. One interface is for control and management of the computer telephony function. This interface will eventually be used to implement the application programming interfaces for each of our modules. This is essentially the industry standard architecture (ISA) bus connectivity of the PC platform to each of the hardware modules. Figure 5–14 shows the interfaces of the T1 hardware module. Shown in the figure are the following three interfaces:

- *ISA bus interface.* Provides the path for the software to monitor and control the functions of the card. This is an I/O mapped implementation that allows the PC's CPU to communicate with the T1 card in order to command it to perform certain operations and also to provide certain statuses.
- *Digital stream interface.* This is a serial bit stream that is synchronized to the serial bit streams of the digital switch card. In this case, the T1 provides the gateway to external calls to be completed through the central office switch. The digital signal traveling over this interface is unipolar.

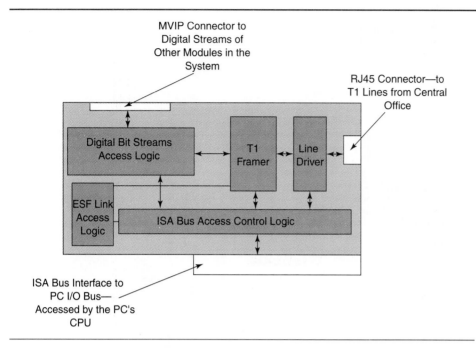

Figure 5–14
Functional block
diagram of the T1
hardware module.
Shown are the three
interface connectors:
ISA, MVIP, and
RJ45.

- *Line interface.* This is the bipolar digital line interface to the central office. In general, this line interface is used to interconnect to other T1 equipment. In this case, the T1 equipment is located at the central office. It is possible that repeater(s) sets are needed between the T1 and the central office equipment.

Also shown in Figure 5–14 are three interface connectors: the ISA bus connector, the RJ45 connector, and the MVIP connector. You are already familiar with the ISA and RJ45 connectors. The multivendor integration protocol (MVIP) bus is probably a new term to you. The connector is essentially a 40-pin ribbon cable connector that carries the digital streams among the telephony resources in our system. It has 32 pins for transmitting and receiving 16 digital bit streams (each digital stream consists of 32 64-kbps PCM channels), one lead for an 8-kHz clock, and one lead for a 4.048-mHz clock. Our T1 uses only one of the digital streams since it supports only 24 channels. The other six channels on the stream assigned to the T1 board will be wasted, but we do not expect a perfect fit all the time.

The cable from the MVIP connector will travel to the MVIP connectors of the other hardware modules (i.e., the DSP DTMF card, the time slot interchanger, and the codec boards).

The other hardware cards have similar interfaces. The ISA bus interface provides similar functions on all the hardware modules. The logic for that interface contains control and status registers that can be written to or read from

Figure 5–15
Connectivity of
telephony resource
modules.

Computer Telephony Integration-Based PBX
(PC Chassis and Telephony Resources)

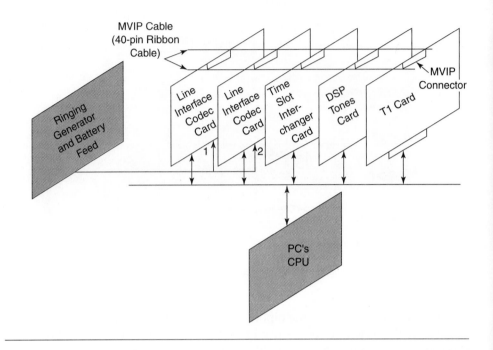

via the PC's CPU. As we already stated, this interface is used to monitor the statuses and control the operations of all the cards in our system. The digital bit stream (MVIP) interface on the digital card is similar to that of the T1 and those of the DSP tones cards. It provides digital connectivity among the collaborating modules to complete a telephone call digitally. The codec cards contain the analog interface connectors (a block of 24 connectors to connect to the telephone sets on each card) and one connector for the power supplies—48 volts and the ringing signals. Shown in Figure 5–15 is the diagram of our system hardware and the relevant interfaces.

5.7.3   Software

Our software is divided into two parts: one that is provided by third parties and the other we will have to develop. The part provided by the third parties includes the device drivers and the API interfaces for the T1, DSP, codec, and time slot interchanger cards. The drivers, once installed, communicate directly with the board's hardware and access to the drivers is accomplished via the computer telephony integration messaging. The messaging, as indicated earlier, is implemented in the CTI API software. Once we familiarize ourselves with the features and functions of the API, we can then architect our software to provide the glue that will hold the components of the system together in order to satisfy the requirements set forth for us.

Functions of Drivers   Once a PC computer telephony module is installed, its driver is responsible for its proper operation. The driver will initialize the various registers with the appropriate setting. For example, if the card uses interrupts, the driver initializes the board with an accurate interrupt number according to the system configuration. If there are special settings, for example if we want to operate our T1 in the ESF mode, the driver will be used to initialize the appropriate registers on the T1 card with that setting. In the case of the DSP tone board, the driver software will be used to command the board to, say, apply a dial-tone to a user who just went off-hook. To accomplish this action, a sequence of software requests might be issued to the board via the driver. In this example, we first need to tell the board to which digital channel on the digital stream the tone must be applied. Then we need to tell it which part of the circuit on the board must be activated to initiate the tone application. Then we may issue a request to find out the status of the command completion: success or failure.

The drivers usually operate on a very low-level basis, which may be cumbersome for software developers. The computer telephony integration software layer insulates the software developer from the intricate details of the driver operation by specifying a well-defined set of messages that deals with the operation of the board in question. In this case, we will have a number of independent APIs, each of which is concerned with one of the boards in the system. We are provided via the third party with one API for the T1, one for DSP tones, one for each of the codec cards, and one for the time slot interchanger. Although we need to know in general terms the workings of the drivers and their installation procedures, we do not need to concern ourselves with the intricate details of their operations. What we really need to know are the operations of the APIs that form the pipes between our software and the drivers. Figure 5–16 shows the layers of our software and the underlying hardware. Let us now examine our software design and its feature specifications.

Software Design   Many of the features of our system will be implemented in software, including the following:

- Call processing.
- System maintenance.
- System management and data reporting.
- Dialing plan.
- System resource management.
- Station and system administration.
- Switching function.

We can design a software module for each of these features. Let us now examine the features and functions of each of these modules, as shown in Figure 5–17.

Figure 5–16
Layers of CTI-based
PBX (software and
hardware).

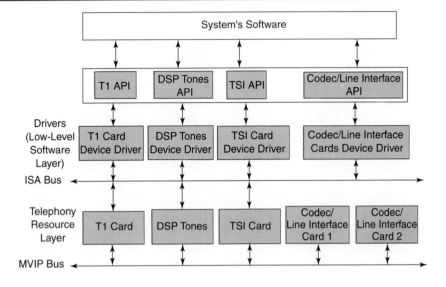

Key:
API: Application Programming Interface
DSP: Digital Signal Processing
Codec: Coder/Decoder
ISA BUS: Industry Standard Architecture Bus
MVIP: Multivendor Integration Protocol Bus
TSI: Time Slot Interchanger

Call Processing This function is the heart of our system. The call processing module is responsible for responding to events related to making the telephone calls, monitoring the telephone calls and controlling the states of telephone calls, determining the routes for telephone calls, forwarding telephone calls, putting calls on hold, and so on. To better understand how a basic call may be completed, let us explore a scenario of making a telephone call.

A user lifts the handset of a telephone (goes off-hook). This action causes the line interface corresponding to that line to complete a current loop indicating an off-hook condition at the hardware level. The line interface card, via the use of interrupts, informs the driver of that condition. The driver in turn, through the API, passes a message to the call processing software module that line X went off-hook. The call processing module interacts with the switching software to enable a path between the DSP tones hardware resource and our telephone line. Then the call processing module requests, via an API message, the tones resource to apply a dial-tone to that line and activate the digit collection circuitry to be ready to collect the digits that the user enters.

The digit collection circuitry passes these collected digits in the form of a string to the call processing software, via yet another API message. The call processing software, based on the digits it receives, in consultation with the dialing

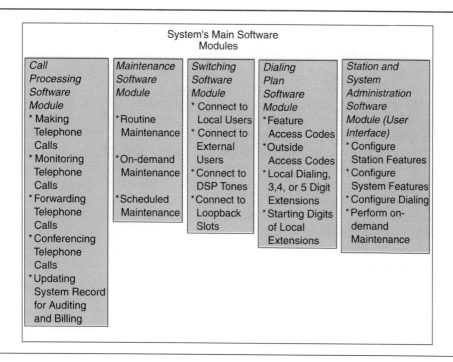

Figure 5–17
System's software main modules and their functions.

plane software module, and based on the access permission allowed for that user, determines how the call will be treated. The following are possible treatments:

- Request from the tone resource to apply an intercept tone if access to the number dialed is not allowed.
- Select a path via the time slot interchanger to connect to another local user if the number is a local number within the premises.
- Select a path via the time slot interchanger and the T1 line interface card to an external user if the number dialed is external. We usually determine that if the first digit dialed is "9." In this case the call processing software will instruct the tones resource to insert the digits into the PCM stream on the outgoing bound of the T1, via the time slot interchanger.

As you can see, many logical decisions need to be made in the call processing software not only for completing a normal telephone call, but also in responding to the calling user properly if the intended hardware is out of order. In this case the call processing software will have to consult with the maintenance software module, which is continually monitoring the health of the hardware and updating its maintenance record with the latest status. In addition, the call processing software module will need to continually monitor the call to detect changes in the state. For example, when one of the users on a call goes on-hook (hangs up) the call must be disconnected and the resource allocated for that call must be released to be used with another call.

Maintenance Software   This module will serve a number of purposes in our system. It will perform the following types of maintenance:

- Routine maintenance is done continually to keep track of the health of the maintenance objects. Maintenance objects are usually ports (telephone lines) or boards (T1, DSP, codec, and time slot interchanger).
- On-demand maintenance is usually performed when the system operator troubleshoots certain problems by issuing maintenance commands via a user interface (in our case by the PC keyboard and the PC monitor). Certain tests will be performed on the maintenance objects. For example, the user may issue a command called *test board T1 long*. This is a long set of tests performed on the T1 board to examine the operation of its various components. Each test will have a number and a pass/fail indication that is displayed on the PC monitor. The test numbers and the meaning of each test will be documented in a system user document to which the operator refers to diagnose the problem.
- Scheduled maintenance is a set of tests that can be performed each night (when use of the system is light) to test stations, boards, and ports, and to take corrective actions if possible.

The maintenance software module reports the health state of each maintenance object to a globally accessible table. The call processing module can then consult this table to, say, find out the state of line number 25 to see if it's normal or out of order.

System Management and Data Reporting   This module is responsible for monitoring telephone calls. It determines how long a call took, it places a time stamp on each call, and it identifies the resources that were associated with that call. This information is then written into a file that can be consulted by the system operator to determine traffic patterns, excess usage, and so forth. The information is obtained from the call processing module.

Switching Function   This module is used to make connections between time slots in the system. It manages the time slot interchanger, allocates a dedicated set of slots for tones, and keeps track of time slots that are being used and time slots that are free to be used in future telephone calls. This module is heavily used by the call processing software to complete calls, forward calls, and conference calls. It is also used by the maintenance software module to perform loopback tests on certain parts of the hardware and to test the operations of the MVIP bus.

Dialing Plan   PBX systems in general use a flexible dial plan. For example, we can decide either to use the digit "9" to make calls outside the office or possibly another digit. We can decide what string a user can dial to forward a call or what string a user can dial to reach the operator. We can also decide the number of digits to be used for local extensions—we may use three, four, or five digits.

Station and System Administration Module   The system administrator uses this software module to configure the features of users' telephones, user permission, users' telephone numbers and their assigned ports on the line interface cards, the dialing plan, and many of the configurable system functions. This interface can be made very fancy using Windows and graphical forms or can be simply a line-text interface. Somewhere in between lies a good solution. We would like to have a form-like interface for our system. The system administrator simply fills the form on the screen with applicable configuration choices and submits the form to update the system with the new entries. This makes it easy because these soft forms resemble printed forms. Also, a printout of these forms can be used to easily access and respond to inquiries.

Computer Telephony Integration API   These components of our system are usually provided along with the boards and drivers we acquire from third parties. However, there are instances where only the drivers are provided with the board. In this case, we must design our own API for those boards to allow the software to use a messaging interface to communicate with the underlying telephony resources. To illustrate this idea, we will examine two examples: one API is the time slot interchanger, provided along with the boards and drivers, and the other is the DSP tones that we have to design ourselves.

Time Slot Interchanger API   This API is provided with the time slot interchanger (TSI) board. Recall that the TSI board has a switch matrix that allows two-way interconnection among any of the time slots traveling over the MVIP bus to any other time slot traveling over the bus. The way the time slots are identified over the MVIP bus is through the use of two values: stream number (1–8) and time slot number (1–32). The device driver takes these two values and commands the switch matrix to make a connection. It is really not necessary for the software to know what stream to use and what time slot within the stream we must use. Therefore, the time slots were numbered (1–256). All the software needs to do is ask the API to connect, say, time slot 7 to time slot 35 and we are done. It is the job of the API to translate to the driver the fact that time slot 35 is really time slot 3 of stream 2. We can say then that the API has two interfaces: one to the system's software side and the other to the device driver of the TSI board. Let us now examine the messages that can be passed to that API interface:

- Connect (Time slot #, Time slot #).
- Disconnect (Time slot #, Time slot #).

As you can see our API messages in this case are limited to two: connect and disconnect. The message identifies one of two values (connect, disconnect), and the message has two fields indicating the time slots to be connected.

When the API receives the message, it will determine what command or function call it should make to accomplish this task of connecting or disconnecting the time slots in question. This API can be accessed from the system software or from other APIs in the system to accomplish other resource management functions, as we will see in the next paragraphs. Keep in mind that this API was provided to us along with the TSI board and its driver. Therefore we did not have to do anything except install it.

DSP Tones API   Unfortunately we were not provided an API for the tones resource board, but we were given a driver that controls many of its functions. The driver can receive digits from the board one at a time. It is used to enable or activate certain tones. It can also be used to send digits to a touch-tone dialer (used for outgoing external calls), and so on. Let us now design an API that will make it easy for the software to send its request and receive events in the form of CTI messages.

Recall that our time slots are numbered from 1 to 256. Since we have only 48 lines, the software needs to handle only 48 local time slots and 24 time slots for the T1 lines (time slots 1–72). We will dedicate time slots 73 to 96 for the use of the tones resource. The software does not need to know anything about them; all it needs to do is ask the API to, say, apply a dial-tone to time slot 15 and the API will take care of the rest.

Let us now define our messages. Listed below are the messages that our API will handle for the system's software:

- Apply (tone type, time slot #).
- Get dialed digits(time slot #).
- Dial (dialing string).
- Test (board).

The above four messages seem very simple; to the software they are simple, but they are not that simple to implement. Each one of these messages performs a sequence of low-level operations with each operation being checked for success or failure. Each message has to be replied to when the actions are completed and so on. Let us take the *Apply* message and examine what takes place to execute it.

When the API receives an API message from the system software, it parses the identifier; in this case the identifier is apply. It then looks at the next field to determine the tone type (dial, busy, intercept, fast busy). If the tone-type field does not match one of the available tones, the message is invalid and the software is informed. If the tone type matches, say, dial-tone, then it determines the time slot given to it by the software (which corresponds to a telephone line in this case). At this point the API performs a number of actions:

- Selects an available time slot from those dedicated for tones resources (recall that time slots 72 to 96 are set aside for that purpose).
- Selects an available dial-tone circuit (the API keeps track of what is being used and what is available).

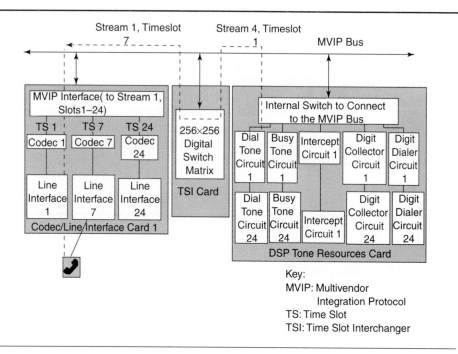

Figure 5–18
A time slot path be-
tween the dial-tone
circuit and telephone
line connected to
line interface 7.

- Sends a command to the driver to connect that dial-tone circuit to one of the tones the time slot selected in step 1.
- Sends an API message to the time slot interchanger API to connect the tones resources time slot to the line time slot.
- Sends a command to the driver to activate the dial-tone circuit.
- Replies to the system's software with the completion of the operation via a CTI message (apply success).

Figure 5–18 illustrates the path used in applying a dial-tone to a user's line. Similar procedures are followed to connect and enable other tones and dialing resources.

5.7.4 System Message Queuing Utility

With that many messages traveling around in the system, it is not practical to implement a private messaging interface between each of the system's software modules and an API. It is more practical to implement a message queuing software utility (similar to a mailbox) with a dispatcher. The system's software simply deposits the API message into the message queue and the dispatcher software routes it to the mailbox of each individual API. The dispatcher also collects messages from the APIs' mailboxes and deposits them into the message queue for the software to retrieve.

5.8 SUMMARY

In this chapter we introduced the concept of computer telephony and computer telephony integration. We saw that we can build a telephony system with most components being acquired from third parties. We have also seen that computer telephony has many useful applications that we interact with in our daily activities. The case study we examined tied many of the components of computer telephony into a switching system with most of its parts acquired from outside vendors. We needed only to develop the system's software and some of the application programming interfaces.

REVIEW QUESTIONS AND PROBLEMS

1. Give two examples of computer telephony application: one for personal use and one for business use.

2. What are the components necessary to implement a telemarketing computer telephony system (in terms of hardware and telephony resources)?

3. What are the advantages of using computer telephony integration messaging interfaces (known as APIs)?

4. In the case study in Section 5.7 we developed an API for the tones resource board. Can you write a sample C code to implement the apply CTI message?

5. What are the disadvantages of CTI from a business perspective?

6. We introduced the concept of message queuing facility. Research the UNIX operating system interprocess communication (IPC) system calls and give an example of how it may be used in a CTI message passing example.

7. In the case study of Section 5.7, what are the functions of:
   a. Call-processing module.
   b. Dialing plan.
   c. Switching-software module.
   d. System maintenance.

8. We briefly described the station and system administration software module. Give an outline of a software design for this module; if possible, implement some of the functions in the C programming language.

Modem Communications

OBJECTIVES
Topics covered in this chapter include the following:

- Role of modem in data communications.
- Modem modulation techniques.
- Signaling and bit rate.
- Huffman code compression algorithm.
- Facsimile compression techniques.
- Asymmetric digital subscriber line (ADSL).

6.1 INTRODUCTION

Computer files are collections of 1s and 0s regardless of the type of code set used to encode them. Therefore, we can say that computer files contain discrete data elements that can be easily translated into a digital signal. If we have two computers connected to each other via a digital network, then transferring a file from one computer to another is a trivial task. One needs to translate the file into a digital signal, put that signal over the transmission facility within the network, and the receiving computer encodes the received signal into a file. However, in the case when we have two computers attempting to transfer the same file over the analog telephone network, more steps are required. A conversion of the digital data into an analog signal must take place, since the telephone network was designed to carry analog signals.

Digital data transmitted as an analog signal means that the discrete data elements are transmitted as a continuous signal. This type of data and signal combination typically requires the use of a modem (modulator/demodulator). The modem modulates (alters) the digital data to an analog form for transmission and demodulates the received analog signal to reconstruct the digital data.

In the vast majority of applications, the user's digital data are first represented as a digital signal by the data terminal equipment (usually a computer and serial communication interface). This digital signal is then supplied to the modem for modulation or received from the modem as a de-modulated signal. The interface between the data terminal equipment and the modem is explored further in Chapter 7.

In this chapter, we present the role of modems in data communications and the modulation techniques. Then we examine some of the data compression techniques used to boost the data rate within the constraints of bandwidth given by the analog telephone networks. We conclude the chapter with a class of high-speed (Mbits per seconds) modems designed for emerging applications such as video-on-demand and interactive television.

6.2   ROLE OF THE MODEM IN DATA COMMUNICATION

Figure 6–1 shows a block diagram of a modem in a communications system. The computers, via their serial communication interfaces, generate and receive digital signals. A computer sends a digital signal to a modem, which transforms the digital input into an analog signal. The analog signal is sent through the analog network (such as the analog telephone network), where it is received by another modem. The receiving modem transforms the analog signal into a digital one for delivery to the serial communication interface, and in turn, to the computer.

The function of the modem is to achieve the transmission of digital data through a network that supplies a user with a bandwidth-limited channel. For example, the filter in the analog telephone network limits the user to a channel

Figure 6–1
Role of the
modem in data
communications.

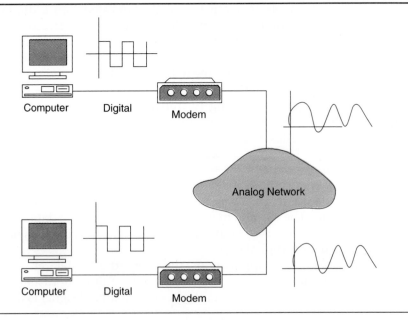

Computer   Digital   Modem

Analog Network

Computer   Digital   Modem

that has a passband of 300 to 3300 Hz. The transmission of a high-speed digital signal through the channel would result in distorted output due to the removal of the high-frequency signal component by the filter. The modem, however, converts the digital signal to an analog signal having only frequency components within the passband of the channel.

6.3 MODULATION TECHNIQUES

As described in Chapter 1, three characteristics of sine waves are of interest: amplitude, frequency, and phase. Indeed, it is these characteristics that will be modified (i.e., modulated) to allow digital data to be transmitted as an analog signal. Three modulation techniques are commonly used to translate a digital signal to an analog form. They are amplitude modulation, frequency modulation, and phase shift modulation, and a combination of amplitude and phase shift modulation called quadrature amplitude modulation (QAM). In the following sections, we will examine how these techniques are used to convert a digital bit stream into an analog signal suitable for transmission over the telephone networks.

6.3.1 Amplitude Modulation

Figure 6–2 illustrates the concept of *amplitude modulation (AM)*, also known as *amplitude shift keying (ASK)*. The figure shows two-level coding; each signal element is a sine wave that may take on one of two allowable amplitude levels and, therefore, carries a single bit, which can be a 0 or a 1. In this example, the higher amplitude signal indicates a 1-bit, and the lower amplitude signal indicates a 0-bit. There is no change in the frequency or phase of the signal. ASK can clearly be expanded to carry additional bits by defining additional amplitude levels. For example, if we allow four amplitude levels but keep the frequency and

amplitude
modulation

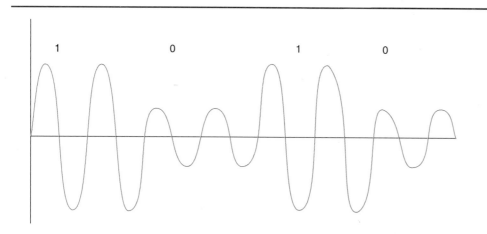

Figure 6–2
Amplitude shift keying (ASK) two-level coding.

phase constant, each level then can encode one of four different values: 00, 01, 10, and 11. If we allow eight amplitude levels, then we can encode eight values: 000, 001, 010, 011, 100, 101, 110, and 111. Therefore, it is obvious that the more amplitude levels are permitted, the more digital values can be encoded. It is worth noting that amplitude shift keying is rarely used in modem communications because of its susceptibility to interference from noise.

6.3.2   Frequency Modulation

frequency modulation

Figure 6–3 shows a two-level coding *frequency modulation (FM)* example, also known as *frequency shift keying (FSK)*. As seen in the figure, a high-frequency signal indicates a 1-bit, and a low-frequency signal is used to represent a 0-bit. In frequency modulation, there is no change in the amplitude or the phase of the sine wave. FSK can also be expanded to encode more than just the 1-bit and 0-bit. This is accomplished by defining additional frequencies. For example, if we define a four-frequency level sine wave, four values would be encoded, 00, 01, 10, and 11. FSK is used in inexpensive and low-speed modems (1200 bits per second and below).

6.3.3   Phase Shift Keying

phase shift modulation, and phase shift keying

Phase shift modulation (PM), commonly called *phase shift keying (PSK),* is often used in modems. In this technique, the amplitude and frequency of the sine wave are held at a constant level. At the beginning of a bit time, the phase of the signal is altered. As shown in Figure 6–4, a 1-bit is indicated by a 180-degree phase shift in the signal, and a 0-bit is indicated by the absence of a phase shift

Figure 6–3
FSK—two-level coding.

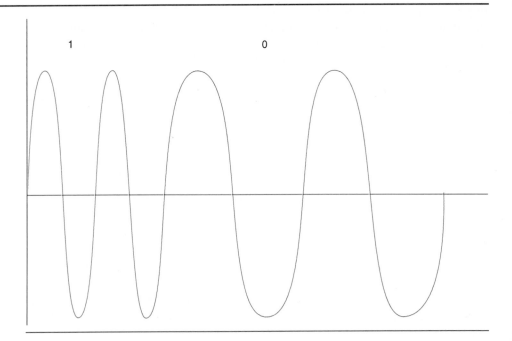

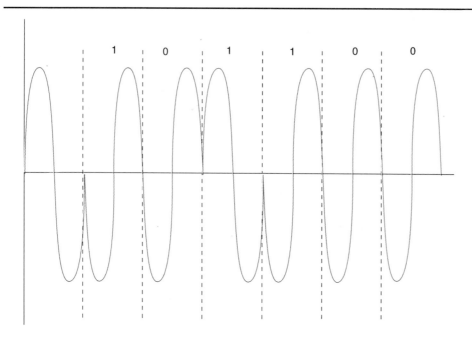

Figure 6–4
Differential phase
shift keying (DPSK).

Because the shift in phase is from the present phase rather than
from an absolute standard, the procedure is called differential phase
shift keying.

(i.e., 0-degree phase shift). As in the examples given for amplitude and frequency modulation, this example presents a two-level encoding carrying a
single bit per signal element.

More precisely, the technique presented here is called *differential phase shift
keying (DPSK)*. The phase shift of the signal is determined by comparing the
phase of the current signal with the phase of the signal sent during the previous bit time. Thus the two bit times are compared and a differential is determined. A phase shift of 180 degrees would indicate the reception of a 1. Because the shift in phase is determined from the present phase compared to the
last bits rather than from an absolute standard, the procedure is referred to as
differential. Phase shift keying is used in moderate-speed modems.

To further explain phase shift keying, let us consider the example in
Figure 6–5 where eight-phase DPSK is employed on a 1600-Hz carrier frequency. We refer to the sine or cosine wave used to encode the binary bits as
the *carrier signal*. Each consecutive three bits, called *tribits*, of the binary serial
input data to the modem are encoded into a single phase change of the carrier
frequency. The encoding of the three bits into a tribit allows the representation
of eight possible phase changes of the carrier frequency. This gives us an effective transmission bit rate of 4800 bits per second (3 × 1600 Hz = 4800 bits per
second).

Figure 6–5
Phasor diagram for a
DPSK 4800-bps
modem, using a
1600-Hz carrier sine
wave.

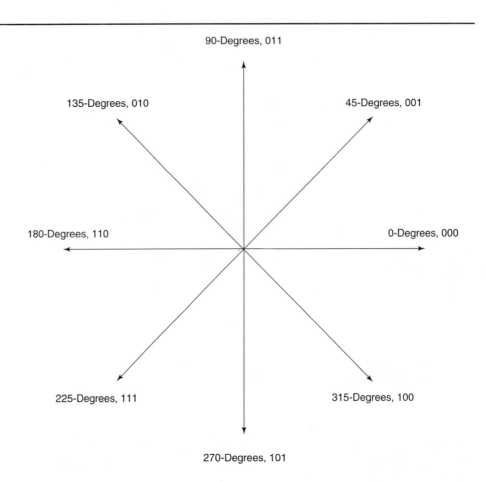

90-Degrees, 011

135-Degrees, 010   45-Degrees, 001

180-Degrees, 110   0-Degrees, 000

225-Degrees, 111   315-Degrees, 100

270-Degrees, 101

Encoding additional bits into a greater number of phase changes will result in increasing the data transfer rates. Four bits, or a *quadbit,* for example, can be encoded into 16 possible phase changes. The phase differential between adjacent phasors is 22.5 degrees (360 degrees ÷ 16 = 22.5 degrees). The problem here, however, is that any phase jitter in excess of 11.25 degrees would result in the detection of erroneous data. This amount of phase shift is not uncommon in long-haul networks that utilize regenerative repeaters and digital multiplexers. The modulation technique, discussed in the next section, allows a more reliable high-data transfer rate.

6.3.4   Quadrature Amplitude Modulation

quadrature
amplitude
modulation

Quadrature amplitude modulation (QAM), pronounced "kwamm," is a modulation technique that varies both phase shift and amplitude, but holds the carrier frequency constant to encode the serial binary data. Figure 6–6 shows an example where 12 different phases and 3 different amplitudes represent 16 possible carrier states. For a 2400-Hz carrier signal, we achieve 9600 bits per

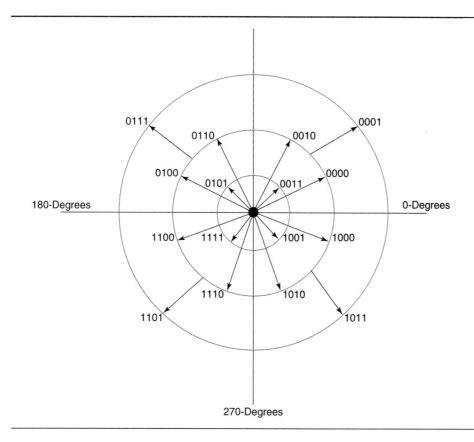

Figure 6–6
Quadrature amplitude modulation (QAM) phasor diagram for a 9600-bps modem using a 2400-Hz carrier signal.

second data transfer rate ($4 \times 2400 = 9600$). In the phasor diagram, the inner circle represents an amplitude of 1 volt, the outer circle represents an amplitude of 3 volts, and the middle circuit represents an amplitude of 2 volts. The encoded quadbits in the first quadrature are shown in Table 6–1.

Data rates of 14,400, 19,200, 28,800, and 33,600 are possible using QAM. This can be accomplished by increasing carrier frequency and the number of states used to encode the input serial data. Combining data compression and QAM can result in even higher throughput reaching, in some cases 128,000 bits per second.

Table 6–1
Quadbits in first quadrature.

Quadbit	Amplitude in Volts	Phase Shift in Degrees
0000	2	22.5
0001	3	22.5
0011	1	22.5
0010	2	22.5

Figure 6–7
Trellis coded modu-
lation (TCM).

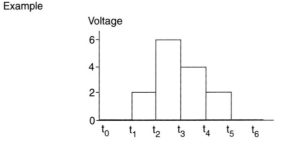

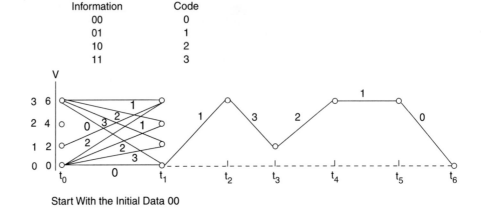

Start With the Initial Data 00

6.3.5 Trellis-Coded Modulation

trellis-coded
modulation

Trellis-coded modulation is used to detect and possibly correct errors when they occur on the transmission lines. Figure 6–7 Illustrates the steps used in this scheme. It combines coding and modulation for improving the reliability of a digital transmission without increasing the transmitted power or the required bandwidth. Trellis-coded modulation is used in high-speed modems (14,400, 28,800, 33,600, and 56,000 bits per second).

6.4 SUMMARY OF MODULATION TECHNIQUES

Table 6–2 shows the main features of the modulation techniques we have described. We can briefly summarize these techniques as follows:

• *Amplitude shift keying.* Modulates the amplitude of the signal. It is simple to build, but it is very susceptible to interference from noise and is not commonly used in data communications.

• *Frequency shift keying.* Modulates the frequency of the signal. It is also simple to build and is not excessively susceptible to noise interference. FSK is used in inexpensive low-speed (1200 bps) modems.

Modulation Techniques	Varying Parameter	Major Properties	Applications
Amplitude shift kying	Amplitude	Simple, but susceptible to noise	None
Frequency shift keying	Frequency	Simple	Low-speed modems
Differential phase shift keying	Phase	Self-clocking, easy to carry multiple bits/ signaling elements	Moderate-speed modems
Quadrature amptitude modulation	Amplitude and phase	Self-clocking, error detection	High-speed error-checking modems
Trellis-coded modulation	Amplitude and phase	Error correction	High-speed error-correcting modems

Table 6–2
Various modulation techniques and their properties.

• *Differential phase shift keying.* Modulates the phase of the sine wave. The phase transition provides timing information, as well as the signal value. Multiple bits can be easily carried with each signal element. This modulation scheme is often used in moderate-speed modems (up to 4800 bits per second).

• *Quadrature amplitude modulation.* Modulates the amplitude and frequency of the carrier signal. It provide high-speed rates, clock synchronization, and the ability to add error-detection capabilities. It is commonly used in high-speed modems (9600 bits per second and above).

• *Trellis-coded modulation.* Combines coding and modulation schemes to achieve improvement in transmission reliability without increasing the required bandwidth and power. It can detect and correct errors. It is used in high-speed modems (14,400–56,000 bps).

6.5 SIGNALING AND BIT RATES

The rate of transmission in data communications systems can be expressed in terms of the number of bits sent per second or the number of signaling elements sent per second. Depending on your perspective and application, both measures are important. The signaling rate, known as *baud*, is expressed as the number of signaling elements per second, or baud. A 1200-baud transmission then is one that sends 1200 signaling elements per second.

baud

The definition of a signaling element will depend on the transmission scheme used. One example, as discussed in the frequency shift keying section, defines the signaling elements as different frequency sine waves. Different signaling elements within a transmission might have different durations. Typically, the baud is calculated based on the signaling element of the shortest length. Thus, if the shortest signaling element lasts X seconds, the baud is $1/X$.

Signaling (baud) rate and bit transfer rate are not always the same. This is because a signaling element might carry more than a single bit. The relationship between signaling rate and bit rate is

Bit rate = (signaling rate) × (number of bits per signaling element).

The bit rate is given as the number of bits per second (bps). Clearly, "bps" = "baud" only if one bit is carried per signal.

Figure 6–8 shows an example using four-level encoding—two bits are carried per one signaling element. In this case the bit rate will be twice the signaling rate. Each signaling element lasts 0.833 milliseconds and carries 2 bits. Therefore

$$Signaling\ rate = \frac{1}{0.833\text{ms}} = 1200 \text{ baud}$$

$$Bit\ rate = \frac{1200 \text{ baud} \times 2 \text{ bits}}{\text{signaling element}} = 2400 \text{ bits per second.}$$

A final point on nomenclature is appropriate. The term *baud rate* is often used incorrectly. The term is redundant since baud is a rate! (Note that we do not speak of the *speed rate* of an element. Similarly, it is common to come across the term *bits/baud* to mean *bits/signal element*. This term, too, is improper since baud is not a synonym for signal element.) In this text, baud will refer strictly to signal elements per second.

Figure 6–8
Bit rate versus sig-
naling rate.

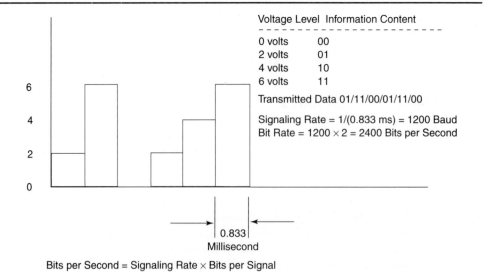

Bits per Second = Signaling Rate × Bits per Signal

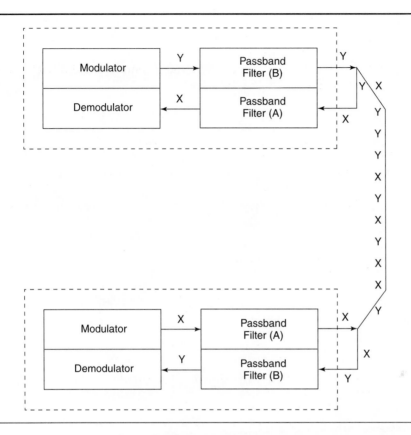

Figure 6–9
Two-wire full-duplex
operation.

6.6   TWO-WIRE FULL-DUPLEX OPERATION

Two-wire asynchronous modems may provide for either full-duplex or half-duplex operation. To operate in full-duplex mode is equivalent to two distinct circuits. This can be accomplished by dividing the bandwidth of the link into two distinct passbands and requiring that the two parties use separate passbands for transmission. Appropriate filtering can ensure that a modem can hear the transmission from its opposite and not its own. Figure 6–9 illustrates this situation when signals from both ends coexist on the telephone loop, but filters ensure that only the correct signals are received.

6.7   COMMON MODEM STANDARDS

As most developments in telecommunications are guided by standards, so is modem design. It is carried out within the framework of a set of standards. Table 6–3 shows some of the common modem standards that are in use today.

Table 6–3
Common modem standards.

Modem Type	Bit Rate (bps)	Signaling Rate (baud)	Modulation Technique	Transmission Scheme
103	300	300	FSK	FDX
V.21	300	300	FSK	FDX
202	1200	300	FSK	HDX
212A	1200	600	DPSK	FDX
V.22	1200	600	DPSK	FDX
201	2400	1200	DPSK	HDX
V.22bis	2400	600	QAM	FDX
V.26ter	2400	1200	DPSK	FDX/EC
209	9600	2400	QAM	HDX
V.29	9600	2400	Modified QAM	HDX
V.32	4800	2400	QAM	FDX/EC
	9600	2400	Modified QAM	
V.32bis	14,400	2400	Trellis coding	FDX/EC
V.32terbo	19,200	2400	QAM	FDX/EC
V.34	28,800	2400	Trellis coding	FDX/EC
		3000		
		3200		
		2743		
		2800		
		3429		
V.34bis	33,600	2400	QAM	FDX/EC
		2749	Trellis coding	
		2800		
		3000		
		3200		
		3429		

Modem type refers to the standard being summarized. The bit rate, signaling rate, and modulation technique are also given. Whenever the modem operates in half-duplex (HDX) or full-duplex (FDX) mode over the two-wire local loop is indicated in the table. Bell standards are also shown (103, 202, 212A, 201, and 209) as well as *International Telecommunications Union—Telecommunications Standardization Sector (ITU-T)* recommendations (V.21, V.22, V.22bis, V.26ter, V.29, V.32, V.34, V.42, and V.42bis). All differential phase shift keying modems carry two bits per signal. In the following subsection, we will selectively examine some of the widely used modem standards.

ITU-T

6.7.1   ITU-T V.22BIS Recommendation

ITU-T V.22bis (*bis* means second revision in French and *ter* means third revision) recommendation provides for 1200- and 2400-bps synchronous full-duplex communication over switched and two-wire lines. Four-phase DPSK is employed as the modulation technique for operation at 1200 bps. QAM is

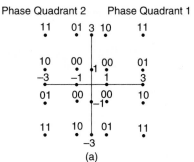

Phase Quadrant 2   Phase Quadrant 1

(a)

Figure 6–10
(a) V.22bis 16-point
constellation, and
(b) V.22bis line en-
coding.

First Two Bits in Quadbit (2400 bps) or dbit Value (1200 bps)	Phase Quadrant Change
00	1 → 2 2 → 3 3 → 4 90° 4 → 1
01	1 → 1 2 → 2 0° 3 → 3 4 → 4
11	1 → 4 2 → 1 270° 3 → 2 4 → 3
10	1 → 3 2 → 4 180° 3 → 1 4 → 2

(b)

used to achieve a data rate of 2400 bps. Signaling rate is specified at 600 baud for both operating speeds. Full-duplex operation is achieved by phase and amplitude shift keying a low-channel carrier frequency of 1200 Hz and a high-channel carrier frequency of 2400 Hz.

Phase and amplitude assignment for dibit (1200 bps) and quadbit (2400 bps) encoding is depicted by ITU-T as a 16-point signal constellation rather than a phasor diagram. They are essentially the same. Figure 6–10 illustrates the signal constellation of the V.22bis recommendation.

6.7.2   ITU-T V.32 Recommendation

The ITU-T V.32 recommendation is intended for the use of 9600-bps synchronous modems connected to switched and leased lines. The recommendation includes rates of 2400 bps and 4800 bps as well as 9600 bps. QAM is the

Figure 6–11
V.32bis signal constellation (32-point signal structure with trellis coding for 9600 and states ZWYX used at 4800 bps).

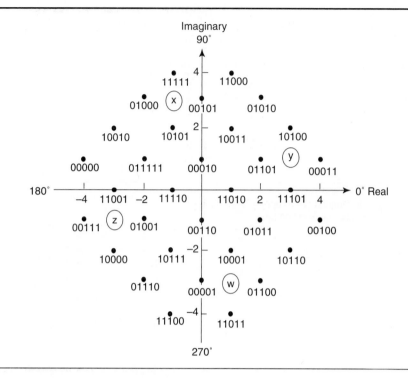

modulation technique employed on a carrier frequency of 1800 Hz. V.32 includes trellis encoding as an option. Trellis encoding divides the data stream to be transmitted into five consecutive bits or *quintbits*. A 32-point signal constellation is achieved. As mentioned earlier, trellis encoding results in a superior signal-to-noise ratio. Figure 6–11 depicts the 32-point signal constellation diagram.

6.7.3 ITU-T V.42 and V.42BIS Recommendations

The main contribution of V.42 revolves around a new protocol called *link access procedure for modems* (LAP M). Link access procedures are discussed in detail in Chapter 7. In an effort to enhance the performance of error-correcting modems that implement the V.42 standard, ITU-T adopted the V.42bis standard. It addresses data compression for data communication equipment using error-correcting procedures. Modems employing data compression have significantly increased throughput performance over their predecessors. The V.42bis can achieve 3:1 to 4:1 data compression ratios for ASCII text. Throughput rates up to 56,000 bps can be achieved by today's modems that employ V.42bis data compression.

6.7.4 The V.34 ITU-T Modem Recommendation

V.fast was adopted in June 1994 and has been designated V.34. The new standard speed is 28,800 bps without data compression. With data compression, the modem is capable of sending data two or three times as fast. V.34 automatically adapts to line conditions and adjusts its speed up or down to ensure data integrity.

Several innovations have enhanced the speed and flexibility of V.34. They include: nonlinear coding, multidimensional coding, constellation shaping, reduced complexity decoding, precoding, and line probing. In 1998, the maximum speed achieved by employing those innovations and data compression was 115,200 bps. Data compression plays an important role in pushing the limits in file transfer throughput. In the next sections, we discuss a number of data compression techniques used in a wide range of applications.

6.8   DATA COMPRESSION TECHNIQUES

Data compression plays an essential role in today's data communication. Either in the transmission of text files or digital motion picture images, data compression provides the tool to achieve the high throughput performance we are witnessing. The term *data compression* refers to the ability to remove redundant elements from the transmitted data stream, while the receiving end is capable of recovering the original information. Two types of data compression techniques, lossy and lossless, depending on the particulars of the application, are in use today.

Lossy compression is widely used to encode still and motion images for transmission over a variety of communication media. With this technique, we may get rid of information elements that are considered unimportant to the overall image. Depending on the number of information elements we remove from the original source, the resolution of the received image is affected. For example, when we transmit motion video over the telephone network, because of bandwidth limitations we are forced to preserve a minimum amount of image data, and the resolution of the received image is very low. On the other hand, if we transmit the same image over a digital network with moderate bandwidth (T1 rate), the compressed transmitted data has more information elements, and thus the received image has a high resolution. We will discuss the types of compression techniques used in video conferencing in detail in Chapter 12.

On the other hand, *lossless compression* preserves all of the information in the original data so that, upon decompression, that data can be fully recovered. For example, consider a transmission of data files that contain financial information. If a single number in a received file is missing one digit, the result can be unpleasant. Another example, in the medical field, is transmitted X-ray images. By the nature of those images, they seem to contain artifacts. As engineers, we are not trained to interpret X-ray images. A doctor can diagnose a medical condition based on the presence or absence of what seems to us to be artifacts. In the following subsections, we will examine some of the most widely used lossless compression techniques, including run-length, facsimile compression techniques, and string-matching algorithms.

data compression

lossy compression

lossless compression

6.8.1   Run-length Encoding

run-length
encoding

Run-length encoding is a simple lossless compression technique that can be quite effective for text compression. It is also used in facsimile compression. We start in this section with the simpler null suppression technique and then describe run-length encoding, which is a generalization of null suppression.

null suppression

Null Suppression   One of the oldest and simplest data compression techniques is known as *null suppression* or *blank suppression*. A long string of blanks, or nulls, are a common occurrence in character streams. With the null suppression scheme, the transmitter scans the data for strings of blanks and substitutes a two-character code for any string that is encountered. The code consists of a special control character followed by a count of the number of blanks. For example, the string of characters

ABCD▯▯▯▯▯▯▯▯▯EFG

is replaced by

ABCDZ9EFG

where Z is a special compression indicator character and ▯ is the symbol for a blank space. This scheme results in a saving for all strings of three or more blank characters.

When null suppression is in use, the receiver of the transmission scans the incoming characters for the special character used to indicate null suppression. Upon detection of that character, the receiver knows the next character contains the count of the number of blanks that were eliminated. From this information, the original data can be reconstructed.

Even though this scheme seems primitive, it is easy to implement. Furthermore, the payoff from this simple technique can be substantial. Some computers that did not previously use any form of data compression achieved a throughput gain of 30 to 50 percent when switched to null suppression.

The Run-length Algorithm   This technique is a generalization of the null suppression scheme. Run-length encoding is used to compress any type of repeating data sequence. Figure 6–12 summarizes the technique when applied to character data and gives some examples. As with null suppression, the transmitter looks for sequences of repeating characters to replace. In this case any sequence of repeating characters is eliminated and replaced by a three-character code. The code consists of a special character that indicates suppression, followed by the character to be suppressed, then followed by a count that indicates the number of occurrences. Therefore, any sequence of four or more identical characters can be suppressed with a net reduction of the total number of characters transmitted.

Run-length encoding efficiency depends on the number of repeated character occurrences in the data to be compressed and the average repeated character length. A measure for compression is the ratio of the length of the

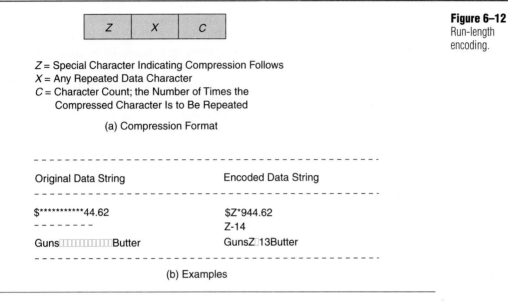

Z = Special Character Indicating Compression Follows
X = Any Repeated Data Character
C = Character Count; the Number of Times the
    Compressed Character Is to Be Repeated

(a) Compression Format

Original Data String	Encoded Data String
$***********44.62	$Z*944.62
- - - - - - - -	Z-14
Guns⬜⬜⬜⬜⬜⬜⬜⬜⬜⬜⬜Butter	GunsZ⬜13Butter

(b) Examples

Figure 6–12
Run-length
encoding.

uncompressed data to the compressed data. On average a compression ratio
of 15 : 1. is possible with this technique. The performance of the technique depends on the input stream.

6.8.2 Huffman Code

Huffman code is another one of the oldest data compression techniques. It has
been in use for more than 25 years and has served as a springboard for many
of the advanced techniques used today. The idea behind Huffman encoding is
to reduce the number of bits that represent those characters that have a high
rate of occurrence. For example, the letter *e* is the most common letter in the
English text and the letter *z* is the least common. Huffman encoding takes this
fact into account. It encodes the most probable characters (those that appear
most frequently in text) with fewer bits and the least probable characters with
more bits.

Huffman code

Consider a hypothetical text file that contains only the set of characters *A, D,
E, H, K, L, M,* and *R*. The characters have the following probabilities of occurrence:

$P(A) = 0.20$ $P(D) = 0.22$
$P(E) = 0.50$ $P(H) = 0.02$
$P(K) = 0.015$ $P(L) = 0.007$
$P(M) = 0.013$ $P(R) = 0.025$

This means that out of all the characters in the text file, *E* will repeat 50 percent
of the time, *D* will repeat 22 percent of the time, and *R* will repeat only 2.5 percent of the time, and so on. Suppose we wish to encode these characters in binary form. One obvious assignment is to use fixed-length three-bit code with

three-bit value for each of the eight characters. A better strategy is to use a variable-length code in which longer code words are assigned to less probable characters and shorter words are assigned to more probable characters. This is the technique used in Morse code, and more efficiently in the Huffman code.

In our hypothetical file, each character is to be uniquely encoded as a binary sequence. What we are interested in is the construction of an optimal code—one that gives the minimum average length for the encoded text file.

Another way to look at the requirement is to consider that we are given a file that is already encoded using fixed-length assignment of binary words to characters. So if there are eight characters, each is encoded in three bits. If there are 16 characters, each will be encoded in four bits, and so on. The requirement then can be stated that we are looking for an optimal variable-length coding scheme that gives the minimum average length for the compressed file.

Now let us see how the Huffman method can be used to construct this optimal code. We begin by arranging all the characters to be encoded in order of decreasing probability so that we have the characters (E, D, A, R, H, K, M, L) with probabilities $P(E)$, $P(D)$, $P(A)$, . . . $P(L)$. Next, we combine the last two characters into an equivalent character with the probabilities of the two combined characters added together (see Figure 6–13). The code for these two characters will be the same except for the last digit. We now have a new set of seven characters. If necessary we reorder the characters in the new set in decreasing probabilities as we did initially.

Figure 6–13 illustrates this process for the set of characters with their probabilities. Consider each character to be a leaf node in a tree to be constructed. Merge the two nodes of the lowest probability into a node whose probability is the sum of the two constituent probabilities. At each step, repeat this process until one node remains. The result is a tree in which each node except the root has one branch from the right and two branches from the left. At each node label the two branches on the left with 0 and 1, respectively. The code word for any character is the string of labels from the root node back to the original character; all of the resulting code words are shown along the left-hand side of the figure.

Some of the properties that one must watch for when working these examples are as follows:

- No two characters can consist of an identical arrangement of bits.
- No code word can be a prefix of another code word.
- Higher-probability characters have shorter code words.
- The two least probable code words have equal length and differ only in their final digit.

Table 6–4 summarizes the properties of the Huffman code for our example. The average length of an encoded file, $E(L)$, is an expected value calculation

$$E(L) = \sum_{i=1}^{i=N} = L_i P_1.$$

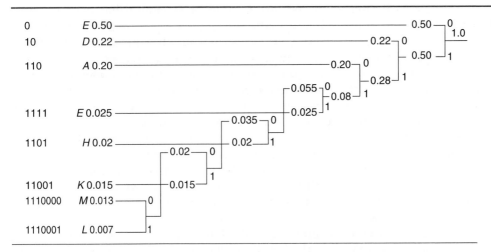

Figure 6–13
Huffman code for eight characters.

Table 6–4
Huffman code for eight characters.

Character	Code	Probability	Length	Probability × length
E	0	0.5	1	0.5
D	10	0.22	2	0.44
A	110	0.20	3	0.6
R	1111	0.025	4	0.1
H	1101	0.02	4	0.08
K	11001	0.015	5	0.075
M	1110000	0.013	7	0.166
L	1110001	0.007	7	0.049
				average length = 2.01

where L_i = length of code i,
         P_i = probability of the code i, and
         N = the number of codes.

So, for our example

$$E(L) = 1 \times 0.5 + 2 \times 0.22 + 3 \times 0.2 + 4 \times 0.025 + 4 \times 0.02 +$$
$$5 \times 0.015 + 7 \times 0.013 + 7 \times 0.007 = 2.01.$$

So if we have a text file that consists of 1000 characters, the average length of the encoded file is 2010 bits. If a straightforward assignment of 3 bits per character is used, then the file length would be 3000 bits.

6.8.3 Facsimile Compression

Data compression techniques are essential to the widespread use of digital facsimile. Consider a typical page with 200 black or white pels resolution. A *pel* is   pel pixel

the smallest discrete scanning-line sample of facsimile that contains only black-white (no gray-scale information). A *pixel* is a picture element that contains gray-scale information. The 200-pel resolution generates 3,740,000 bits (8.5 inches × 11 inches × 40,000 pels per square inch). At the basic ISDN rate 64 kbps, a page would take about one minute to transmit. Ultimately, users will expect their systems to operate at a rate similar to that of a copier, or one page every few seconds. To meet this requirement without extraordinarily high channel bandwidth requires the use of data compression.

modified Huffman code

modified READ

ITU-T met this requirement by specifying two lossless data compression techniques for facsimile. They are the *modified Huffman code,* and the *modified READ.* Modified Huffman was adopted for facsimile data compression by ITU-T Group-3 in 1988 with modified READ as an option. Group-4 facsimile specifies modified READ for use in facsimile. In the following paragraphs we briefly discuss these two facsimile standards.

Group-3 is the first digital facsimile standard. This system provides only black-and-white values, with sampling densities of 200 spots per inch horizontally across the paper and 100 or 200 lines per inch vertically down the page. Group-3 uses a digital encoding technique and incorporates a means of reducing the redundant information in the document signal prior to modulation. The assumption here is that transmission takes place via modems over analog telephone lines. Transmission rate is enhanced by a factor of three or more compared to prior ITU-T standards.

Group-4 is also a black-and-white digital facsimile standard. It is intended for use over digital networks at speeds of up to 64 kbps and with provision for error-free reception. Resolution between 200 to 400 pels per inch is specified. As with Group-3, compression techniques are used to reduce the number of transmitted bits. When using Group-4, transmission time drops to a few seconds rather than the minutes of the earlier standard. In the following subsections, we discuss the modified Huffman and modified READ compression techniques used in both Group-3 and Group-4 standards.

Modified Huffman Code Black-and-white areas tend to cluster in typical documents. We observe that there are long runs of black (B) and white (W) pels if we view the document as a sequence of lines. This property suggests the value of compression based on run-length encoding. The two-level (black or white) input is converted to many run-length units, each consisting of the black or white symbol and the count for each symbol as discussed earlier. This run-length input is subsequently coded for transmission. In addition, because longer runs of black or white are, in general, less probable than shorter runs, variable-length coding can be employed advantageously. The Huffman code, described in the previous section, could be used for the purpose of facsimile encoding. It could be applied to an image one horizontal line at a time to encode the sequence of black-and-white pels. For example, suppose that the scan of a single line of an image produces a sequence of W7, B7, W4, B8, W4, B7, and W10. If we consider each of these elements as a symbol in a source alphabet, then Huffman encoding can be used to encode the source. However, since ITU-T standards require at least 1728 pels per line, the number of different codes and hence the average length of a code is very large.

An alternative is the modified Huffman encoding technique. Modified Huffman regards a run-length N as the sum of two terms

$$N = 64m + n; m = 1, 2, \ldots; n = 0, 1, 2, \ldots 63.$$

More precisely, each run of black or white pels consist of a multiple of 64 points plus a remainder.

Each run length can now be represented by two values, one for m and one for n. These values can then be encoded using Huffman code. For example, a string of 200 black points in a row can be expressed as $64 \times 3 + 8$. For this purpose, ITU-T has defined eight representative documents and calculated the probabilities of the different run-length occurrences. Since these probabilities are different for black and white, two sets of probabilities were calculated and two tables were developed from this information, as shown in Table 6–5. The terminating code is used for run lengths less than 64. For run lengths greater than 64, a combination of terminating code (n) and a makeup code (m) is needed. Table 6–5 lists examples of run lengths and their corresponding terminating code words.

Modified READ Code   Applying modified Huffman encoding to facsimile transmission significantly reduces the number of bits that must be transmitted compared to a straightforward bit-map image. However, further gain can be achieved by recognizing that a strong correlation exists between the black-white patterns of two adjacent lines. In fact, for a typical facsimile document, approximately 50 percent of all black-white and white-black transitions are directly underneath a transition on the previous line. Additionally 25 percent of the adjacent lines differ by only one pel. Therefore, approximately 75 percent of transitions can be defined by a relationship that is plus or minus at most one pel from the line above it. This principle is the underlying basis for the modified Relative Element Address Designate (modified READ) code.

In modified READ encoding, run lengths are encoded based on the position of changing elements. A changing element is defined as a pel of a different color from the immediately preceding pel on the same line. A changing element $W1$ is coded in terms of its distance to one of two references: either a preceding changing element $W0$ on the same line or a changing element $X1$ on the previous line. The selection of $W0$ or $X1$ depends on the exact configuration, as explained in the next paragraphs.

Figure 6–14 illustrates the five changing elements that are defined for this scheme, which are as follows: (Note that W always refers to the coding for the current line and X always refers to the reference or previous line.)

$W0$:   The reference or starting changing element on the coding line. At the start of the coding line, it is set to an imaginary white changing element just to the left of the first element on the line. During the coding of a line, it is redefined after every encoding.

Table 6–5
Modified Huffman code table.

Run Length	White	Black	Run Length	White	Black
		Terminating Code Words			
0	00110101	0000110111	32	00011011	000001101010
1	000111	010	33	00010010	000001101011
2	011	11	34	00010011	000011010010
3	1000	10	35	00010100	000011010011
4	1011	011	36	00010101	000011010100
5	1100	0011	37	00010110	000011010101
6	1110	0010	38	00010111	000011010110
7	1111	00011	39	00101000	000011010111
8	10011	000101	40	00101001	000001101100
9	10011	000100	41	00101010	000001101101
10	10100	0000100	42	00101011	000011011010
11	00111	0000101	43	00101100	000011011011
12	01000	0000111	44	00101101	000001010100
13	001000	00000100	45	00000100	000001010101
14	110100	00000111	46	00000101	000001010110
15	110101	000011000	47	00001010	000001010111
16	101010	0000010111	48	00001011	000001100100
17	1-1-11	0000011000	49	01010010	000001100101
18	0100111	0000001000	50	01010011	000001010010
19	0001100	00001100111	51	01010100	000001010011
20	0001000	00001101000	52	01010101	000000100100
21	0010111	00001101100	53	00100100	000000110111
22	0000011	00000110111	54	00100101	000000111000
23	0000100	00000101000	55	01011000	000000100111
24	0101000	00000010111	56	01011001	000000101000
25	0101011	00000011000	57	01011010	000001011000
26	0010011	000011001010	58	01011011	000001011001
27	0100100	000011001011	59	01001010	000000101011
28	0011000	000011001100	60	01001011	000000101100
29	00000010	000011001101	61	00110010	000001011010
30	00000011	000001101000	62	00110011	000001100110
31	00011010	000001101001	63	00110100	000001100111
		Make-up Code Words			
64	11011	0000001111	960	011010100	0000000110011
128	10010	000011001000	1024	011010101	0000001110100
192	010111	000011001001	1088	011010110	0000001110101
256	0110111	000001011011	1152	011010111	0000001110110
320	00110110	000000110011	1216	011011000	0000001110111
384	00110111	000000110100	1280	011011001	0000001010010
448	01100100	000000110101	1344	011011010	0000001010011
512	01100101	0000001101100	1408	011011011	0000001010100
576	01101000	0000001101101	1472	010011000	0000001010101
640	01100111	0000001001010	1536	010011001	0000001011010
704	011001100	0000001001011	1600	010011010	0000001011011
768	011001101	0000001001100	1664	011000	0000001100100
832	011010010	0000001001101	1728	010011011	0000001100101
896	011010011	0000001110010	EOL	000000000001	000000000001

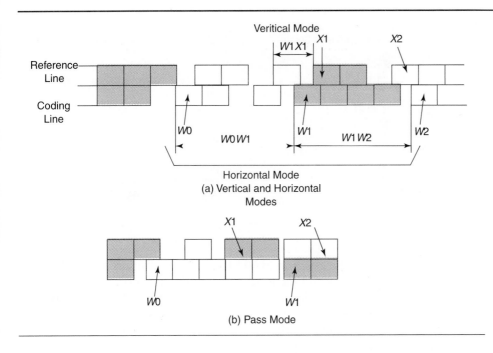

Figure 6–14
Changing picture elements for the modified READ technique.

W1:   The next changing element to the right of W0 on the coding line.
W2:   The next changing element to the right of a1 on the coding line.
X1:   The first changing element on the reference line to the right of A0 and of opposite color of A0.
X2:   The next changing element to the right of X1 on the reference line.

The encoding procedure is summarized in Table 6–6 and can be defined as follows:

- Step1:
  a. If the position of X2 lies to the left of W1, this is coded using word 0001. After this encoding, the position of W1 is shifted to lie under X2. This is referred to as pass mode. The algorithm resumes at Step 1.
  b. If the condition in (a) is not satisfied, go to Step 2.
- Step 2:
  a. If the position of W1 is within three of the position of X1 ($|X1W1|3$), then W1 is coded in the vertical mode, after which the old position W1 becomes the new position W1, W2 becomes W1, and so on.
  b. If the position of W1 is not within three of the position of X1, then W1 is coded in the horizontal mode. Following the horizontal mode code 001, W0W1, and W1W2 are encoded by one-dimensional modified Huffman coding. After this, the old position W2 becomes the new position W0.

Table 6–6
Modified READ code table.

Mode	Element to be Coded		Notation	Code Word
Pass	$X1, X2$		P	0001
Horizontal	$W1\,W1$, $W1\,W2$		H	$001 + M(W0\,W1) + M(W1\,W2)$
	$W1$ is just under $X1$	$W1\,X1 = 0$	$V(0)$	
Vertical	$W1$ to the right of $X1$	$W1\,X1 = 1$	$Vr(1)$	011
		$W1\,X1 = 2$	$Vr(2)$	000011
		$W1\,X1 = 3$	$Vr(3)$	0000011
	$W1$ to the left of $X1$	$W1\,X1 = 1$	$Vl(1)$	010
		$W1\,X1 = 2$	$Vl(2)$	000010
		$W1\,X1 = 3$	$Vl(3)$	0000010

Step 1 is used to move the position of b1 along after execution of Step 2. Also, Step 1 has the effect of avoiding long run length. In Step 2, if the current changing element to be encoded is within three positions of the same transition in the previous line, then the position is encoded with one of seven possible values using modified Huffman. This situation will hold most of the time. In the few cases in which a transition in the current line is not within three positions of the same transition in the previous line, the next two runs are encoded using modified Huffman.

6.9   ASYMMETRIC DIGITAL SUBSCRIBER LINE

In the above sections we have seen the techniques used to transmit digital data over the analog local loop (one wire pair from the subscriber premises to the local office). The speed limitations were imposed by the fact that the maximum frequency spectrum allowed for modem operations is the 4 kHz, imposed by the audio filters in the telephone system. Recall that most of the intelligent speech lay between 300–3300 Hz. This is not to say that the single-wire pair between the subscriber's premise and the local office could not handle higher bandwidth, in fact, it can handle bandwidth in the Mbps range. A new class of modems designed for applications that require high bandwidth between the serving network and the service subscriber have appeared.

An example of such an application is the interactive video-on-demand system. The goal of this application is to compete with video tape rentals. Instead of going to the video rental store, the subscriber uses the remote control to operate a modem connected to the TV. In real time, the user can play a movie, fast forward, stop, rewind, and resume playing of a movie at his or her convenience. The movie images reside in a server owned and operated by the service provider (in this case the telephone company). For quality comparable to that of cable TV, a high downstream (from the network to the subscriber) bandwidth is required. The upstream bandwidth, however, may not need to be large since only control functions are carried over it. One class of modems used to implement these functions, and keep the existing telephone services intact is called an *asymmetric digital subscriber loop (ADSL)*. Asymmetric in this context means that the downstream bandwidth is quite large compared to the upstream bandwidth. In the next subsection we examine the technology and implementations of ADSL modems.

6.9.1   Basics of DMT

The *discrete multitone (DMT)* is a multilevel encoding scheme that achieves its high bit rate by encoding groups of bits in discrete subchannels, commonly referred to as *bins*. DMT aims for a maximum of 15 bits per Hz inside, although outside of perfect conditions (which do not exist), this number is normally never approached. DMT is a standard developed by the American National Standards Institute (ANSI) in 1993 in a recommendation designated T1.413.

DMT

At a very basic level, DMT asymmetric digital subscriber loop (ADSL) systems can be thought of as employing 256 mini-modems, 4 kHz each, that run simultaneously in a single chipset. Figure 6–15 shows how DMT utilizes the frequency spectrum above 4 kHz. In a DMT system, bits are assigned to bins on the basis of each bin's signal-to-noise ratio—the bins that have the highest ratio taking the most bits. If noise increases above a certain level on a given bin, bits from that bin are moved to bins with lower noise (i.e., higher signal-to-noise ratio).

In T1.413, DMT's 256 bins are grouped into larger channels: four unidirectional, downstream and three bidirectional, downstream and upstream. Most

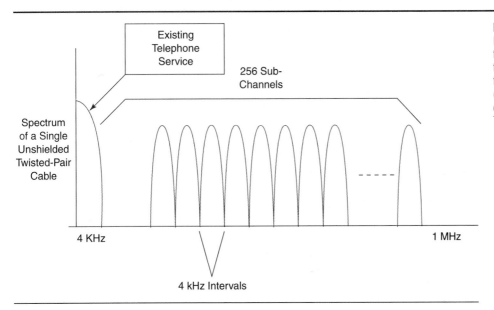

Figure 6–15
DMT's utilization of the frequency spectrum on a single twisted-pair wire (i.e., telephone cable) for ADSL T1.413.

DMT systems use echo cancellation techniques (similar to the one discussed in the two-wire full duplex operations in the previous sections) to separate these channels. Note that in the ADSL systems, telephone service is maintained by transporting digitized voice signals between the subscriber's digital telephone set and the local office.

6.10 SUMMARY

In this chapter we covered modem communications and modem encoding techniques used in high- and low-speed communications. We also covered compression techniques that are commonly used in modem and facsimile communications. New breeds of modems are emerging for use with the digital subscriber loops known as ADSL. These high-speed modems enable communications speeds in the tens and hundreds of Mbps. These modems use a multilevel encoding scheme known as DMT.

REVIEW QUESTIONS AND PROBLEMS

1. A low-speed modem uses the phase shift keying modulation technique. It encodes five bits in each signal element. Assume that the modem uses a sine wave with an 1800-kHz frequency.
   a. Calculate the signaling rate.
   b. Calculate the bit rate.
   c. Assuming equal phase shift angles, how many phase shift angles are required?

2. What attributes of the sine wave are kept constant in:
   a. Phase shift keying?
   b. Frequency modulation?
   c. Quadrature amplitude modulation?

3. In a QAM modem, if there are 5 amplitude levels and 12 phases:
   a. How many signal elements can be encoded?

   b. If the carrier frequency is 4 kHz what is the bit rate and the signaling rate?

4. Given a text file that contains 2000 characters, the characters used in the following are K, H, A, D, E, and R with the following probabilities of occurrence:

   $P(K) = 0.12$ $P(H) = 0.28$
   $P(A) = 0.2$ $P(D) = 0.6$
   $P(E) = 0.49$ $P(R) = 0.01$

   a. Find the Huffman code for each character.
   b. Find the average length of the file.
   c. If you use straightforward assignment, three bits for each character, what would the length of the file be? Compare this answer to that of part b.

5. Assume we have the text in Figure 6–16.

Figure 6–16
Text for encoding.

```
$*****************************************************$
    START OF PROGRAM
$*****************************************************$
    MOV AX, CX        ;store
    ADD DX, BX        ;add
```

Using the run-length data compression technique, encode this file. What is the compression ratio?

6. Assume the black-and-white image in Figure 6–17 is to be faxed. Give the encoded data that will be transmitted using modified Huffman codes listed in Table 6–6.

7. In DMT ADSL systems, what is the upstream channel and what is the downstream channel? Can you think of the design difficulties in implementing ADSL systems? List three problems with this technique.

Figure 6–17
Image to be faxed.

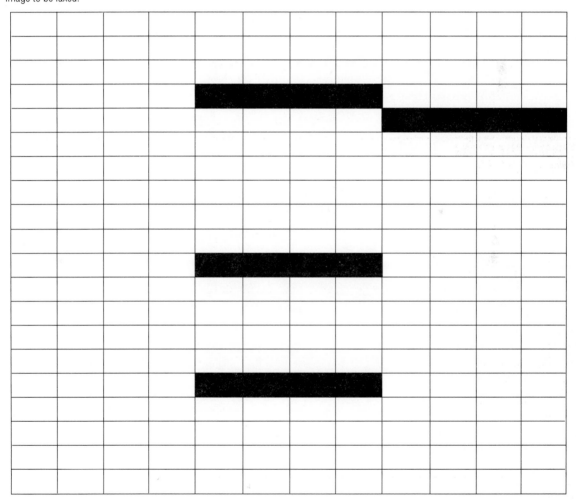

7

Protocols and Standards

7.1 INTRODUCTION

In the early 1970s, the rapid proliferation of modems, signaling methods, network architectures, and the methods of communication among devices created many problems because of a lack of standardization by the various manufacturers. Users were obliged to support extensive and redundant equipment. This situation prompted standards organizations to define common structures and integrated networks of data processing machines to alleviate some of the problems. In 1978 the *International Standards Organization (ISO)* proposed a layered network architecture called the *open systems interconnection (OSI) reference model*. This model, approved in 1983, makes it possible for data processing machines from different vendors to freely communicate and interchange data.

In the OSI environment the complicated communication task is divided into a set of manageable functions called *layers*. Each *layer* performs a subset of the tasks required to communicate with another system. The telecommunication group of the *International Telecommunications Union (ITU)* defined the basic reference model of the open system interconnections in a recommendation referred to as the X.200.

The open systems interconnection (OSI) is concerned with the exchange of information between systems, in fact, between "open" systems. Within OSI, a distinction is made between "real" systems and "open" systems. A real system is a computer system together with its associated software, peripherals, terminals, human operators, physical processes, and even subsystems that are responsible for information transfer. On the other hand, an open system is only a representation of a real system that is known to comply with the architecture and protocols as defined by OSI. Put differently, an open system is that real portion of a real system that is visible to other open systems in their attempts to transfer and process information jointly. The OSI is not only concerned with interconnection and exchange of information among open systems (i.e., transmission), but also with interconnection aspects of cooperation among systems. Cooperation among systems includes activities such as interprocess communication (information exchange and synchronization among activities).

7.2 PROTOCOLS AND LAYERS

Figure 7–1 depicts a collection of real systems that are interconnected by using physical transmission media. The overall objective of software or hardware that forms the open system is to enable interconnection among application processes. Application processes are an abstraction of user programs that implement certain user applications. However, only certain aspects of application programs are of interest—those that concern communication or cooperation among open systems. To design and implement complex mechanisms for information transfer and cooperation, the entire network must be viewed as a series of layers. Each open system is logically divided into a collection of *subsystems*, each implementing a certain function. A layer is a collection of all subsystems from various open systems that perform similar functions.

subsystem

A network of computers may then be viewed as a series of layers, each layer cutting across several systems. The design of layers across computer networks is called *network architecture.* Each layer provides a collection of services to the subsystems in the higher layer—the ability to move data across the network, possibly even in an unreliable manner, or the ability to enable users to exchange data without concern for its representation, for example. As we go up the layers the kind of services made available at each layer changes. In particular, the range of services is either enlarged or improved. Each layer must implement certain functions over and above the service made available to it by the lower layers, thereby enlarging the service or improving its quality.

network
architecture

The concept of layering is akin to solving a large problem in a series of smaller manageable problems. As an example, one has to provide a service by which a user transfers data reliably across a network without concern for representation of information. The solution involves implementing specific functions in different layers to solve each aspect of the problem. This process is essentially a top-down approach, which leads to modularity as well as independence from how an underlying function is implemented. The design of a layer may be changed to incorporate new technology without having to change the other layers. This concept parallels that of transparency, where a given layer transfers user data without concern for what that data is. Furthermore, data in any form may be transferred by the supporting layer.

In some ways the layered architecture approach is very similar to the top-down design of hardware and software systems. One breaks down the problem

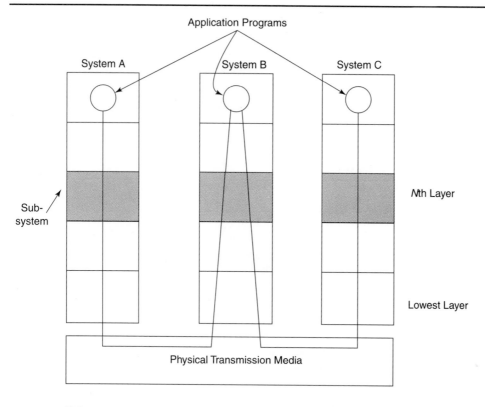

Figure 7–1
Concepts of layered architecture.

Application Programs

System A
System B
System C

Sub-
system

Nth Layer

Lowest Layer

Physical Transmission Media

Notes:
• Each system is divided into a
  collection of subsystems.
• A subsystem implements a collection of functions.
• Subsystems performing similar functions,
  but from different real systems, belong to
  the same layer.

of interconnection among systems into a succession of smaller problems, each problem being solved within a layer. This approach has a number of advantages, including modular construction, independence from implementation of the underlying layer, maintenance, and standardization. Once the service made available by each layer is defined, then it is easy to see the functionality of each layer. A layer bridges the gap between the services made available by it and those available to it. Figure 7–2 illustrates the concept of layers, services, and functions.

7.2.1 Entities, Protocols, and Interfaces

Having defined the concept of layers, and that of subsystems within them, next we describe what a subsystem looks like. A subsystem consists of one or more *entities*. An *entity* is a hardware and/or a software module. Only that aspect of a module that is significant or visible from the viewpoint of communication

entity

Figure 7–2
Layers, services, and functions.

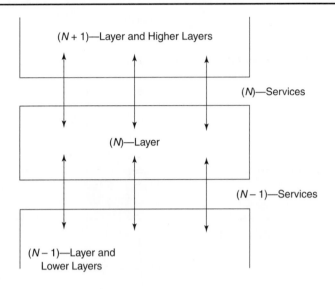

$(N + 1)$—Layer and Higher Layers

(N)—Services

(N)—Layer

$(N - 1)$—Services

$(N - 1)$—Layer and Lower Layers

*Each Layer:
    - Provides a set of services to its higher layer.
    - Uses services made available by its lower layer.
    - Implements certain functions.

*Functions implemented within a layer close the gap between services used and those to be made available.

*This concept of layering results in:
    - Modular construction.
    - Independence from underlying implementation.
    - Transparent data transfer.

between entities is considered part of the entity. For example, four entities are required to handle four transmission lines in each system. The four entities together form the subsystem that uses the physical transmission media.

    A network may be viewed as a collection of entities. The entities in two different systems may interact or communicate with each other, provided they belong to the same layer. Such entities are called *peer entities,* and the interaction between them is governed by what is known as a *protocol.* A protocol specifies the nature or the format of information that may be exchanged by two peer entities, both with respect to the contents (i.e., semantics) and the manner in which information is to be represented (i.e., syntax). It also specifies under what conditions an entity may send or receive information. A *protocol stack* is a list of protocols used by a system.

    Similarly, two entities within a network may interact with each other provided they belong to two *adjacent* layers. This form of interaction across a layer is consistent with service made available by an entity in the next higher layer. Entities' specifications are in terms of the collection of primitive operations

protocol

protocol stack

that either an entity is allowed or capable of supporting. The basic operations are called *primitives*. These operations are primitive in the sense that they may not be broken down any further. Associated with each primitive is a collection of parameters that may be exchanged.

primitive

Aspects pertaining to the implementation of service primitives are not part of the architectural specification and neither is the representation of parameters. An implementation of services is referred to as an *interface*.

Services are made available to an entity in the Nth layer at service access points (SAPs). An entity makes service available at one or more SAPs, but an entity in the higher layer may access services at one or more SAPs. The only restriction is that two entities within the same subsystem may not access services at a common SAP. In some ways, the concept of an SAP is very similar to that of a mailbox. A mail carrier may serve a number of mailboxes, but no more than one individual may use a mailbox. One may now model the interaction among entities in a higher layer and the layers below. Figure 7–3 shows that a layer, together with those below, may be viewed as a service provider, and entities in the higher layer are service users.

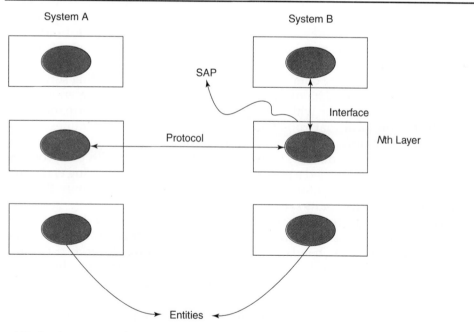

Figure 7–3
Entities, protocols, and interfaces.

* Each subsystem is composed of a number of "entities."
* An entity is hardware and/or software module.
* There are two forms of interactions between entities:
    - Access services across an interface (within the same system, but across a layer boundary)
    - Exchange data and control information according to a protocol (within a layer, but across systems).
* Service access points (SAPs): Services are made available by an entity in Nth layer.

In communication protocols design, layer N provides services to the layer above it $(N + 1)$. Layer $N - 1$ provides services to layer N and so on. This example is similar to real-life situations. In a work environment you usually provide services to the person above you. If you have someone working for you, he or she provides services to you. As illustrated in Figure 7–3, we need to remember the following:

- Each subsystem is composed of a number of entities.
- An entity is a hardware and/or software module that provides specific protocol services.
- There are two forms of interactions between entities:
  a. Access services across an interface (within the same system, but across a layer boundary).
  b. Exchange data and control information according to a protocol (within the same layer but across systems).
- Services are made available to the layer above via the service access points (SAPs).

7.2.2   Transmission of User Data

Figure 7–4 shows how user data gets transmitted across a layer and delivered to the remote user. User A, wishing to transmit data to User B, hands it over to the supporting layer. It may do this by using a service primitive that provides data transfer service. It may also supply the identity of User B, or its address. It then becomes the responsibility of the layer to deliver the contents of the user data to the receiving User B.

Entity C, supporting data transfer service at the corresponding service access point, sends user data out to the corresponding supporting entity D, at the remote end, together with control information. Control information pertains to the protocol that runs between the two supporting entities and is called *protocol control information*. It may include, for example, a sequence number, error checksum, etc. The two—user data together with protocol control information—are called a *protocol data unit (PDU)*. The receiving end strips off the protocol control information, interprets it, and delivers only the user data to the $N + 1$ layer of its system.

A related question is how an entity transfers the protocol data unit. The entity uses the services made available to it by the lower layer to transmit the PDU across. It typically hands the PDU over to the lower layer as user data, but this time it corresponds to the lower data. If all goes well, it is then delivered to the entity at the remote end. This lower layer may have to append its own protocol control information, and so on.

Transparency of data transfer is an important concept in all layered architectures. It basically suggests the following:

- The supporting layer does not in any way constrain the kind of data that are to be transferred across the layer, both in terms of its contents or its encoding.
- The supporting layer is not required to examine user data in order to carry out its functions. For instance, a layer may not compare the contents of two user messages to determine whether they are duplicates or not.

PDU

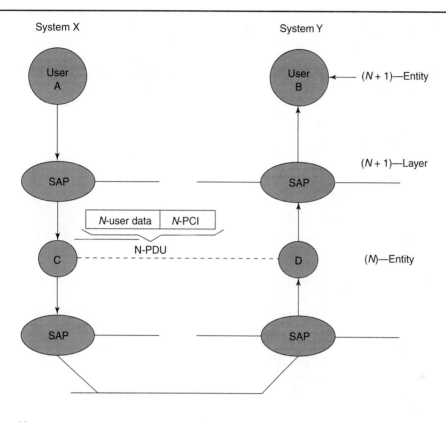

Figure 7–4
Transmission of user data.

Key:
* PCI: Protocol Control Information
* PDU: Protocol Data Unit
* SAP: Service Access Point

7.2.3 Communication over Computer Networks

Figure 7–5 shows the structure of a computer network composed of several data stations, A, B, C, etc., that can communicate with one another over the subnet in a store-and-forward fashion. The data stations are host computers that host the application program's processes (e.g., x, x', x'', y, y', z, z'). The computers do not have anything to talk about. It is an application program in one host (e.g., x in A) communicating with an application program in another host (e.g., y in B) connected by high-speed trunks. Each node is a specialized computer whose purpose is to:

- Accept data from a link and temporarily store the data.
- Determine which link the data should be sent over so that it will eventually be delivered to the destination data station.
- Send the data on the chosen link.

Figure 7–5
Communication over
computer networks.

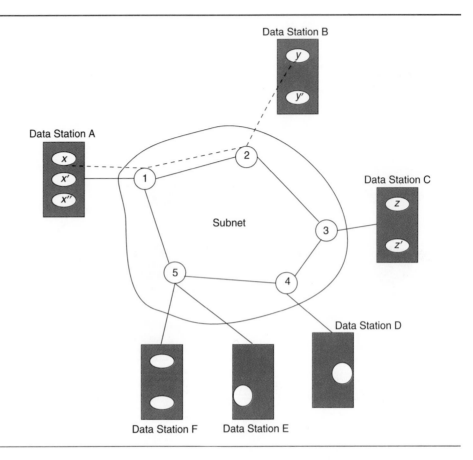

7.3   OPEN SYSTEMS INTERCONNECTION (OSI) REFERENCE MODEL

Figure 7–6 shows the open systems interconnection reference model allowing
program *x* in host A to communicate with application program *y* in host B. The
connection between *x* and *y* is shown by the dotted line. The model indicates
seven layers as shown. The lower layers, one through three (called *chained
layers*), exist in both the node and the host computers and are responsible for
transporting data packets across the subnet. Layers four through seven (called
end-to-end layers), exist in the host computer only. The upper layers, five
through seven, include information processing-oriented capabilities and deal
with the way the application programs communicate. The transport layer is a
liaison between the lower layers and the upper layers and provides an end-to-
end reliable path between the applications.

Each layer provides services to the layer above it and receives some services from the layer below. A layer requests the services of the layer beneath it
through an interface (shown by vertical arrows).

Figure 7–6
The open systems intrconnection reference model.

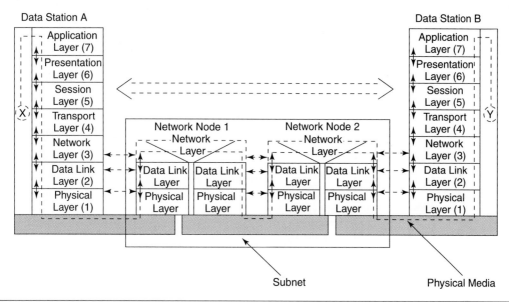

Each entity converses with its peer entity, which resides in a network node or another data station. The peer communication is governed by a protocol (shown by dotted horizontal arrows). The services and the protocols at each layer are specified by the OSI and ITU-T. The peer entities communicate with one another over a virtual connection (i.e., logical connection). The physical connection exists only at the physical media (i.e., twisted pair, coaxial cable, digital radio, optical fiber), over which the actual physical movement of information occurs.

7.3.1   Data Transfer among Users

Figure 7–7 shows how user x in System A may transfer data to user y in System B in an OSI environment. User x presents its data to the application layer. The application layer will add protocol control information in the form of a header to it. This header may contain information such as the message number, date, time, etc.; it is only meaningful to the application entities and is treated as data by the presentation layer. The combination of the user data plus the protocol control information make up the protocol data unit . The application layer sends its protocol data unit to the presentation layer; this layer adds its protocol control information to it, and sends the resulting data unit to the session layer.

So, as the protocol data unit at each layer in the transmitter is presented to the lower layer, an appropriate header will be added to it and sent down. Finally

Figure 7–7
Data transfer among users in OSI environment.

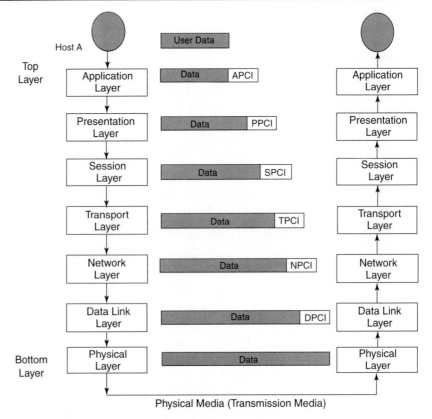

Physical Media (Transmission Media)

PCI= Protocol Control Information

the resulting message at the physical layer is transmitted as a series of bits via the transmission media. At the receiver each layer strips off and interprets its header (protocol control information), and the data are sent to the next higher layer. Eventually user *y* will receive the data via the application layer.

Now let us briefly describe the functions of each layer in the open systems interconnection (OSI) reference model:

physical layer

• *The physical layer* provides for transmission of bits over a physical medium, which can be twisted pair, coaxial cable, optical fiber, or air. It defines procedures to activate, maintain, and deactivate a physical circuit. It receives the bits and passes them up to the data link layer. It may also provide for multiplexing of several data links over a single physical link (as we saw in T1 transmission). It can be point to point or multipoint. The data unit at the physical layer is a bit. The standards for this layer include EIA RS-232-C, RS-232-D, RS-449, EIA-530, V.35, and X.21 (used in Europe).

- *The data link layer* is responsible for providing a relatively error-free path between adjacent stations to higher layers. It delimits the data transmission units (frames) and performs an error and sequence check on the frames. The contents of the information is not relevant. Acknowledgments from the receiver guarantee the transmitter that the data frames are received. If the data frames are not acknowledged, the transmitter will retransmit the frames. *Flow control* is also provided at this layer. The data unit at the data link layer is a frame. Some of the existing standards at this layer are as follows:

data link layer

flow control

a. International Standards Organization (ISO) high-level data link controller (HDLC), which is a superset of several classes of data link protocols. We will examine the HDLC later in this chapter.
b. Link-layer access protocol for the B channel (LAPB), which is layer 2 of an ITU-X.25 packet-switching interface. B channel can be thought of as a single T1 channel (64 kbps).
c. Link-layer access protocol for the D channel (LAPD), which is layer 2 of ITU-T ISDN.
d. ANSI X3.28.
e. IEEE 802.2 for local area networks.
f. IBM's BISYNC.

- *The network layer* provides for transfer of network transmission data units (packets) among network layer service users across the subnet. It is therefore required to route the packets among switching nodes. The services of this layer include connection establishment, normal and expedited data transfer, flow control, and reset. The data unit at the network layer is a packet. The following are among the existing standards:

network layer

a. ITU-T X.25, the packet layer protocol.
b. ITU-T Q.931, the third layer of ISDN.
c. ITU-T X.213, the network layer protocol.
d. ISO 8473, the connectionless network protocol.
e. DOD Internet Protocol (IP).

- *The transport layer* provides reliable transport of messages from the upper layers among the end system entities. It is an end-to-end protocol, and it corrects for possible deficiencies in quality of service provided by the supporting network layer. It may provide for multiplexing several transport connections over a network connection to optimize data transfer or it may split a transport connection over several network connections for higher throughput or reliability. It also provides flow control and recovery mechanisms for error and abnormal conditions. The data unit at the transport layer and the layer above it is a message. Some of the existing standards are as follows:

transport layer

a. DOD transmission control protocol (TCP) used in Internet.
b. ISO 8072 and ITU-T X.214 for transport layer service definitions for transport protocol classes 0, 1, 2, 3, and 4.
c. ISO 8073 and ITU-T x.224 for transport layer protocol specification.

session layer

• *The session layer* establishes, releases, and manages the transport connections for the presentation layer entities, and provides for negotiation of the dialogue and communication parameters (e.g., half duplex versus full duplex, who should talk next) for the higher layers. It also provides for synchronization points so that users may synchronize and retransmit only that part of the data that needs retransmission. The exception reporting service permits the presentation entities to be notified of exceptional situations that cannot be recovered from by the session layer.

presentation layer

• *The presentation layer* performs services that pertain to representation of user data in a form that can be used and understood. The meaning of information is preserved, but the format and the language difference (syntax) is resolved. This layer selects the syntax to be used for transfer of information during communication. Some examples are: code conversion of ASCII to EBCDIC, data encryption, and data compression. It is generally easier, more cost effective, and more reliable to provide these services centrally than to make each user develop individual techniques. Among the standards is the ISO abstract syntax notation (ASN.1) or ITU-T X.490.

application layer

• *The application layer* provides all services that are directly accessible by the application programs and the ultimate user. These services reflect accumulation of all the services performed by lower layers. This layer establishes the association between the application entities and allows for identification and authorization for the communicating users. Among the existing standards are ISO file transfer, access, and management (FTAM); ITU-T message handling system (MHS); airline flight reservation protocols; and credit check protocols.

One final point worth mentioning regarding the seven layers of the open systems interconnection (OSI) reference model is that layers 1 through 3 usually reside in the nodes in a network. Layers 1 through 3 and 4 through 7 reside in the host computers that communicate over the network using the OSI model. In the next few sections, we will examine in more detail the first three layers in the OSI reference model.

7.4   THE PHYSICAL LAYER

In this section, we will examine the serial transmission of user information between data terminal equipment (DTE). Two important aspects of a DTE will guide and focus our discussion. First, a DTE has an internal parallel architecture, and second, the transmission distance of a DTE is limited. To interface the parallel architecture of a DTE to the serial transmission medium requires the DTE to perform a parallel-to-serial conversion in the transmit direction and a serial-to-parallel conversion in the receive direction (for distances greater than a few score feet, the transmission medium is serial).

DCE

The distance across which the DTE can transmit information can be extended with addition of *data circuit terminating equipment (DCE)*. One example

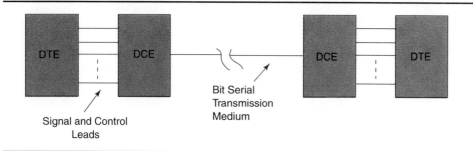

Figure 7–8
DTE–DCE interface.

of a DCE is a modem that allows data transmission through the analog telephone network. Thus, you could consider your PC a DTE and the modem card installed in it a DCE. Standards defining the interface between the DTE and DCE have been developed. These standards allow consumers to purchase data terminal equipment and data circuit equipment from different vendors. These devices will work together as long as both manufacturers implement the same interface standard. Figure 7–8 shows the basic configuration in a DTE–DCE interface.

The central elements of the parallel-to-serial and serial-to-parallel interface in data terminal equipment (DTE) are the shift registers, as shown in Figure 7–9. The top illustration in the figure shows a single-buffered interface. In such a configuration the receiver must obtain the current character before the arrival of the first bit in the next one. Frequently, the receiver is overlapping processing with input-output activities and cannot respond to an incoming character in time to prevent a data overrun. To prevent this type of error, most interfaces are designed using the double-buffered approach shown in the bottom illustration of Figure 7–9. In this type of interface, a second register (the holding register) is provided. This allows the received character to be transferred in parallel to this register as soon as the last bit has been sampled. The receiver now has the entire time of the assembly of the second character to read the first one. The flag that is raised when the parallel transfer occurs alerts the receiver that a character is in the holding register. Interfaces with higher degrees of buffering are sometimes used in situations where bursts of characters can be expected.

Transmitter interfaces are also implemented using shift registers. When a character has been shifted out onto the line, a flag is raised signaling the transmitting DTE that the interface is ready for a new character. The interface must hold the line at the mark state for the stop interval required before transmitting the start of a new character. If the transmitter cannot provide a character during the stop interval, no real difficulty occurs, the line is just not being used as efficiently as possible. A double-buffering procedure may be employed to keep the line running at maximum speed. Figure 7–10 shows a simplified, double-buffered transmitter interface.

Figure 7–9
DTE parallel <—>
serial conversion: (a)
single-buffered re-
ceiver interface, and
(b) double-buffered
receiver interface.

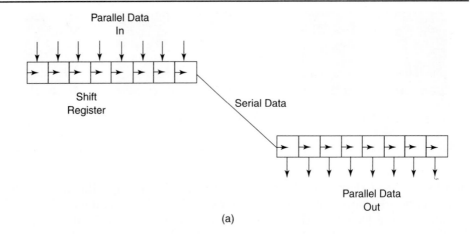

Parallel Data
In

Shift
Register

Serial Data

Parallel Data
Out

(a)

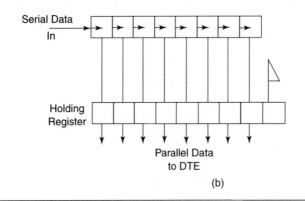

Serial Data
In

Holding
Register

Parallel Data
to DTE

(b)

Figure 7–10
The asynchronous
transmitter interface.

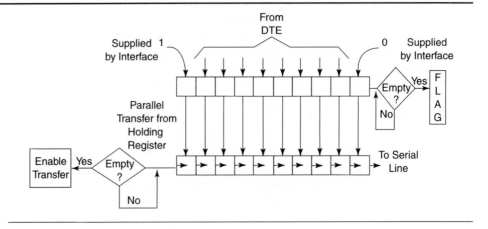

From
DTE

Supplied   1
by Interface

0   Supplied
by Interface

Empty
?

Yes

F
L
A
G

No

Parallel
Transfer from
Holding
Register

Enable
Transfer

Yes

Empty
?

No

To Serial
Line

7.4.1 Universal Asynchronous Receiver/Transmitter (UART)

It is possible to buy a single large-scale integration (LSI) chip called a *universal asynchronous receiver/transmitter (UART)* that performs buffering and error indication. The UART is mounted on a 28- or 40-pin dual in-line package (DIP). Pins are provided for data, error indication, character length selection, and stop bit length selection, along with other possibilities depending upon the device. Clock signals are typically externally generated to establish the bit rate for transmission and reception, and may be independent.

UART

A UART has two internal status registers that may be read or written to by an attached computer. These registers contain indicators for overrun errors, framing errors, and parity errors; that is, a test performed on the incoming character indicates that an error occurred during its transmission. The registers also contain interrupt enable bits and flag bits to signal that a character has been received or that the UART is ready for a new character to transmit.

The UART can be used in either program-controlled or interrupt-driven I/O systems. In program-controlled systems, the attached computer reads the status registers to determine whether a character has arrived or a character can be transmitted. In interrupt-driven systems, the UART notifies the processor, via an interrupt signal, that action is required. Figure 7–11 illustrates a functional block diagram of the universal asynchronous receiver/transmitter.

7.4.2 Synchronous DTE–DCE Connection

So far in this section we have examined the asynchronous DTE–DCE interface. A similar approach is followed to implement the synchronous interface. Recall

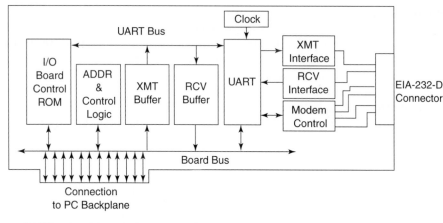

Figure 7–11
UART PC board block diagram.

• UART and serial connector are generic.

• Control ROM, board bus, board layout, and connector to PC backplane are specific to a given device (or class of devices).

Figure 7–12
Synchronous
DTE–DCE data format.

SYNC Character(s)	Start-of-text Character	User Data (Sometimes Referred to as a Packet)	End-of-text Character

USRT

from Chapter 1 that one example of synchronous communication is to transfer user data framed by a synchronization (SYNC) character and an end-of-test character. The interface circuit used for this type of synchronous communication is a *universal synchronous receiver/transmitter (USRT)*. USRT performs the serial-to-parallel conversion and the double-buffering techniques we discussed earlier plus appends and removes headers and trailers; it also locates incoming SYNC characters. A SYNC character register in the USRT allows the user to define this character. The Intel 8251 and Signetic 2657 are typical USRT devices. Figure 7–12 shows the data format in a DTE–DCE synchronous connection.

7.4.3   Physical Interface Standards

Physical interface standards typically specify a number of characteristics about the connection between data terminal equipment (DTE) and data circuit equipment (DCE):

• Electrical Properties—Include the type of signaling used, voltages and their meaning, and whether the interface is electrically balanced or unbalanced. (In an electrically unbalanced line, one wire is at ground potential; in a balanced transmission, both wires are at above-ground potential and each wire carries signals with equal amplitudes but opposite phase.)

• Mechanical Properties—Include specifying the types of connector to be used, pin size, and cable type.

• Function of Interface Circuits—As alluded to earlier, a serial interface is really composed of many leads for data transfer, electrical grounding, timing, and other control functions. Interchange circuits perform each of these functions, and they must be defined.

• Interaction Among Devices—The interchange circuits allow the exchange of information between the DTE and DCE. The function of each circuit must be defined, as well as the action that devices should take when certain events occur.

• Subset of Leads for Specific Uses—Most interface standards have many options and features to allow their use in many application environments. Subsets of the numerous circuits may be defined for specific applications. For example, a full-duplex application would not need to implement the leads that control the line in half-duplex applications.

7.4.4   Overview of the EIA-232-D Electrical Specification

EIA–232 standard

The most commonly used serial DTE–DCE interface standard in the United States is the *Electronic Industries Association (EIA) EIA-232 standard*. EIA-232 was introduced in the early 1960s as RS-232. The third revision, RS-232-C, was rewritten in August 1969 and was the standard until early 1987 when EIA-232-D was introduced. These two versions are very similar. In this subsection we will describe the EIA-232-D and will point out the major differences between the C and D versions. In summary, the following are the electrical specifications of the EIA-232-D:

- Maximum open-circuit voltage: +/–25 volts.
- Maximum short-circuit current: +/–0.5 amps.
- Driver slew rate: $dv/dt < 20 \mu v/\mu s$.
- Maximum capacitance: 2500 pf.
- Maximum distance: 50 feet.
- Maximum data rate: 20,000 bps.
- Polar signaling.

EIA uses polar signaling (positive voltage value represents a 0 and negative voltage value represents a 1). Figure 7–13 shows the meaning of the various

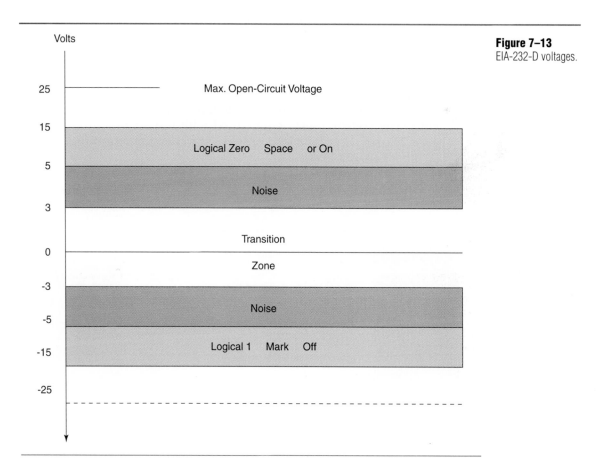

Figure 7–13
EIA-232-D voltages.

voltages levels on the interchange circuits. On a data transmission circuit, a logical 1 bit will be represented by a signal between –5 and –15 volts, while a logical 0 bit will be represented by a signal between 5 and 15c volts. A noise margin of 2 volts is introduced since the reception of a signal 3 volts or greater is interpreted as a logical 0, while the reception of –3 to –15 volts is a logical 1. On control circuits, on is a positive signal and off is a negative signal.

The EIA-232-D standard specifies that one 25-pin D-shaped subminiature (DB-25) connector will be used. The standard defines the use of 24 of the pins, including two reserved for test circuits (one pin is unassigned). Not all of the 24 pins are required for all applications; the minimum subset is shown in Figure 7–14. Although almost universally used, the DB-25 connector was never specified in RS-232-C. Furthermore, only 22 pins were defined by RS-232-C, including the two for test points.

To exchange data between two devices, three circuits are required, namely, the transmitted data, received data, and signal ground circuits. An additional circuit, called *shield,* is usually used as well although it is not essential. (The shield lead was called *protective ground* in RS-232-C.) The functions of these leads are as follows:

- Transmitted Data—Serial data output by the DTE that is to be transmitted by the DCE.
- Receive Data—Serial data that is being sent to the DTE from the DCE.
- Signal Ground—The voltages on the leads must be between + and –25 volts. The voltage must be measured with respect to a ground reference. The signal ground lead provides the common return function for all other EIA-232-D circuits.
- Shield—The shield circuit electrically isolates the interface to protect it from abnormal electrical events.

Figure 7–14
EIA-232-D interface circuits (minimum system).

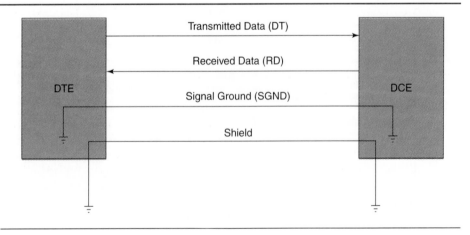

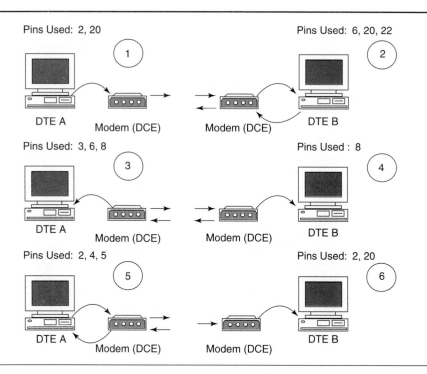

Figure 7–15
EIA-232-D timing
operation.

Let us briefly discuss the sequence of events required to establish a data communication link between two computer systems labeled A and B in Figure 7–15.

1. The sequence begins with a handshake between the sending terminal A (DTE)—the computer in this case—and its modem (DCE). Terminal A turns on the data terminal ready pin (pin 20) to tell its modem that it wants to begin a data exchange. This pin will be held on throughout the communication. Terminal A then transmits a phone number via the transmitted data pin (pin 2) for the modem to dial.

2. Next, modem B alerts its terminal to the incoming call via the ring indicator (pin 22). Terminal B turns on its data terminal ready pin (pin 20). Modem B then generates a carrier signal, to be used by system B in the exchange, and turns on its data set ready pin (pin 6) indicating its readiness to receive data.

3. When modem A detects a carrier signal, it alerts terminal A via the received line signal detector pin (pin 8), and also tells the terminal that a circuit has been established via pin 6. If the modem has been so programmed, it will also send an online message to the computer screen via the terminal's received data pin (pin 3).

4. Modem A then generates its own carrier signal to modem B, which detects it via pin 8, the received line signal detector.

5. When a terminal (say, terminal A) wishes to send its data, it activates pin 4, the request-to-send pin. Modem A then responds via the clear-to-send pin (pin 5). Terminal A will then send its data in the form of pulses representing binary pulses 1s and 0s to modem A via pin 2, the transmit data pin. Modem A will modulate these pulses to send the data over its analog carrier signal.

6. Modem B reconverts the signal to a digital form and sends it to terminal B via the terminal's received data pin (pin 3).

Two additional EIA-232-D leads are required for half-duplex operations. These are the request-to-send (RTS) and clear-to-send (CTS) leads. The request-to-send lead is controlled by the DTE. It is on when the DTE is ready to transmit. The clear-to-send lead is controlled by the DCE and it is on when the DCE is granting permission for the DTE to transmit.

7.4.5   Other Physical Interfaces

V.24, and V.28

There are a number of other important physical interface standards. We briefly list these standards in this section. ITU-T recommendations *V.24* and *V.28* provide mechanical and electrical definitions, respectively, that are essentially identical to EIA-232-D. They define the common serial interface used in Europe. The only difference between V.24/V.28 and EIA-232-D is that the ITU-T recommendation is more specific in its handling of some circuits between calls. This is easily solved by implementing EIA-232-D following the V.24/28 rules.

V.235

ITU-T recommendation *V.235* is commonly used in Europe for 48-kbps transmission service and in the U.S. for 56-kbps services. It is electrically similar to V.28, except that its data and clock signals are electrically balanced. Rather than use the DB-25 pin connector, it uses a 34-pin (square) connector.

ITU-T recommendation X.21 is used in digital-switched networks and provides more functionality than EIA-232-D. X.21 may be used at United States rates up to 10 Mpbs or distances up to 3333 feet. This standard is not commonly used in the United States.

EIA-449 is a mechanical specification defining many more functions than EIA-232-D. Its electrical characteristics are described in two other standards: EIA-422-A and EIA-423-A. EIA-422-A is an electrical specification only and defines an electrically balanced interface. Because of protection from noise, this interface may be used for data rates up to 10 Mpbs or distances up to 4000 feet. EIA-422-A is equivalent to ITU-T recommendations V.11 and X.27.

EIA-423-A is also an electrical specification and defines an electrical unbalanced interface. It may operate at speeds up to 20,000 bits per second in EIA-232-D compatibility mode, or may operate at speeds up to 100 kbps or distances up to 4000 feet (EIA-423-A is equivalent to ITU-T recommendation V.10 and X.26).

7.5   DATA LINK LAYER PROTOCOLS

There are many causes of errors in a communication facility. To ensure relatively error-free data communications, a method of detecting and correcting errors in transmission must be developed. Two general approaches to the problem are as follows:

- Send enough extra bits with a message to allow the receiver to detect and correct most errors with high probability. This may lower the efficiency of transmission because a relatively large number of noninformation bits are sent with every transmission.

or

- Send only enough extra bits to allow detection of errors by the receiver. If an error is detected, the receiver then requests a retransmission of the message.

The efficiency of the second approach depends on the characteristics of the communication facility. If the probability of errors is low enough, the efficiency of this method exceeds that of the error-correcting method, because relatively few noninformation bits are sent. If the probability of errors is high, the number of retransmissions will also be high and efficiency will be low. As a practical matter, error detection is the preferred choice since the communication channels are reasonably error-free and bandwidth costs for correction overhead are not always economical. In this section, we will discuss the second approach—the error detection method.

7.5.1 The Automatic Repeat/Request (ARQ) System

When using the detect-only approach, the receiver will typically signal the acceptance or rejection of a block of user information. If a block is rejected, the transmitter is to retransmit the block in question. Thus, an automatic repeat request is generated by a reject signal. This type of error control system is referred to as an *automatic repeat/request (ARQ)* system. ARQ systems may be stop-and-wait or continuous. In stop-and-wait ARQ, the transmitter waits after transmitted blocks for an acknowledgment or a reject signal (ACK or NACK). In continuous ARQ, the transmitter continues to send blocks while acknowledgments arrive on a return channel. Therefore, some means of identifying the block being acknowledged or rejected must be provided.

Continuous ARQ strategies may be "selective" or "go-back-N." In selective ARQ, the transmitter only transmits blocks reported to be in error. This approach may be optimal in minimizing the amount of data retransmitted; however, it forces the receiver to accept data in an out-of-sequence order. This problem can be avoided with the go-back-N scheme. In this case the transmitter repeats all blocks starting with the offending block in the original sequence sent.

Choosing between these two strategies makes it necessary to weigh the consequences of retransmitting a potentially large number of blocks against the complexities of rearranging all the blocks after they have been received. In the go-back-N case, the transmitter has to be able to buffer as many blocks as can be outstanding (i.e., unacknowledged), but the receiver has to hold only the current block. In the selective ARQ case, the transmitter still has to buffer all allowable outstanding blocks, but the receiver must also be able to buffer all blocks

Figure 7–16
Automatic repeat
request protocols.

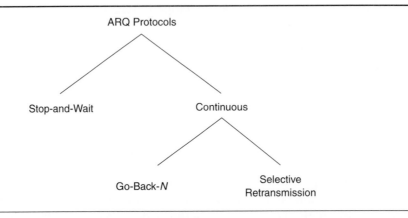

because they are liable to be received out of order. The go-back-N strategy is, in fact, more common. However, observe that in very noisy environments, go-back-N may fail miserably as each retransmission is likely to be several blocks long and will probably incur a hit. This can lead to very high overheads with progress occurring in jumps of only one or two blocks despite each retransmission containing several times that many blocks. Figure 7–16 summarizes the relationship among ARQ protocols discussed in this section.

To gain a deeper understanding of the operation of ARQ protocols, the concept of a *sliding window* is now introduced. Both the transmitter and receiver are considered to have a window each. The maximum allowable size of the transmitter window refers to the maximum number of outstanding, or unacknowledged, blocks that the transmitter can have at any time. The size of the receiver window is the number of blocks the receiver is prepared to accept at any time. Each time the receiver acknowledges a block, the transmitter window rotates by one and the transmitter is permitted to send one more block. This rotating of the window results in the concept of the sliding window.

Let us examine an example of the sliding window, so we can define both windows a little better. Figure 7–17 shows the example we are about to study. Let us assume that every outbound block contains a sequence number in the range (0, 3). The sender is required to keep a list of all blocks that have been sent but not yet acknowledged. This structure is called *transmitter's window.* Likewise, the receiver keeps a list of blocks that it can accept. The *receiver's window,* as it is called, is of constant size, but the transmitter's window may grow to a maximum size agreed to by both parties. Note that the window size permitted to the sender is a flow control parameter. If, for example, the transmitter window size is restricted to 1, we actually have a stop-and-wait protocol; if the transmitter window size is large, a substantial number of blocks can be sent in short order. Thus, by adjusting the window size, the receiver can protect itself from being overrun by a zealous transmitter.

If the receiver gets a block that is not in its window, it discards it without comment. If the block is the one at the lower edge of the window, it is

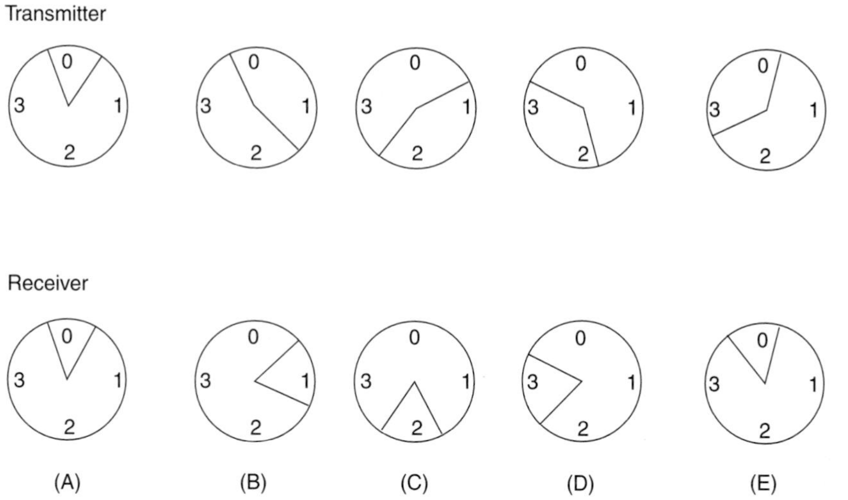

Figure 7–17
ARQ sliding window
protocol.

acknowledged and the window is rotated. Figure 7–17 illustrates this idea with a transmitter's window size of 2 and a receiver's window size of 1.

The concept of the sliding window can now be used to classify ARQ protocols. In the stop-and-wait ARQ protocol, both the transmitter and receiver have a window size of 1. In go-back-N protocol, the receiver must accept data blocks in sequence. Thus, the size of the receiver's window is 1. The transmitter may have more than $(N - 1)$ outstanding data blocks. In selective retransmission protocols, the receiver and transmitter window sizes are identical. In fact, the window must be larger than one-half of the sequencing module, assuming all sequencing is performed by module N.

7.5.2 Protocols for the Data Link

The basic function of a *data link protocol* is to provide reliable transfer of data between adjacent machines. Reliability in this context means:

data link protocol

- Error-free.
- Without loss of data.
- Without duplication of data.
- In sequence.

In this section, we will discuss three types of data link protocol approaches:

- Character-oriented protocol, in which special characters frame a message.
- Byte count–oriented protocols, in which a count field at the start of a message indicates how many characters constitute the message.
- Bit-oriented protocols, in which a special bit pattern frames the bits that constitute the message.

In bit-oriented protocols, as opposed to character-oriented protocols, the flag (or framing pattern) is never permitted as part of the message itself. This is accomplished without loss of transparency by a technique known as *bit stuffing*, which will be discussed with this type of protocol. In principle, bit-oriented protocols treat all messages as bit strings. Thus, you can theoretically implement the same bit-oriented protocol in different machines, even though they may have different code sets.

The classic character-oriented protocol is IBM's binary synchronous communication (BISYNC). Digital's digital data communication message protocol (DDCMP) is an example of a byte count–oriented protocol, whereas IBM's synchronous data link control (SDLC), ISO's high-level access procedures on the D channel in ISDN (LAPD), logical link control (LLC) and medium access control in LANs, and frame relay are all bit-oriented protocols.

In this section we will study the high-level data link control (HDLC), the current ISO bit-oriented data link protocol standard. It is a superset of many bit-oriented data link protocols including link access procedures for the D channel used in defining ISDN layer 2.

7.5.3 Bit-Oriented Data Link Control Protocol, HDLC

HDLC

High-level data link control (HDLC) is the most publicized and implemented data link control protocol. It is a bit-oriented protocol, capable of transmitting bit streams in lengths that are not necessarily multiples of character length. Most implementations, however, use bit streams that are multiples of character length. It is the ISO standard for the data link layer of the open systems inter-connection (OSI) reference model. As shown in Figure 7–18, it can operate in point-to-point or multipoint modes. All frames are exchanged in either direction between a primary and a secondary station. Frames from primary to secondary are called *commands* (whether they command the secondary to do anything or not) and frames from secondary to primary are called *responses* (whether they respond to a command or not).

HDLC Frame Format   The normal format of HDLC frames is given in Figure 7–19; options allow modified format. The flag field contains an 8-bit pattern, 01111110, that is a 0, six 1s and another 0. The bit pattern is not allowed to occur in a transmitted frame between leading and trailing flags. A 0 insertion algorithm inserts a 0 at the transmitter after five consecutive 1s in the address, control, information, and the frame check sequence of each frame, so a flag cannot appear in a transmitted bit stream in this portion of a frame. The receiver counts successive 1s; after five 1s, it looks at the next bit and throws it out if it is a 0. If it is a sixth 1, the following bit is examined to verify whether a flag has been found. If a 0 comes next, a flag has been found; otherwise the frame is aborted.

The *address field* always contains the address of the secondary station. There may be several secondary stations on a multipoint link, but only one primary station is allowed. All transmission occurs between a primary and a secondary, so this identifies the secondary station intended for communication

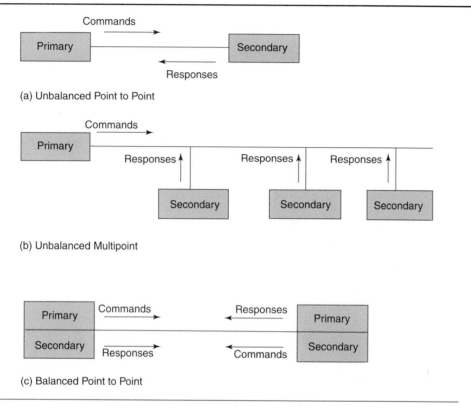

Figure 7–18
HDLC link structure.

(a) Unbalanced Point to Point

(b) Unbalanced Multipoint

(c) Balanced Point to Point

with the primary. Group addressing can be used to address several stations, with an all-1s address (defined to be a broadcast address) addressing all stations. An option to use extended address fields is available. If it is used, the address field is $8n$ bits long, with n representing any integer. The first bit of the final address field byte is 1, and the first bit of each preceding address field byte is 0, so the extended address field is recursively extendible.

The *control field* identifies the type of frames and contains frame sequence counts. Extended control fields are optional. Control field formats are given in Figure 7–20.

The *information field* contains the user data. Its length is arbitrary; it is of zero length for some frames, including supervisory frames, but may be longer for user data frames. Any bit pattern is allowed, so transparent text is handled naturally.

Flag	Address	Control	Information	FCS	Flag

Figure 7–19
HDLC frame format.

Figure 7–20
HDLC control field
formats.

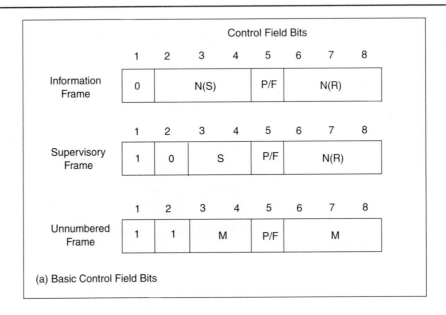

(a) Basic Control Field Bits

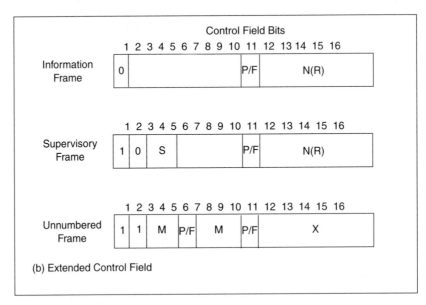

(b) Extended Control Field

The *FCS field* contains a frame check sequence. It is obtained from the polynomial

$$X^{16} + X^5 + X^2 + 1$$

and checks the contents of the address, control, and information fields, excluding 0s inserted by the 0 insertion algorithm. This method is known as the cyclic redundancy check (CRC), which was briefly introduced in Chapter 2. It is one of the most widely used methods of error checking in synchronous serial data communications. It is also used for error checking of disk storage. CRC is used for error checking a stream of bits, in contrast to checksum, which is byte-oriented, but both are used for error checking blocks of data. In the CRC method, one byte called the CRC byte is appended to the stream of data and transmitted with the data. At the destination, by hardware or software, the data and the CRC byte are tested for data integrity. The CRC bytes are calculated using the formula

$$\frac{M(X) * X^n}{G(X)}.$$

The polynomial $M(X)$ represents the bit stream. This is multiplied by

$$X^n$$

where n is the number of bits in the stream. The product of these two terms is divided by a polynomial called the *generator polynomial*. This division will result in a quotient, $Q(X)$, and a remainder, $R(X)$. It is the remainder that forms the CRC byte(s). The following example shows the CRC method used for a 16-bit data stream, 4D92H. In binary this is 0100 1101 1001 0010.

To get $M(X)$, first the bits are reversed: 0100 1001 1011 0010. This series of bits is interpreted as a series of coefficients of a polynomial, as shown below.

$$M(0100100110110010) = 0X^{15} + 1X^{14} + 0X^{13} + 0X^{12} + 1X^{11} + 0X^{10} + OX^9 + 1X^8 + 1X^7 + 0X^6 + 1X^5 + 1X^4 + X^1.$$

Removing the terms with 0 coefficients yields

$$M(X) = X^{14} + X^{11} X^8 + X^8 + X^7 + X^5 + X^4 + X^1.$$

Now $M(X)$ is multiplied by

$$X^{16}$$

since the data is a 16-bit stream.

$$M(X) * X^n = (X^{14} + X^{11} + X^8 + X^7 + X^5 + X^4 + X^1) * X^{16} = X^{30} + X^{27} + X^{24} + X^{23} + X^{21} + X^{20} + X^{17}.$$

The above term is the dividend of the equation. This will be divided by $G(X)$. We will use

$$G(X) = X^{16} + X^5 + X^2 + 1.$$

This is called the HDLC protocol generator. Other data link protocols may use a different generator polynomial. For example, BISYNC uses

$$G(X) = X^{16} + X^{15} + X^2 + 1.$$

Notice that both begin with

$$X^{16}.$$

This ensures that the remainder will be less than

$$X^{16}.$$

The remainder of the division is remainder $R(X)$, which is the CRC or the FCS in the case of the HDLC protocol.

Information frames (I-frames) are used for normal data transfer and are identified by a 0 in the first bit of the control byte. The send count, N(S), gives the frame's number being transmitted, and the receive count, N(R), gives the number of the next frame the transmitter is looking for.

The P/F-bit stands for poll/final; its interpretation depends on whether the frame is a command or a response. In a command (sent from a primary to a secondary), a poll bit is used to request transmission from the secondary. In a response (sent from a secondary to a primary), a final bit indicates the end of a sequence of frames. P- and F-bits are always exchanged in pairs—that is, a primary cannot send a second P-bit to the same secondary before it has received a returned F-bit (except in error recovery situations with timeouts). A second F-bit cannot be sent from a secondary before it receives a new P-bit. The ordinary use of the P/F-bit is when a computer is polling a group of terminals. When used as P, the computer is inviting the terminal to send data. All frames sent by the terminal, except the final one, have the P/F-bit set to P. The final one is set to F. Another use of the P/F-bit is to force one machine to send a supervisory frame immediately.

Supervisory frames are identified by "10" in the first two bits of the control byte. They are used for positive and negative acknowledgments in flow and error control. Both go-back-N and selective-reject ARQ (automatic repeat request) are allowed. The two S bits identify the type of supervisory frame. There are four possible combinations and four types of supervisory frames. These are receiver ready (RR), receiver not ready (RNR), reject (REJ), and selective reject (SREJ); REJ and SREJ are optional.

The RR (receiver ready) frame is most often used for positive acknowledgment when an I-frame is not available for piggybacking the acknowledgment. A receiver not ready (RNR) frame is used to accept a frame but request that no more information frames be sent until a subsequent RR frame is used. For the RR, RNR, and REJ frames, N(R) indicates the sequence number of the next expected information frame.

Unnumbered frames are used for a variety of control functions. As the name indicates, these frames do not carry sequence numbers and do not alter the sequencing or flow of numbered I-frames. Unnumbered frames can be categorized as follows:

- Mode-setting commands and responses.
- Information transfer commands and responses.
- Recovery commands and responses.
- Miscellaneous commands and responses.

Mode-setting commands are sent by the primary/combined station to initialize or change the data link mode of operation. The secondary/combined station acknowledges acceptance by responding with an unnumbered acknowledgment (UA) frame. The UA frame has the F-bit set to the same value as the received P-bit. Once the link is established, the mode of operation remains in effect until the next mode-setting command is either accepted or transmitted and acknowledged.

Set asynchronous balanced mode (SABM) and set asynchronous balanced mode, extended (SABME) are two examples of mode-setting commands. Upon acceptance of the SABME command, the I-frame sequence number in both stations on the link is set to 0. The disconnect command (DISC) is sent to inform the other station on the link that data link operation is suspended. While in the initialization phase, information is exchanged between a primary/combined and a secondary/combined using unnumbered information (UI) frames. Examples of UI frames are higher-level status, operational interruption, time of day, and link initialization parameters.

Recovery commands and responses are used when the normal ARQ mechanism does not apply or will not work. The frame reject (FRMR) response is used to report an error in a received frame, such as:

- Invalid control field.
- Data field too long.
- Data field not allowed with received frame type.
- Invalid receive count (i.e., a frame is acknowledged that has not yet been sent).

The RST command is used to clear the FRMR condition. RST announces that the sending station is resetting its send sequence number, and the receiving station is expected to reset its receive sequence number.

Typical Operation   Figure 7–21 illustrates several examples of HDLC operations. In these example diagrams, each arrow includes a legend that specifies the frame name, the setting of the P/F-bit, where appropriate, and the values of N(R) and N(S), respectively. The setting of the P- or F-bit is 1 if the designation is present and 0 if it is absent.

Figure 7–21a shows a link setup and disconnect sequence. The HDLC entity for one side issues an SABM command to the other side and starts a timer. The other side, upon receiving the SABM, returns an unnumbered acknowledgment (UA) response and sets its local variables and counters to their initial values. Upon receipt of the UA response, the initiating entity then sets its variables and counters and stops the timer. The logical connection is now active and both sides may begin transmitting frames. Should the timer expire without a response, the originator will repeat the SABM, as illustrated. The originating station will continue sending the SABM until a UA is received, or for a given number of tries. If a UA is not received, the originating station reports an error to a management entity. In such a case, higher level intervention is necessary.

Figure 7–21
HDLC operation
examples.

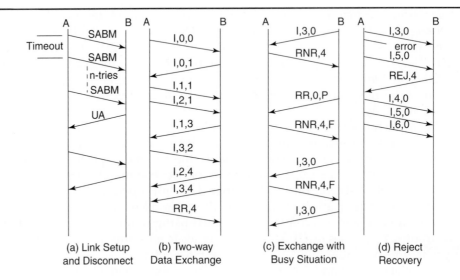

Key

I = I-frame
SABM = Set Asynchronous Balanced Mode
RR = Receiver Ready
RNR = Receiver Not Ready
REJ = Reject
P/F = Poll or Final bit
N(S) = Sequence Number of Transmitted Frame
N(R) = Sequence Number Expected to Be Received

Figure 7–21b depicts a full-duplex exchange of information I-frames. When an entity sends a number of I-frames in a row with no incoming data, the receive sequence number is simply repeated (e.g., I,1,1; I,2,1 in the A-to-B direction). Remember N(R) is the next frame to be received. When an entity receives a number of I-frames in a row without outgoing frames, then the receive sequence number in the next outgoing frame must reflect the cumulative activity (e.g., I,1,3 in the B-to-A direction). In addition to I-frames, data exchange may involve supervisory frames (RR 4 from the B-to-A direction).

Figure 7–21c shows an operation involving a busy condition. Such a condition may arise because an HDLC entity is not able to process I-frames as fast as they are arriving, or the intended user is not able to accept data as fast as they arrive in I-frames. In either case, the entity's receive buffer fills up, and it must halt the incoming flow of I-frames using the RNR command. In this example, the station issues RNR, which requires the station on the other side to halt transmission of I-frames. The station receiving the RNR will usually poll the busy station at periodic intervals by sending RR with the P-bit set. When the busy condition has cleared, B returns an RR, and I-frames exchange may resume.

An example of error recovery using the REJ command is shown in Figure 7–21d. In this example, A transmits I-frames numbered 3,4, and 5. I-frame number 4 suffers an error. B detects the error and discards the frame. When B receives I-frame number 5, it discards this frame too because it is out of sequence and sends a REJ with N(R) of 4. This causes A to initiate retransmission of all frames sent, beginning with frame 4. It may continue to send additional frames after the retransmitted frames.

7.5.4 Data Link Layer in the Internet

The Internet consists of individual machines (hosts and routers) and the communication infrastructure that connects them. Within a single building, LANs are widely used for interconnection, but most of the wide area infrastructure is built up from point-to-point leased lines. In Chapter 10, we will look at LANs; here we will examine only the data link protocols used on point-to-point lines in the Internet. Figure 7–22 shows a home PC acting as an Internet host. In this section we will examine two data link layer protocols used for Internet access over dial-up lines: *serial line internet protocol (SLIP)* and *point-to-point protocol (PPP)*.   SLIP, and PPP

Serial Line IP (SLIP)   SLIP is the older of the two protocols. It was developed to connect Sun workstations to the Internet over a dial-up line using a modem. The protocol is very simple—workstations send raw IP packets over the line, with a special flag byte 11000000 at the end for framing.

   More recent versions of SLIP perform some header compression. They take advantage of the fact that consecutive packets often have many header fields in common. These are compressed by omitting those fields that are the

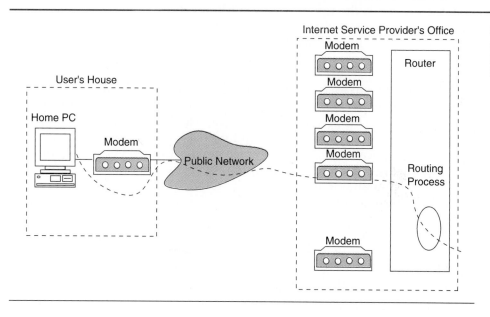

Figure 7–22
A home personal computer acting as an Internet host.

same as the corresponding fields in the previous packet. Furthermore, the fields that do differ are not sent in their entirety, but as increments to the previous value.

Although it is still widely used, SLIP has some serious problems. It does not perform any error detection and correction, it does not provide any form of authentication, and it is not an approved Internet standard.

Point-to-Point Protocol (PPP)   PPP was devised to overcome the deficiencies of the SLIP data link layer protocol. It handles error detection, supports multiple protocols, allows IP addresses to be negotiated at connection time, permits authentication, and has many other improvements over SLIP. While many Internet service providers still support both SLIP and PPP, the future clearly lies with PPP, not only for dial-up lines, but also for leased router-router lines.

PPP provides three functions:

1. A framing method that unambiguously delineates the end of one frame and the start of the next one.

LCP
2. A *link control protocol (LCP)* for bringing lines up, testing them, negotiating options, and bringing them down again gracefully when they are no longer needed.
3. A way to negotiate network-layer options in a way that is independent of the network layer protocol used. Therefore, in this method we may have a different network control protocol for each network layer supported.

router
To see how these pieces fit together, consider the typical scenario of a home user calling up an Internet service provider to make a home PC a temporary Internet host. The PC first calls the provider's *router* via a modem. After the router's modem has answered the phone and established a physical connection, the PC sends the router a series of link control protocol packets in the user's data field of one or more PPP frames. These packets, and their responses, select the PPP parameters to be used.

TCP/IP
Once the PPP parameters have been agreed upon, a series of network control protocol packets are sent to configure the network layer. Typically, the PC wants to run a *TCP/IP* protocol stack. The protocol stack in this case refers to the *Internet protocol* (IP, layer 3) and the *transmission control protocol* (TCP, layer 4). The PC needs an IP address to use. There are not enough IP addresses to go around, so normally each Internet provider gets a block of them and then dynamically assigns one to each newly attached PC for the duration of its login session. If a provider owns n IP addresses, it can have up to n machines logged in simultaneously, but its total customer base may be many times that. The network control protocol for IP is used to create the IP address assignment.

At this point, the PC is now an Internet host and can send and receive IP packets, just as a hardwired host can. When the user is finished, the network control protocol is used to tear down the network layer connection and free up the IP address. Then the link control protocol is used to shut down the data link layer connection. Finally, the computer tells the modem to hang up the phone, releasing the physical layer connection.

The PPP frame format was chosen to closely resemble the HDLC frame format, since there was no reason to reinvent the wheel. The major difference between PPP and HDLC is that the former is character-oriented rather than bit-oriented. In particular, PPP, similar to SLIP, uses character stuffing on dial-up modem lines, so all frames are integral numbers of bytes. Not only can PPP frames be sent over dial-up telephone lines, but they can also be sent over SONET or true bit-oriented HDLC lines (e.g., router-router connections). The PPP frame format is shown in Figure 7–23.

All PPP frames begin with the standard HDLC flag byte (01111110). Next comes the *address* field, which is always set to the binary value 11111111 to indicate that all stations are to accept the frame. Using this value avoids the issue of having to assign data link addresses.

The *address* field is followed by the *control* field, which has the value 00000011 as default. This value indicates an unnumbered frame. In other words, PPP does not provide reliable transmission using sequence numbers and acknowledgment as the default. In noisy environments, such as wireless networks, reliable transmission using a numbered mode can be employed.

Since the *address* and *control* fields are always constant in the default configuration, the link control protocol provides the necessary mechanism for the two parties to negotiate an option to omit them altogether and save two bytes per frame, providing a reduction in overhead.

The fourth PPP field is the *protocol* field. Its job is to tell what kind of packet is in the user's data (payload) field. Codes are defined for link control protocol (LCP), network control protocol (NCP), Internet protocol (IP), AppleTalk, and other protocols.

The *payload* field is of variable length, up to some negotiated maximum. If the length is not negotiated using the link control protocol during the line setup, a default length of 1500 bytes is used. Padding may follow the payload if needed.

After the *payload* field comes the *checksum* field, which is normally two bytes, but a four-byte checksum can be negotiated.

In summary, PPP is a multiprotocol framing mechanism suitable for use over modem, HDLC bit-serial lines, SONET, and other physical layers. It supports error detection, option negotiation, header compression, and optionally, reliable transmission using HDLC framing.

Let us now turn from the PPP frame format to the way the lines are brought up and down. The simplified diagram of Figure 7–24 shows the phases that a line

Bytes	1	1	1	1 or 2	Variable	2 or 4	1
	Flag 01111110	Address 11111111	Control 00000011	Protocol	User's Data (Payload)	Checksum	Flag 01111110

Figure 7–23
The PPP full-frame format for unnumbered mode operation.

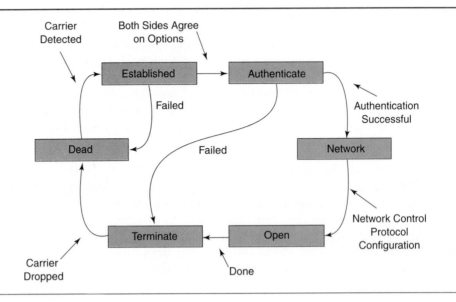

Figure 7–24
A simplified phase
diagram for bringing
a line up and down.

goes through when it is brought up, used, and taken down again. This sequence applies both to modem connections and router-router connections.

When the line is dead, no physical carrier is present and no physical layer connection exists. After the physical connection is established, the line moves to the "establish" state. At this point the link control protocol option negotiation begins, which, if successful, leads to the "authenticate" state. Now the two parties can check on each other's identities, if desired. When the "network" phase is entered, the appropriate network control protocol is invoked to configure the network layer. If the configuration is successful, the "open" state is reached and data transport can begin. When data transport is finished, the line moves to the "terminate" phase, and from there back to "dead" when the carrier is dropped.

The link control protocol (LCP) is used to negotiate data link protocol options during the "establish" phase. The LCP is not actually concerned with the options themselves but with the mechanism for negotiation. It provides a way for the initiating side to make a proposal and for the responding side to accept or reject it, in whole or in part. It also provides a way for the two sides to test the line quality to see if they consider it good enough to set up a connection. Finally, the LCP also allows lines to be taken down when they are no longer needed.

Eleven types of link control protocol (LCP) packets are defined. These packets are listed in Table 7–1. The four *configure* types allow the initiating side (I) to propose option values and the responding side (R) to accept or reject them. In the latter case, the responding side can make an alternative proposal or announce that it is not willing to negotiate certain options at all. The options being negotiated and their proposed values are part of the LCP packets.

Name	Direction	Description
Configure-request	I → R	List of proposed options and values
Configure-ack	I ← R	All options are accepted
Configure-nack	I ← R	Some options are not accepted
Configure-reject	I ← R	Some options are not negotiable
Terminate-request	I → R	Request to shut the line down
Terminate-ack	I → R	OK, line shut down
Code-reject	I ← R	Unknown request received
Protocol-reject	I ← R	Unknown protocol request
Echo-request	I → R	Please send this frame back
Echo-reply	I ← R	Here is the frame back
Discard-request	I → R	Just discard this frame (for testing)

Table 7–1
The LCP packet types.

7.6   SAMPLE NETWORK INTERFACE

In this section, we will bring together many of the concepts we discussed in this chapter so far by describing network interfaces, namely ITU-T recommendation X.25 and the integrated services digital network (ISDN). Before continuing, it is necessary to define a *network interface*. As the diagram in Figure 7–25 suggests, a network interface is the access point for a user into a network. It allows the internal architecture of the network to become unimportant to the user. The network interface provides a point-to-point link between the user and the network, and one interface is required on each side of the network.

network interface

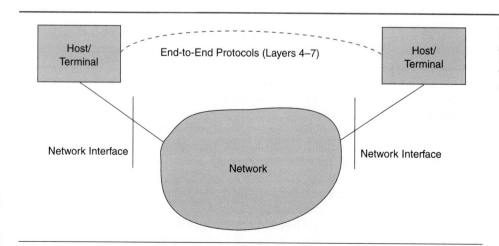

Figure 7–25
A network interface is the access point for a user into a network.

Thus, the network is viewed only as a carrier of information between two end-communicating hosts.

A network interface does not define end-to-end procedures. Therefore, only the chained layers from the OSI reference model (layers 1–3) are defined. The choice of end-to-end protocol is up to the user and is transparent to the network.

7.6.1   ITU-T X.25 Recommendation

During the 1960s, many common carriers throughout the world saw the need to provide public data network service as a companion to the public telephone network service. A public data network can provide local and long-haul data-carrying services just as a public telephone network provides local and long-haul voice-carrying services.

Also during the 1960s, packet switching was proving itself as a viable technology for data transmission. These two concepts came together in the 1970s with implementation of a number of packet-switched public data networks. The common carriers also saw the need to have compatible packet-switched public data networks. Since the compatibility of the network internals could not be guaranteed (or agreed upon), the ITU-T defined recommendation X.25 as the interface between a user and a packet-switched public data network. Thus, manufacturers could build standard devices to connect with any packet-switched public data network.

ITU-T recommendation X.25 is a commonly implemented protocol throughout the world. Some of the notable implementations include U.S. Sprintnet (U.S.), Tymnet (U.S.), AT&T Accunet packet-switched service (U.S.), Datapac (Canada), and Transpac (France).

X.25 defines three layers (previously called levels) of protocols:

- Physical layer corresponding to the physical layer of OSI.
- Link layer corresponding to the data link layer of OSI.
- Packet layer corresponding to the network layer of OSI.

Figure 7–26 shows a schematic of an X.25 connection to a packet-switched public data network. An end user (host) is considered data terminal equipment (DTE), and the packet-switched public data network boundary node is considered data circuit terminating equipment (DCE). Recommendation X.25 defines the DTE–DCE interface. A DTE needs to operate in a synchronous packet mode, and should include the lower and upper layers of software (1–7). The DCE, on the other hand, includes the lower layers of software and hardware (1–3).

No mention is made in X.25 about the subnetwork connection between the DCEs. The network internal architecture is at the option of the packet-switched public data network provider and is totally transparent to the DTE. X.25 also makes no statements about end-to-end protocols, which are themselves transparent to the subnetwork.

As just stated, the X.25 DTE–DCE interface is composed of three layers of protocol. Layer 1 is the physical layer (OSI physical layer), which defines the

X.25

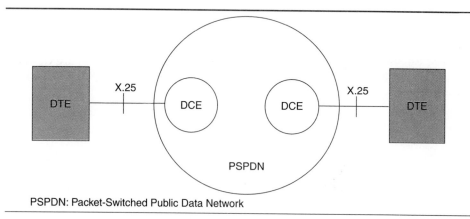

Figure 7–26
X.25 defines the interface between the DTE and DCE.

PSPDN: Packet-Switched Public Data Network

procedure for the physical connection establishment and termination and specifies the physical, mechanical, and electrical aspects of the interface. The physical interface is serial, synchronous, full duplex, and point to point. Recommendation X.25 specifies that ITU-T recommendation X.21 or X.21bis be used as the physical interface. This requirement is not commonly adhered to in the United States. ANSI recommends EIA-232-C, EIA-449, and V.35.

Layer 2 is the link layer (OSI data link layer), which provides a data link layer, namely error-free communication over the physical path between two adjacent data stations. The framing and operational procedures are a subset of the HDLC described above. The link access procedure provides the data link establishment terminating and allows full-duplex, point-to-point transmission of frames with flow control.

Layer 3 is the packet layer (OSI network layer), which fragments messages into packets; reassembles packets into messages; assigns logical channel identifiers; and establishes, resets, and clears virtual calls. It can multplex logical channels over the data link. Only virtual circuit service is provided; datagrams are not supported in X.25 (although datagram service was added in the 1980 version of X.25, it was dropped from the 1984 version). The OSI network layer includes routing functions, but the X.25 packet layer protocol does not, because routing is not necessary across a point-to-point link. Table 7–2 gives a summary of the three layers defined in the X.25 recommendation.

X.25 Call Setup   Figure 7–27 demonstrates the steps involved in the establishment of an X.25 virtual call. For the purpose of terminology, a virtual circuit is a full-duplex, logical, end-to-end connection between two DTEs; a logical channel is the identifier of a virtual circuit across a DTE–DCE interface. The steps in an X.25 call setup are as follows:

1. The calling DTE 1234 sends a call request packet, requesting a virtual circuit to DTE 987. The virtual circuit will have a logical channel identifier of 18 (chosen by the calling DTE) across this DTE–DCE interface.

Table 7–2
Summary of X.25 layers.

Layer	1-(Physical)	2-(Link)	3-(Packet)
Responsibility	Transmission of bits	Reliable path of frames	Transfer of packets (virtual circuit service)
Protocol	X.21 bits, X.21, EIA232-C, EIA-449, V.35	Link access procedure balanced (LAPB)	Packet layer protocol
Functions: Establish/Terminate	Physical connection	Data link	Virtual circuit
Data transfer	Point-to-point, full-duplex, synchronous, serial	Point-to-point, full-duplex, error-free, sequential delivery	Point-to-point, full-duplex, sequential delivery
Others	Physical, electrical, mechanical specification	Frame structure, error control, flow control	Packet format, fragment/recombine, flow control, multiplex logical channels

Figure 7–27
X.25 call setup.

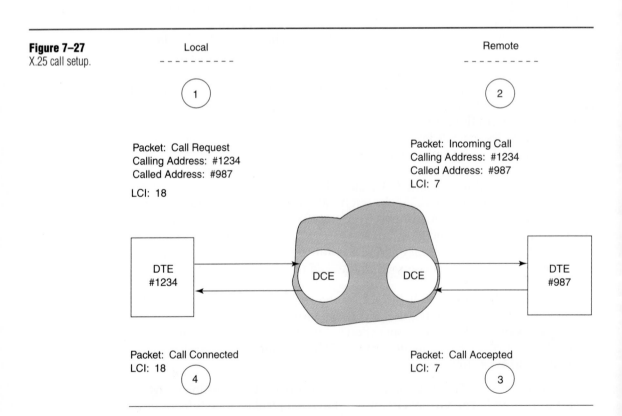

Local

Remote

1

2

Packet: Call Request
Calling Address: #1234
Called Address: #987

LCI: 18

Packet: Incoming Call
Calling Address: #1234
Called Address: #987
LCI: 7

DTE #1234

DCE

DCE

DTE #987

Packet: Call Connected
LCI: 18

Packet: Call Accepted
LCI: 7

4

3

2. At the remote interface, the called DTE (987) receives an incoming call packet, indicating a call from DTE 1234 on logical channel 7 (chosen by remote DCE).

3. Deciding to accept the call, DTE 987 returns a call accepted packet on logical channel 7. Note that full DTE addresses are not necessary; the logical channel numbers provide unambiguous address information at this point.

4. The calling addresses are not mandatory in the call accepted and call connected packets; they are placed in these packets for redundancy and error checking only.

X.25 Call Request/Incoming Call Packet Figure 7–28 shows the format of the X.25 call request (sent by a DTE) and incoming call (sent by a DCE) packets. The first three *octets* (bytes) of all packets are the same: octet

- General format identifier (octet 1, bits <8:5>): usually set to "0001" as shown.
- Logical channel identifier (octet 1, bits <4:1> and octet 2): a 12-bit quantity identifying the logical channel. This will be a number between 1 and 4095.
- Packet type identifier (octet 3): indicates the type of packet.

The following additional fields are specific to the call request/incoming call packet:

- Calling DTE address length (octet 4, bits <8:5>): indicates the number of digits in the calling DTE address. This will be a value between 0 and 15.
- Called DTE address length (octet 4, bits <4:1>): indicates the number of digits in the called DTE address. This will be a number between 0 and 15.
- Called and/or calling DTE addresses: the called DTE address (if present), followed by the calling DTE address. Each address digit is coded in binary-coded decimal (BCD), two BCD digits per octet. This field is between 0 and 15 octets in length.
- Facility length: facilities are optional capabilities that a user may want associated with a given call (e.g., large packet sizes or reverse charging). The value of this field indicates the number of octets in the facility field and will be a number between 0 and 109.
- Facility field: contains optional user facility requests, if any. This field will be between 0 and 109 octets in length.
- Call user data field: contains any additional information that might be needed to set up this call, such as a password. This field may be absent or between 1 and 16 octets in length.

The call accepted/call connected packets have a similar format to that of the call request packet.

X.25 Data Packet Figure 7–29 shows the format of an X.25 data packet. Note that there is very little overhead with the data packet; once the call is established,

Figure 7–28
X.25 call request/
incoming call
packet.

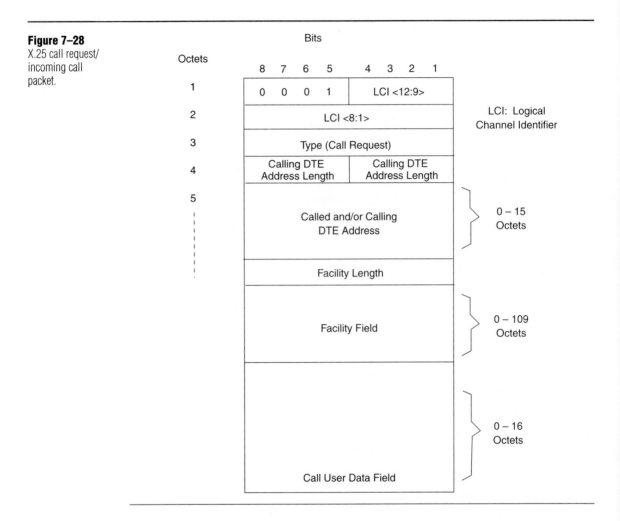

only the logical channel number (LCI) is needed. The first two octets in the data packet are the same as in the call request/incoming call packet. The third octet indicates the packet type plus additional information:

- Packet type (octet 3, bit <1>): this bit is set to 0, indicating that this is a data packet. This bit is set to 1 in all other X.25 packets.
- Packet sequence number (octet 3, bits <4:2>).
- More indicator (octet 3, bit <5>): indicates whether more data packets follow this one that are logically associated with this one to form a message. Remember that layer 3 fragments and reassembles messages into packets and packets into messages.
- Acknowledgment number (octet 3, bits <8:6>): indicates the sequence number of the next expected data packet.

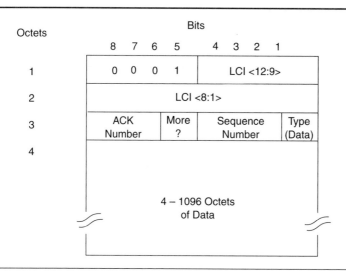

Figure 7–29
X.25 data packet.

- User data field: between 1 and 4096 octets of user data. The maximum size of this field is 128 octets by default, although X.25 supports up to 4096 octets. The maximum number is chosen depending upon the buffer size of the data station, the channel characteristics, and the error-detection capability of the data link layer.

X.25 Call Termination   At the conclusion of the data exchange, the DTE must clear the call. This action is required to remove the logical channel number assignment at both interfaces, free up buffer space, and remove the routing table entries that may have been created by the packet-switched public data network for this virtual circuit. This procedure is called *call clearing*. The call request/clear indication packets are used to clear the virtual circuit. these packets are confirmed by the clear confirmation packet as will be seen later in this section.

X.25 Clear Request/Indication Packet   Figure 7–30 shows the format of the clear request (sent by a DTE) and clear indication (sent by a DCE) packets. The first three octets have the same format as the first three octets of the call request/incoming call packet. In addition, there are two more octets:

- Clear cause field (octet 4): the reason that this call was cleared (e.g., remote DTE busy, network is congested).
- Diagnostic code field (octet 5): additional information about the clearing cause.

Complete X.25 Call Scenario   Figure 7–31 shows all of the steps involved in the setup of an X.25 virtual call, the exchange of data, and the clearing of the

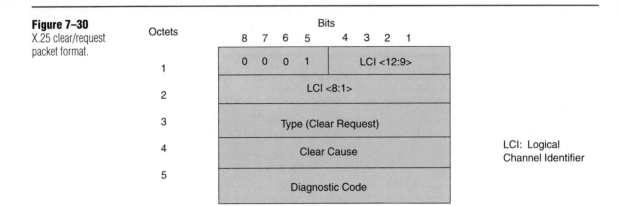

Figure 7–30
X.25 clear/request
packet format.

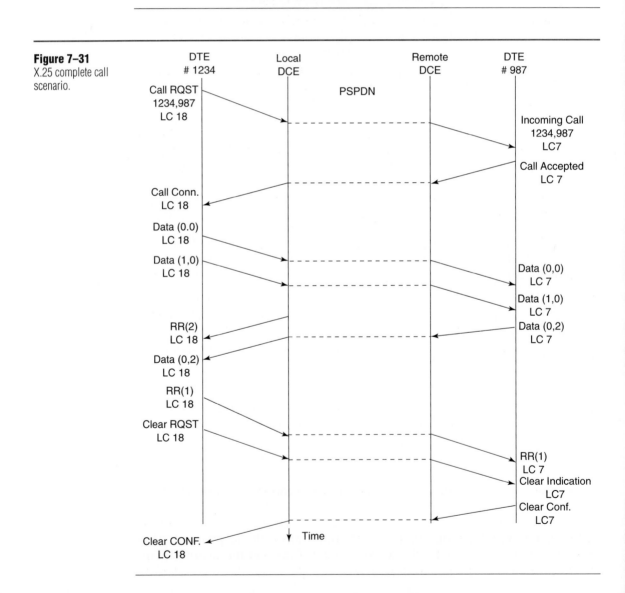

Figure 7–31
X.25 complete call
scenario.

virtual call in time. The slope of the arrows indicates the propagation and the transmission delay over the DTE–DCE interface. The following is the sequence of events that takes place in an X.25 virtual call:

1. DTE 1234 sends a call request packet on logical channel (LC) 18 to establish a virtual circuit to DTE 987. An incoming call packet from DTE 1234 is delivered to DTE 987 on LC 7.
2. DTE 987 accepts the call by sending a call accepted packet on LC 7; this is delivered as a call connected packet on LC 18.
3. DTE 1234 sends two data packets, numbers 0 and 1 consecutively. (Both indicate that data packet 0 is expected from DTE 987.)
4. Local DCE acknowledges data packet 1 by RR(2) (receiver ready packet) since it does not have a data packet to send and piggyback the acknowledgment.
5. DTE 987 sends back its own data packet 0, indicating receipt of DTE 1234's data packet 0 and 1 (next expected data packet is 2).
6. Remote DCE acknowledges this packet by RR(1).
7. DTE 1234 acknowledges DCE's data packet by returning an RR(1).
8. Finished with the call, DTE 1234 sends a clear request packet on logical channel (LC) 18. The packet-switched data network (PSDN) clears the call at the local DTE–DCE interface and sends DTE 1234 a clear confirmation packet.
9. The packet-switched public network is responsible for clearing the entire call through the network and at the remote DTE–DCE interface. DTE 987 receives a clear indication packet on logical channel 7 and responds with a clear confirmation packet. The call is completely cleared at this point.

7.6.2   Integrated Services Digital Network (ISDN)

Integrated services digital networks (ISDN) is a natural evolution of the telephone network. It specifies a standard for interfacing the user to the network. The ISDN standard, similar to the X.25 standard, specifies the lower three layers (1 through 3) for interfacing user's equipment to network equipment. An ISDN is a public network, a digital network, and integrates voice and data services. ISDN also supports circuit-switched data and/or voice services as well as packet-switched data services. We could devote a whole book to ISDN, but we have to cover other topics in this book; therefore, in this section we will present an overview of ISDN and its three layers. We will also provide the scenario of making data and voice calls over ISDN networks.

ISDN Evolution Figure 7–32 shows a logical evolution of network access toward the ISDN model. Scenario A shows an older model where users need to access several types of networks to exchange voice, data, and signaling information. In that model, each type of information transfer requires a different network, and each different network requires a different user-network interface.

The interim model of scenario B is the first step toward ISDN. Although the signaling, voice, and data networks are still separate entities, the user gains

Figure 7–32
ISDN evolution.

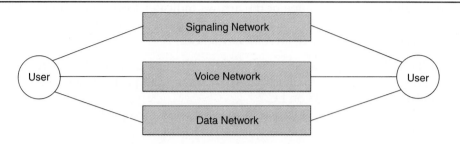

A. Older model: User needs an interface for
    every different network accessed.

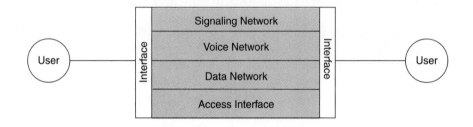

B. Interim model: User has
    "integrated access" to all networks.

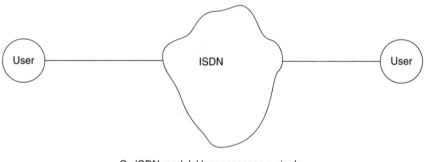

C. ISDN model: User accesses a single
    network for many purposes.

access via a single interface. Thus, the interim model provides integrated access to the network.

Scenario C shows the true ISDN model. The user has the same integrated access to a single network, and that single ISDN provides all of the necessary services. An important part of this function is transparency. The move from scenario B to scenario C should progress without negatively affecting the user. In fact, the user should not even know (or care!) that the evolution has occurred.

This transparency results because the standard integrated access interface is the same in both scenarios and the internal network structure is invisible to the user.

An example of this transparency in today's world is with respect to central office equipment. A user's telephone will work the same way after the central office converts from a step-by-step switch to a 5ESS switch. The available service from 5ESS, of course, will be superior to that of the step-by-step switch.

The evolution from scenario A to B will not be transparent. This is because in-house equipment will be digital rather than analog, in-house wiring will have to be modified, and the local subscriber loop must be modified to carry digital signals.

ISDN Services Figure 7–33 shows the services that an ISDN network may provide. A large number of devices may access this type of network, including the following:

- The PBX acts as an intermediate device for the connection of telephones, personal computers, and other ISDN and non-ISDN devices in a business environment.
- A home or business security or energy management system may send data packets to some monitoring station through ISDN.
- A multimedia terminal, incorporating voice, data, and video features may also interface to the ISDN from a business environment.
- Non-ISDN devices may interface via an ISDN-compatible station on a LAN.
- DTEs supporting X.25, or other non-ISDN equipment, may access the ISDN via a device called a terminal adapter (TA). A TA makes a non-ISDN device present itself to the network as an ISDN-compatible device.

ISDN Reference Points A large amount of new terminology has been introduced with respect to ISDN. Figure 7–34 introduces a number of terms dealing with two important ISDN concepts. The first concept deals with the names of the different pieces of equipment on an ISDN. The local exchange (LE) is the provider of the ISDN service (i.e., the central office). Network termination 1 (NT1) equipment is the termination of the network channel from the LE (this is analogous to the DCE in X.25). Network termination 2 (NT2) equipment, if present, is customer premises equipment (CPE) that provides on-site distribution, such as a PBX.

User equipment falls into two categories: terminal equipment 1 (TE1) is an ISDN device and terminal equipment 2 (TE2) is a non-ISDN device. TE2 devices may access the ISDN via a terminal adapter (TA). Clearly, the TA is specific to a TE2 device and, in fact, could be integrated into TE2 devices.

The second ISDN concept in Figure 7–34 deals with reference points:

- The U reference point is the link between the LE and the NT1 (i.e., the network transmission line). The ITU-T ISDN recommendations do not describe this interface since they view it as part of the network. The American

Figure 7–33
ISDN services.

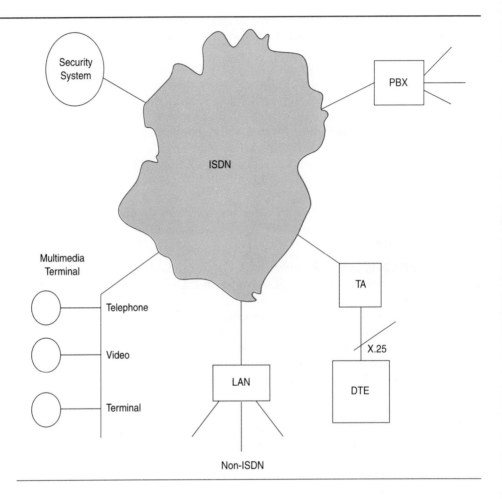

National Standards Institute (ANSI) is actively defining this interface since the FCC views this point as the network boundary.

• The T reference point is the network boundary according to the ITU-T and is analogous to the X.25 DTE/DCE interface.

• The S reference point is between CPE (i.e., it falls between the NT2 and either the TE1 or the TA). In the absence of NT2 equipment, the S/T reference point is the network boundary.

• The R reference point is specific to a particular TE2 device and its TA. Every manufacturer of TE2 equipment must create the appropriate R interface so that the TA-TE2 equipment presents itself to the ISDN as a TE1 device.

ISDN B- and D-Channels   Two main types of channels will be used in ISDNs, normally B- and D-channels. The D-channel is primarily used for user-network signaling. When a user needs any type of service, the service request is sent to the network on the D-channel. For example, when the handset from

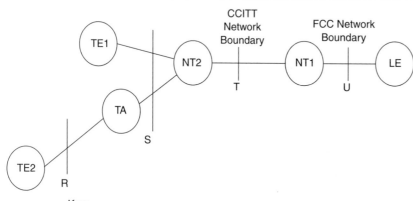

Figure 7–34
ISDN reference
points.

Key:

LE = Local Exchange
NT1 = Network Termination 1 (e.g., Network Channel Termination Equipment)
NT2 = Network Termination 2 (e.g., PBX)
TE1 = Terminal Equipment 1 (ISDN Terminal)
TE2 = Terminal Equipment 2 (Non-ISDN Terminal)
TA = Terminal Adapter
U Reference Point = Network Transmission Line; Not Specified in ITU-T ISDN Standards
T Reference Point = Analogous to X.25 DTE-DTE Interface
S Reference Point = ISDN TE to NT
R Reference Point = Vendor-specific

an ISDN telephone is picked up, the telephone will automatically make a request on the D-channel for an outgoing voice line. Since the bandwidth of the D-channel is greater than that needed for signaling alone, packet-mode data may also be sent on this channel.

The B-channel may be used for all forms of telecommunication services. A B-channel, if available, is granted in response to a request for service over the D-channel. Since ISDN carries only digital information, data, voice, or other transmissions may be carried transparently through the network. All B-channels operate at 64 kbps "clear." That is, all bits on the B-channel are dedicated to carrying user information and not for signaling on the physical channel (as we saw the robbed bit signaling in T1, where one bit was stolen to transmit signaling information, this is not the case with ISDN). In summary, the B- and D-channels carry the following types of information:

- B-channel (bearer channel)
    a. User information.
    b. Voice, data, image.
    c. Clear channel (no signaling).
- D-channel
    a. User-network signaling (primarily).
    b. Request for service.
    c. Data packet exchange.

ISDN Rate Interfaces   The final set of new ISDN terms relates to the basic and primary rate interfaces. The basic rate interface (BRI) is composed of two 64-kbps B-channels and one 16-kbps D-channel (2B + D). The user data rate is 144 kbps, and the overall bit rate (including signaling information on the transmission facility) is 192 kbps. The 2B + D interface is designed for residential customers and small users.

The primary rate interface (PRI) is composed of 23 B-channels and a 64-kbps D-channel (23 B + D) or 24 B-channels (24B). This interface is designed for business customers and large users. The data rate of the primary rate interface is 1.536 Mbps, but the overall bit rate (including signaling) is 1.544 Mbps (the T1 rate). There are actually two PRI recommendations. The 23B + D PRI is for ISDN in the U.S., Canada, and Japan. The European ISDN has defined a different primary rate interface, called 30B + D or 31B, which operates at 2.048 Mbps.

ISDN Standards   ISDN standardization is one of the most active areas of current standards creation. In particular, the user-network interface (S/T) is defined in the ITU-T I-series recommendations, first published formally in 1984. Other ISDN-related recommendations are the E-, G-, Q-, and X-series

In the United States, the ANSI T1D1 task group is working on U.S. contributions to the ITU-T recommendations and on the U.S. ISDN standards. Lucent Technologies, among others, has its own ISDN standards as well.

The R reference point is specific to given TE2 devices. For this reason, there are no R reference point standards; instead, the TE2 manufacturer is responsible for this interface. The R interface may use public standards, such as X.21, V.35, EIA-232-D, or EIA-530.

The ITU-T views the U reference point as internal to the network, thus beyond the scope of its recommendations. The FCC views the U interface as the network boundary, however, and many organizations in the United States (and around the world) are proposing U interface standards to the ANSI T1 D1 committee. United States organizations proposing standards include Lucent Technologies, AT&T, Bell Communication Research (BELLCORE), Northern Telecom, and Siemens. In the following three sections, we present an overview of the three layers of protocols included in the ISDN user-network interface.

ISDN Layer 1   The ISDN layer 1 (OSI physical layer) is defined in ITU-T recommendations I.430 (basic rate interface) and I.431 (primary rate interface). The BRI uses pseudo-ternary signaling, as shown in Figure 7–35. Two observations can be made about the pseudo-ternary signals. The signal pulses occur with the 0-bits instead of 1-bits, and the signal pulse lasts for 100 percent of the bit-time.

Figure 7–36 shows the TDM scheme used by the physical layer. The basic rate frame is shown without any of the layer 1 signaling information. A frame lasts for 250 microseconds, yielding a rate of 4000 frames per second. Each frame is composed of two 8-bit groups from the B1-channel, one bit from the

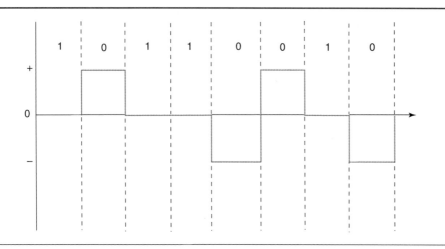

Figure 7–35
Pseudo-ternary signal formats.

D-channel, and two 8-bit groups from the B2-channel. Thus, each frame consists of 16 bits from each B-channel (yielding a 64-kbps bit rate for each B channel) and four bits from the D-channel (yielding a bit rate of 16 kbps). A frame is actually composed of 64 bits; the remaining 12 bits per frame are overhead used for maintaining the physical link (48 kbps).

The North American and Japanese primary rate (23B + D or 24B) frame is identical to a T1 carrier frame. Each frame lasts for 125 microseconds, yielding 8000 frames per second. Each frame has a framing bit, followed by 24 octets, each representing one of the 64 kbps channels.

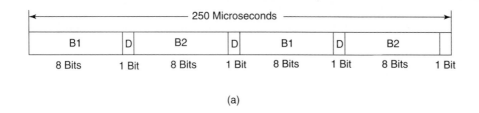

Figure 7–36
ISDN layer 1 frame formats: (a) ITU-T recommendation I.430—basic rate interface, (b) ITU-T I.431—primary rate interface.

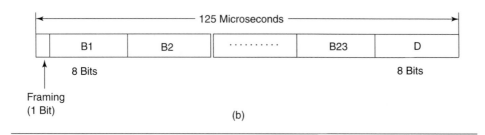

A PRI multiframe is composed of 24 frames. The 24 framing bits of a multiframe are used in a fashion similar to the 24 framing bits of a T1 ESF.

ISDN Link Layer (Layer 2)   The ISDN link layer (layer 2) definition is contained in ITU-T recommendation I.440 (Q.920) and I.441 (Q.921). The ISDN data link protocol is called the link access procedures on the D-channel (LAPD). LAPD is a go-back-N bit-oriented protocol that is a subset of HDLC, described in previous sections of this chapter.

LAPD can provide three types of services to layer 3. First, using the call control procedure, the D-channel can provide circuit-switched services for voice or data on the B-channels. The D-channel can provide packet-switched services on the B- or D-channels. The D-channel can also provide operation, administration, and management service on the D-channel.

No link layer procedures are specified for the B-channel. This is because the user may send any type of information over the B-channel. The frame format (if any) of the B-channel transmission is determined by the service and/or protocol that the user has requested at call setup.

To clarify the role of the D-channel, say, when we make voice calls over any of the B-channels, the signaling information messages (analogous to off-hook and on-hook signals in the plain old telephone service) are carried over the D-channel. Thus, the D-channel acts as the signaling conduit for all of the 23 B-channels in the primary rate interface, and as the signaling conduit for the 2 B-channels in the basic rate interface.

ISDN Layer 3   The ISDN layer 3 specifications are contained in ITU-T recommendations I.450 (Q.930) and I.451 (Q.931). In particular, I.451 defines the set of network signaling messages used on the D-channel. Table 7–3 shows the analogy between ISDN signaling versus signaling used in plain old telephone service (POTS) networks:

• When a user lifts a telephone handset, today's network sends an off-hook signal to the central office (CO); an ISDN telephone will send an I.451 (Q.931) setup message on the D-channel.

• The CO in the current telephone network replies to the off-hook signal with a dial-tone; this action signals the user to dial a telephone number. The

Table 7–3
Plain old telephone
service (POTS) –
ISDN analogy.

POTS	ISDN
Phone sends off-hook signal	Phone sends a setup message
Network replies with dial tone	Network sends setup ack message
User dials telephone number	Phone sends info message
User hears ringing	Network sends alerting message

ISDN will send a setup acknowledgment message on the D-channel; the ISDN telephone will generate a dial-tone for the user.

• The user dials a telephone number, generating pulses or dual-tone, multi-frequency (DTMF) tones for the central office. The ISDN telephone will buffer the dialed digits and send the telephone number to the CO in an information message on the D-channel.

• Today's network lets the user know that the call is going through by generating a ringing sound. The CO will send an alerting message to the ISDN telephone, which, in turn, will generate a ringing sound for the user.

Take notice that the ISDN telephone, not the network, will generate a dial-tone and a ringing sound for the user. These sounds are not really for the network; however, if they are not present, ISDN providers would have to retrain the entire population in how to use their telephones! Continuing to use these sounds is part of the transparency in the ISDN evolution.

There is no ISDN layer 3 for the B-channel for essentially the same reason that there is no data link definition on the B-channel; the B-channel can carry any information in any format.

ISDN Voice Call Scenario   Figure 7–37 shows the steps involved in establishing a voice call through an ISDN. Signaling information will be sent on the D-channel, and the voice conversation will occupy a B-channel. The steps are as follows:

1. The user picks up the telephone handset. The ISDN telephone will send an I.451 setup message. The message will request a B-channel for a voice call; the request is automatic since the telephone can only use B-channel setup for voice!

2. The network replies with an I.451 setup ack message.

3. The setup ack causes the telephone to generate a dial-tone, directing the user to dial the desired telephone number. This number is sent in an info message.

4. The local exchange places the call through the ISDN. The local exchange sends an I.451 alerting message to cause the ISDN telephone to make a ringing sound for the user.

5. The other party picks up his or her telephone. The network sends an I.451 connect message to indicate that the B-channel is ready for the conversation.

6. The two parties talk to each other over the B-channel. Eventually, they finish talking and . . .

7. . . . the user hangs up the telephone. An I.451 disconnect message is sent over the D-channel.

8. The network responds with an I.451 release message and the telephone replies with an I.451 release complete message on the D-channel. At this point, the B-channel is free for another call.

Figure 7–37
ISDN voice call scenario.

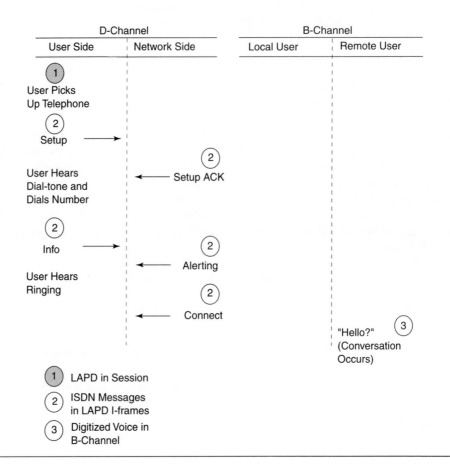

Voice on the B-Channel

7.7 SUMMARY

When network devices and equipment communicate, they have to follow certain rules. These rules govern the methods of communication among devices in the network. Standards bodies and organizations define the rules of communication among devices. These rules are referred to as protocols. The OSI reference model defines a framework within which devices in networks can communicate with one another. This reference model specifies 7 layers of protocols:

- Layer 1—the physical layer.
- Layer 2—the data link layer.
- Layer 3—the network layer.

- Layer 4—the transport layer.
- Layer 5—the session layer.
- Layer 6—the presentation layer.
- Layer 7—the application layer.

Computers and equipment that are not part of a network, but that use a network to communicate, have to implement all of the seven layers. Network devices, however, need to implement only the first three layers in the OSI reference model (1 to 3).

In this chapter, we examined many examples of standards protocols that are in compliance with the OSI reference model. Particular examples include the integrated services digital network (ISDN) designed for voice and data communications and X.25, an earlier standard that defined packet-switched communications.

REVIEW QUESTIONS AND PROBLEMS

1. Why is the idea of layering so important in protocol design?

2. Describe the OSI reference model.

3. Give the protocol data unit for each of the layers in the OSI reference model.

4. Give a scenario of communication between two devices, each using HDLC for layer 2 of the OSI model. In your scenario, show the establishment of a data link layer and transfer of information between the two devices when no transmission errors occur.

5. Give the same scenario when errors occur in the transmission of I-frames.

6. What are I-frames, supervisory frames, and unnumbered frames in the HDLC standards?

7. How are the P- and F-bits used in the HDLC standards?

8. Give the frame formats for the HDLC, describing each field in brief detail.

9. What is X.25?

10. Give a scenario of communication over an X.25 network, including setting a connection.

11. List the different types of packets used in X.25.

12. What are the advantages of ISDN over X.25?

13. Give a voice call scenario using ISDN.

14. What is LAPD?

Broadband ISDN and ATM

8.1 INTRODUCTION

The highest bit rate an ISDN primary rate interface can offer the user is 1.5 Mbit/s or 2 Mbit/s. Connection of local area networks (LANs), however, or transmission of moving images with good resolution may, in many ways, require considerably higher bit rates. Consequently, the conception and realization of a *broadband ISDN (B-ISDN)* was needed.

ITU-T recommendation I.113 (vocabulary of terms for broadband aspects of ISDN) defines *broadband* as:

> . . . a service or system requiring transmission channels capable of supporting rates greater than the primary rate.

B-ISDN thus includes 64 kbit/s ISDN capabilities but in addition opens the door to applications using bit rates above the 1.5 Mbit/s or 2 Mbit/s defined for the primary rate ISDN interface. The bit rate available to a broadband user is typically from about 50 Mbit/s up to hundreds of Mbit/s. The first concrete idea of B-ISDN was simply to:

- Add new high-speed channels to the existing channel spectrum.
- Define new broadband user-network interfaces.
- Rely on existing 64 kbit/s ISDN protocols and only to modify or enhance them when absolutely necessary.

B-ISDN

B-ISDN development can be justified and will be successful only if it meets the needs of potential customers. Therefore, a brief outline of foreseeable broadband applications will be given before we enter into the discussion of the technical details. In principle, B-ISDN should be suitable for both business and residential customers. Thus, as well as data communication, the provision of TV program distribution and other entertainment facilities has to be considered. B-ISDN supports both variable and fixed bit rates; data; voice; still and moving picture transmission; and of particular note, multimedia applications, which may combine data, voice, and picture service components.

In the business arena, videoconferencing is already a well-established but still not commonly used method that facilitates the rapid exchange of information between people. As traveling can be avoided, videoconferencing helps to save time and costs. B-ISDN may considerably improve the current situation and allow videoconferencing to become a widespread telecommunication tool, as it allows for high-picture quality (at least today's TV quality or even better), which is critical for its acceptance.

Another important feature of B-ISDN is the (cost-effective) provision of high-speed links with flexible bit rate allocation for interconnection of customer networks. The residential B-ISDN user may appreciate the combined offer of text, graphics, sound, still images, and films, giving information about topics such as vacation resorts, shops or cultural events, as well as interactive video services and video on demand.

8.2   B-ISDN SERVICES

The motivation behind incorporating the broadband feature into ISDN is documented in ITU-T recommendation I.121 (broadband aspects of ISDN). The B-ISDN recommendations were written taking into account the following:

- The emerging demand for broadband services.
- The improved data and image processing capability available to the user.
- The advances in software application processing in the computer and telecommunication fields.
- The need to integrate interactive distribution services and circuit and packet transfer modes into a universal broadband network.
- The need to provide flexibility in satisfying the requirements of both user and operator (bit rates, quality of service, etc.).

As a result of those considerations, B-ISDN supports switched, semipermanent, permanent, point-to-point, and multipoint connections and provides on-demand, reserved, and permanent services. Connections in B-ISDN support both circuit mode and packet mode services of a mono and/or multimedia type, of a connectionless or connection-oriented nature, and in a bidirectional or unidirectional configuration. A B-ISDN contains intelligent capabilities for the purpose of providing advanced service characteristics, supporting powerful operation and administration management (OAM) tools. By now, you are probably impressed by the intended capabilities of B-ISDN, which is designed to be *the* universal future network.

8.3   MOTIVATION FOR ATM

In order to understand ATM, we have to look at what technology was used before and why ATM came into existence. *Synchronous transfer mode (STM)* is   STM used to transfer packetized voice and data across long distances. It is a circuit-switched networking mechanism, where a connection is established between two endpoints before data transfer can take place, and is torn down when the two endpoints are done. The two endpoints allocate and reserve the connection bandwidth for the entire duration, even when they are not actually transmitting the data. Data is transmitted in an STM network by dividing the bandwidth of the STM links, which can be T1 or T3 links, into fundamental units of transmission called time slots or buckets as we saw in Chapter 2.

   Data transfer over STM can be compared to a train. The train is organized into a number of buckets labeled from 1 to N. The train repeats periodically every T time period, with the buckets in the train always in the same position with the same label. There can be up to M different trains labeled from 1 to M, also repeating with the same time period T, and all arriving within the same time period T. For example, 8 T1s are labeled 1 to 8, and each T1 has 24 buckets (time slots) labeled 1 to 24.

   On a given STM link, a connection between two endpoints is assigned a fixed bucket number between 1 and N on a fixed train between 1 and M. Data from that connection are always carried in that bucket number on the assigned train. This is analogous to saying the connection uses time slot 19 on T1 number 3. If there are intermediate nodes, it is possible that a different bucket number on a different train is assigned on each STM link in the route for the connection. In other words, once a time slot is assigned to a connection, it generally remains allotted for the connection's sole use throughout the life of that connection.

   To understand this concept better, picture the same train arriving at a station every T period. Then if a connection has any data to transmit, it drops its data into the assigned bucket (time slot) and the train departs. If the connection does not have any data to transmit, that bucket in the train remains empty. No passengers waiting in line can get on that empty bucket. If there are a large number of trains, and a large number of total buckets are remaining empty most of the time (although during rush hours the trains may get quite full), this is a significant waste of bandwidth, and limits the number of connections that can be supported simultaneously. Furthermore, the number of connections can never exceed the total number of buckets on all different trains ($N \times M$). This is the reason that technology shifted to ATM.

8.3.1   Birth of ATM

The telecommunications companies are investigating fiber optic cross-country and cross-oceanic links with gigabit/sec speeds. They would like to integrate real-time traffic such as voice and high-resolution video, which can tolerate some losses but no delays, as well as non-real-time traffic such as computer data and file transfer, which may tolerate some delays but no losses. The problem with carrying these different characteristics of traffic on the same medium in an integrated fashion is that the peak bandwidth requirement for these traffic sources may be quite high, as in high-resolution full-motion video. The duration for which the data is actually transmitted, however, may be quite small. In other words, the data comes in bursts and must be transmitted at the peak rate of the burst, but the average arrival time between bursts may be large and random. For such bursty connections, it would be a considerable waste of bandwidth to reserve them a bucket at their peak bandwidth rate for all times, when on average only one in ten buckets may actually carry the data.

It would be nice if that bucket could be reused for another pending connection. Thus, using the STM mode of transfer becomes inefficient as the peak bandwidth of the link, peak transfer rate of the traffic, and overall burstiness of the traffic, expressed as a ratio of peak to average, all go up. Hence ATM was conceived. It was independently proposed by Bellcore, the research arm of the regional telephone operating companies in the U.S. The main idea here is that instead of always identifying a connection by bucket number, the connection identifier can be carried along with the data in any bucket. The size of the bucket should be small enough so that if a bucket is dropped en route due to congestion, not too much data is lost. This sounds very much like the packet switching discussed in Chapter 4, so they called it "fast-packet switching with short-length packets." The fixed size of packets arose out of a motivation to sustain the same transmitted voice quality as in STM networks, but in the presence of some lost packets on ATM networks.

The two endpoints in an ATM network are associated with each other via an identifier called the *virtual circuit identifier (VCI label)* instead of by a time slot or bucket number as in an STM network. The VCI, discussed in more detail later in this chapter, is carried in the header portion of the fast packet. The fast packet itself is carried in the same type of bucket as before, but there is no label or designation for the bucket anymore. The terms *fast packet, cell,* and *bucket* are used interchangeably in ATM literature.

8.4   STATISTICAL MULTIPLEXING

Fast-packet switching attempts to solve the unused bucket problem of the synchronous transfer mode (STM) by statistically multiplexing several connections on the same link based on their traffic characteristics. In other words, if a large number of connections are very bursty (i.e., their peak-to-average ratio is 10:1 or higher), then all of them may be assigned to the same link in the hope that statistically they will not all burst at the same time. If some of them burst simultaneously, there is sufficient elasticity for bursts to be buffered up and put in subsequently available free buckets. This process is called *statistical multiplexing,* and it allows the sum of the peak bandwidth requirements of all connections on a link to even *exceed* the aggregate available bandwidth of the link under certain conditions. This achievement was impossible on an STM network, and it is the main distinction of an ATM network.

statistical
multiplexing

8.5   TYPES OF USER NETWORK INTERFACES (UNI) FOR ATM

It is envisioned that ATM network service providers may offer several types of interfaces to their networks. One interface that is likely to be popular with companies that build routers and bridges for local area networks (discussed in later chapters of this book), is a frame-based interface. One or more of the IEEE8.2.X frames may be supported at the user network interface (UNI), with frame-to-ATM cell conversion and reassembly done inside the UNI at the source and destination endpoints, respectively. Thus, a gateway host on a local area network might directly connect its *Ethernet,* token ring, or other LAN interface to the UNI, and thus bridge two widely separated LANs with an ATM network. This arrangement will preserve the existing investment in these standards and equipment and enable a gradual transition of the ATM networks into the marketplace.

Ethernet

An alternate interface likely to be more popular in the long run, and for which the concept of broadband-ISDN really makes sense, is direct interface at the UNI with standard ATM cells. Such streaming interfaces can hook telecommunication subscribers, data communication, and computer equipment directly to the network, and can allow orders-of-magnitude greater performance and bandwidth utilization for integrated multimedia traffic of the future. Thus, it is no accident that the IEEE 802.6 packet for the media access control layer of the metropolitan area network protocol looks very much like an ATM cell (packet).

Companies with investments in existing STM networks such as T1 and T3 backbones are likely to want a direct T3 interface to the user network interface (UNI), thus allowing them to slowly integrate the newer ATM technology into their existing system. That is why we see a flurry of small startup companies rushing to make large T3 multiplexers for connecting T3 pipes into one large ATM pipe at the UNI.

8.6   BUILDING BLOCKS OF B-ISDN: ATM

ATM is considered the ground on which B-ISDN is to be built, as stated in ITU-T recommendation I.120: "Asynchronous transfer mode (ATM) is the transfer mode for implementing B-ISDN . . ."

The term *transfer* comprises both transmission and switching aspects, so a *transfer mode* is a specific way of transmitting and switching information in a

Figure 8–1
ATM cell structure.

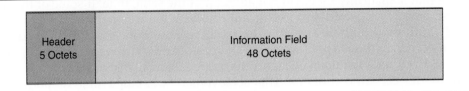

network. In ATM, all information to be transferred is packed into a fixed-size slot called a *cell*. These cells have a 48-octet information field and a 5-octet header. The information field is available to the user, but the header field carries information that pertains to the ATM layer functionality itself, mainly, the identification of cells by means of a label, as shown in Figure 8–1. A detailed description of the ATM layer functions and ATM header structure and coding will be given in following sections.

cell

ATM uses a label field inside each cell header to define and recognize individual communications. In that respect, ATM resembles conventional packet transfer modes. Similar to packet-switching techniques, ATM can provide communications with bit rates that are individually tailored to actual needs, including time-variant bit rates.

The term *asynchronous* in the name of the new transfer mode refers to the fact that, in the context of multiplexed transmission, cells allocated to the same connection may exhibit an irregular recurrence pattern as they are filled according to actual demand. A cell associated with a virtual channel may occur at essentially any position in the bit stream.

In ATM-based networks, the multiplexing and switching of cells are independent of the application. Thus, the same piece of equipment can, in principle, handle a low bit-rate connection as well as a high bit-rate connection, be it of stream or bursty nature. Dynamic bandwidth allocation on demand with a fine degree of granularity is provided. The overall protocol architecture of ATM networks comprises:

- A single link-by-link cell transfer capability common to all services.
- Service-specific adaptation functions for mapping higher layer information into ATM cells on an end-to-end basis. Examples are packetization/depacketization of continuous bit streams into and from ATM cells, and segmentation/reassembly of larger blocks of user information into and from ATM cells (core-and-edge concept).

8.7 ATM BASICS

The information transfer and signaling capabilities of B-ISDN include:

- Broadband capabilities.
- 64 kbit/s capabilities.
- User-to-network signaling.

- Interexchange signaling.
- User-to-user signaling.

This process is depicted in Figure 8–2.

ATM provides broadband information transfer. The ATM data unit is the fixed-size cell (a block of 53 octets). The 5-octet cell header carries the necessary information to identify cells belonging to the same virtual channel. Cells are assigned on demand, depending on source activity and the available resources. ATM guarantees (under normal, fault-free conditions) cell sequence integrity. This means that nowhere in the network can a cell belonging to a specific virtual channel connection overtake another cell of the same virtual channel connection that has been sent earlier.

ATM is a connection-oriented technique. A connection within the ATM layer consists of one or more links, each of which is assigned an identifier. These identifiers remain unchanged for the duration of the connection. Signaling information for a given connection is conveyed using a separate identifier (*out-of-band signaling*). Although ATM is a connection-oriented technique, it offers a flexible transfer capability common to all services, including connectionless data services. We will discuss the connectionless services via ATM-based B-ISDN later in this chapter.

out-of-band signaling

8.7.1   ATM Channels

In a connection-oriented service over a virtual circuit, the data stream from origin to destination follows the same path. Data from connections is distinguished by means of a *virtual channel identifier (VCI)*. A connection over a virtual circuit is called a virtual channel in ATM.

VCI

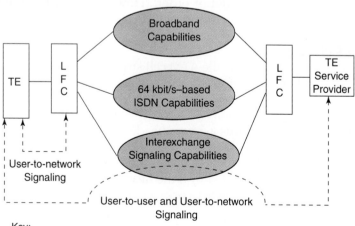

Figure 8–2
Information transfer and signaling capabilities.

Figure 8–3
Relationship between virtual channels, virtual path, and transmission path.

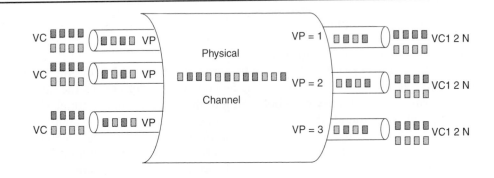

Figure 8–3 illustrates the concept of the virtual channel and the virtual path in ATM. A virtual channel (VC) is a unidirectional communication capability for the transport of ATM cells. A virtual path (VP) is a bundle of virtual channels. All of the virtual channels within a virtual path have the same endpoints. Every virtual channel (within a virtual path) and every virtual path (within a physical channel) has a unique identifier, such as the *virtual path identifier (VPI)*, that is assigned at the originating node and removed at the terminating node. VCIs may retain their values through the network for permanent virtual circuits. A transmission path may carry several virtual channels. The virtual path concept allows the grouping of several virtual channels.

Each cell incurs an overhead corresponding to the length (number of bits) of the VCI, which is generally much smaller than the length of a full source/ destination address needed in a datagram service. Also in the same connection, cells reach the destination in the same order they are sent from the source, thus eliminating the need for sequence numbers and for buffering packets at the destination if they arrive out of order. (However, sequence numbers are necessary if the application layer must detect cell loss.)

8.7.2 ATM Cell Header Structure

In this section, we will examine the ATM cell structure and we will see how an ATM switch can obtain the information it needs from the cell header. As indicated in Figure 8–4, the *ATM adaptation layer (AAL)* produces a data stream of 48-byte cells or protocol data units (PDUs). We will see later the function of the AAL layer. The ATM layer adds a 5-byte header and forwards the 53-byte cell to the physical layer. The 53-byte cell is converted into a serial bit stream by reading the cell from right to left (most significant bit first) and from top to bottom (byte 1 first). Figure 8–4 shows the header structure for both the user-network interface (UNI) and the network-network interface (NNI). The abbreviations used in Figure 8–4 are:

- *GFC, generic flow control*, is used at the customer site only (not end-to-end). It is overwritten by ATM switches and is currently undefined (0000 default).

VPI

AAL

GFC

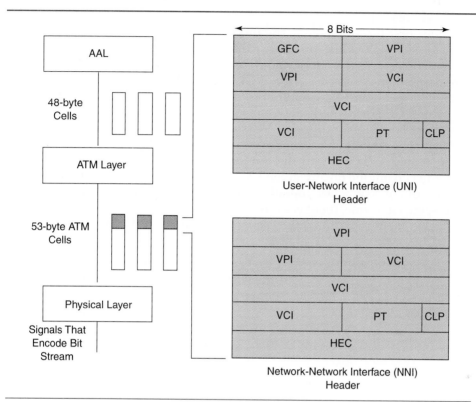

Figure 8–4
Header of ATM cells across the user-network interface (UNI) and across the network-network interface (NNI).

• *VPI, virtual path identifier,* consists of 8 bits and is used for routing. The VPI at the network-network interface (NNI) comprises the first 12 bits of the cell header, thus providing enhanced routing capabilities.

• *VCI, virtual channel identifier,* together with the virtual path identifier, constitute the routing field of a cell. This is a 16-bit field for both user-network and network-network interfaces.

• *PT, payload type,* is a three-bit field that indicates eight payload types, including user data and operation, administration, and management payloads.

• *CLP, cell loss priority,* is one bit that is used explicitly to indicate the cell loss priority. If the value is one, the cell is subject to be discarded, depending on the network condition. However, the agreed-upon quality-of-service parameters will not be violated.

• *HEC, header error control,* is not used by the ATM layer. It contains the header control sequence and is processed by the physical layer.

The only difference between the two headers is that the UNI header has a 4-bit field that can be used to indicate to the user the need for flow control. The NNI uses those bits to expand the virtual path identifier (VPI) field.

8.7.3   VPI and VCI

Consider first the 16-bit VCI. The most important feature is that the VCI is local to each link. More precisely, different simultaneous connections that share a common link must have different VCIs, but connections that have no common link may have the same VCI. Because VCIs are local to a link, connections coming into a switch from different switches (or sources) may have the same VCI. However, if these connections share the same outgoing link, they must be assigned different VCIs on that link. As shown in Figure 8–5, ATM switches maintain routing tables that include the relationship between the incoming VPI/VCI and the outgoing VPI/VCI. Virtual channel identifiers must be unique on any given link. Figure 8–6 shows four user nodes (A, B, C, D) and two switches. Initially, there is a connection over VCI 1 from A to C. At some later time, there is a request to establish a connection from B to D. This assignment is carried out by separate control cells during the connection

Figure 8–5
ATM connection establishment: purpose of VPI/VCI.

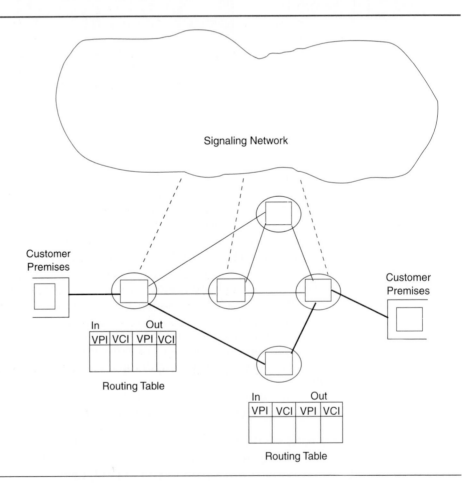

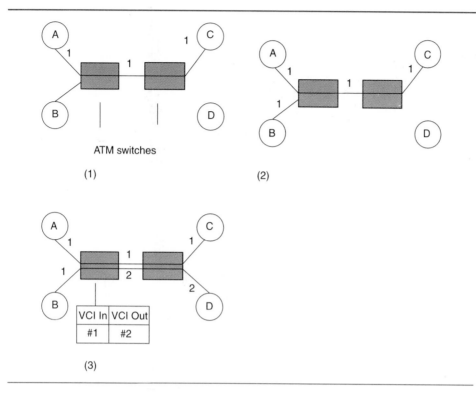

Figure 8–6
Virtual channel (VC) switching: Four user nodes and two ATM switches, the switch creates and updates the VPI/VCI translation tables.

ATM switches

(1)

(2)

VCI In	VCI Out
#1	#2

(3)

establishment phase. Since B is not currently using VCI 1, it assigns that VCI to that connection. When the packet establishing this connection reaches the first switch, that switch knows that VCI 1 is assigned to another connection that shares a link with the proposed connection. The switch therefore changes the VCI from 1 to 2, noting the change in the table. During the data transfer phase, this switch must change the VCI on each cell on the B–D connection from 1 to 2.

The 16-bit VCI field permits 64,000 simultaneous connections through each link, and since the same identifier may be reused by connections with disjointed paths, the network can support orders-of-magnitude more simultaneous connections. Some purposes may be better served by treating several virtual channels together as a group. Suppose, for example, as shown in Figure 8–7, that a user wishes to establish three virtual channels from A to B. It would then make sense to give these channels the same path and permit the user to arbitrarily assign VCIs to those channels. That is the function of the VPI. Virtual channels with the same VPI form a group. They are assigned the same path and are switched together (i.e., routing and switching decisions are based only on the VPI). More interestingly, the bandwidth and buffer resources allotted to a VPI may be done statically (at the time of service subscription) and shared only by virtual channels with that VPI. This use of VPIs permits the

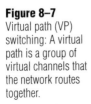

Figure 8–7
Virtual path (VP)
switching: A virtual
path is a group of
virtual channels that
the network routes
together.

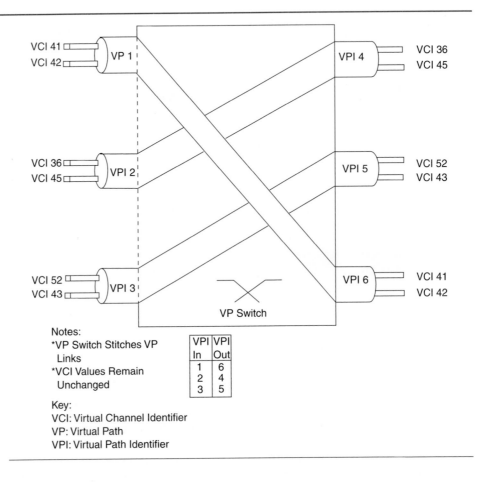

Notes:
*VP Switch Stitches VP
 Links
*VCI Values Remain
 Unchanged

VPI In	VPI Out
1	6
2	4
3	5

Key:
VCI: Virtual Channel Identifier
VP: Virtual Path
VPI: Virtual Path Identifier

creation of virtual private networks: a multilocation firm can rent several virtual paths to form its own private network whose resources are then shared by its virtual channels.

From Figure 8–4, we note that the user-network interface (UNI) header has an 8-bit VPI field and the network-network interface (NNI) has an additional 4 bits. The network may use these additional bits to create certain fixed routes, similar to what is done in today's long-distance telephone networks. This system may allow resources allocated to a virtual path to be statically assigned and shared dynamically among component virtual channels. The use of VPIs may speed up processing, since switches need only to consult a table with entries indexed by shorter VPIs.

Another important feature of ATM switches is that they potentially can discriminate among different connections according to quality-of-service levels assigned to a particular virtual connection. This potential can be used in many ways: admission control (refusing certain connections if sufficient network resources are unavailable), congestion control (limiting the amount of traffic accepted from a connection), resource allocation (negotiating the bandwidth and buffers allocated to a connection), and policing (monitoring the burstiness and average rate of traffic in a connection).

8.7.4 ATM Connection Subscription: ATM Bearer Services

An ATM bearer service at a public UNI is defined to offer point-to-point, bidirectional or point-to-multipoint unidirectional virtual connections at either a virtual path (VP) level and/or a virtual channel (VC) level. Networks can provide either VP or VC (or combined VP and VC) level service. For ATM users who desire only VP service from the network, the user will be able to allocate individual VCs within the VP connection (VPC) as long as none of the VCs are required to have a higher quality of service than the VP connection. Customers can subscribe to permanent virtual connection service or switched virtual connection service.

8.7.5 Quality-of-Service Parameters for an ATM Connection

Listed below are the seven parameters that define the quality of service for an ATM connection:

- Cell loss ratio = *number of lost cells ÷ number of (lost + successfully delivered) cells.*
- Cell misinsertion rate = *number of misinserted cells ÷ second.*
- Cell error ratio = *number of cells with errored information field ÷ number of delivered cells.*
- Cell transfer delay = *coding delay + packetization delay + transmission delay + switching delay + queuing delay + reassembly delay.*
- Cell delay variation (jitter).
- Cell transfer capacity (throughput).
- Skew, which is defined as the difference in presentation time of two related streams, for example video stream and audio stream.

Table 8–1 and Table 8–2 contain the quality-of-service objectives for some ATM-based services.

Service	Bit Error Rate	Cell Loss Ratio	Delay (ms)
Telephony	10 E-3	10 E-3	
Without echo cancelers			<25
With echo cancelers			<500
Data transmission	10 E-7	10 E-6	1000
Distributive computing	10 E-7	10 E-6	50
Hi-fi sound	10 E-5	10 E-7	1000

Table 8–1
Examples of quality-of-service parameters for ATM services.

Application	Delay (ms)	Jitter (ms)
64-kbps video conference	300	130
1.5-mbps MPEG NTSC video	5	6.5
20-mbps high-definition TV (HDTV)	0.8	1
16-kbps compressed voice	30	130
256-kbps MPEG voice	7	9.1

8.7.6 Other Fields in the ATM Cell Header

The 4-bit generic flow control (GFC) may be used by the network to signal to a user the need for momentary changes in the instantaneous cell stream rate. The functionality of the GFC has not yet been established.

The 3-bit payload type (PT) permits networks to distinguish between different types of information such as user data (PT = 000) and maintenance (PT = 100 or 101). This field permits the network equipment to introduce and remove special cells that are routed as ordinary cells but that carry special information for control purposes.

The 1-bit cell loss priority (CLP) distinguishes among cells that the network may not discard (CLP = 0) and cells that it may discard (CLP = 1) if necessary. Of course, the network may introduce a low-priority service at the level of a connection—all cells in such a connection are subject to be discarded. The CLP field is needed if different priority cells are present in the same connection. One example would be if voice is encoded into an equal number of high- and low-order bits and sent over cells that alternately carry the high- and low-order bits. The low-order bit cells would be assigned a CLP = 1, and be subject to be discarded, whereas the cells with the high-order bits would be assigned CLP = 0 and would not be discarded. By encoding voice in this way, the statistical multiplexing gain can be increased considerably. Also, if a switch notices that cells do not conform to the service parameters, then that switch can set CLP = 1 in those nonconforming cells so they are eligible for discarding.

Finally, the 8-bit header error control (HEC) field is equal to the sum of the byte 01010101 and the cyclic redundancy check (CRC) calculated over the rest of the header. The HEC is also used to delineate the cell boundary. The HEC can be used to correct single-bit errors and to detect multiple-bit errors. The error-control algorithm has two states: error detection (D) and error correction (C). The algorithm is initially in state C. If the algorithm is in state C and detects a single-bit error, it corrects the error and moves to state D. If it detects a multiple-bit error when in state C, the algorithm discards the cell and moves to state D. The algorithm discards cells that contain errors when in state D and remains in state D. The algorithm moves from state D to state C when it gets a cell without error. Figure 8–8 depicts a state diagram of the error detection/correction mechanism at an ATM receiver.

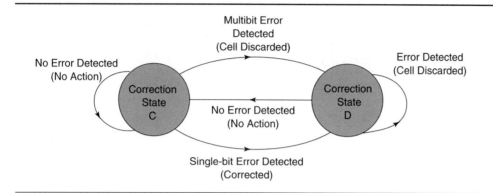

Figure 8–8
Header error correction, receiver mode operation.

8.8   ATM/B-ISDN LAYERS

Figure 8–9 illustrates the layers and sublayers of the B-ISDN/ATM protocol reference model. It indicates the functions of the physical layer, the ATM layer, the AAL, and associated sublayers.

8.8.1   The Physical Layer

The physical layer consists of two sublayers: the physical medium and the transmission convergence. The physical medium sublayer includes physical medium-dependent functions. It provides bit transmission capability, including bit transfer and bit alignment, as well as line coding and electrical optical transformation. Of course, the principle function is the generation and reception of waveforms suitable for the medium, the insertion and extraction of timing information, and line coding where required.

The transmission convergence sublayer converts the flow of cells into a flow of bits. In other words, it takes each ATM cell and puts it on the transmission medium bit by bit. It also forms ATM cells from the received bits. Among the important functions of the transmission convergence sublayer is the generation and recovery of transmission frames. Another function is transmission frame adaptation, which includes the actions necessary to structure the cell flow according to the payload structure of the transmission frame and to extract the cell flow out of the transmission frame. The transmission frame may be SONET envelopes, E1/T1 envelopes, and so on. In the transmit direction, the header error control (HEC) sequence is calculated and inserted in the header. In the receive direction, cell headers are checked for errors, and errors are corrected where possible. Cells are discarded where it is determined that header errors are not correctable.

8.8.2   The ATM Layer

The ATM layer is completely independent of the physical medium. One important function of this layer is *encapsulation*. This function includes header   encapsulation

Figure 8–9
Layers of B-ISDN/
ATM reference
model.

Higher Layer Functions		Higher Layers	
Convergence		CS	AAL
Segmentation and Reassembly		SAR	
Generic Flow Control Cell Header Generation/Extraction Cell VPI/VCI Translation Cell Multiplex/Demultiplex		ATM	
Cell Rate Decoding HEC Header Sequence Generation/Verification Cell Delineation Transmission Frame Adaptation Transmission Frame Generation/Recovery		TC	Physical Layer
Bit Timing Physical Medium		PM	

Key:

CS: Convergence Sublayer
PM: Physical Media
SAR: Segmentation and Reassembly Sublayer
TC: Transmission Convergence

generation and extraction. In the transmit direction, the cell header generation function receives a cell information field from a higher layer and generates an appropriate ATM cell header except for the header error control (HEC) sequence. (HEC is handled by the physical layer.) In the receive direction, the cell header extraction function removes the ATM cell header and passes the cell information field to the higher layer.

In a switch, the ATM layer determines where the incoming cells should be forwarded. The ATM layer also handles traffic management functions and buffers incoming and outgoing cells. It indicates to the next higher layer (the AAL, described in the next section) whether or not congestion has occurred during transmission. The ATM layer monitors both transmission rates and conformance to the agreed-upon quality of service—called *traffic shaping* and *traffic policing*. Many other functions of the ATM layer are beyond the scope of this book.

8.8.3 The ATM Adaptation Layer

The basic function of the ATM adaptation layer (AAL) is to isolate the applications and higher layers from the specific characteristics of the ATM layer. It maps the application messages and data units into the information field of the ATM cell and vice versa.

Services offered by the ATM adaptation layer are classified according to the following criteria:

- Real-time relation between the source and destination.
- Whether the transmission bit rate is constant or variable.
- Whether the transport connection is connection-oriented (virtual circuit) or connectionless (datagram).

When those parameters are combined, four service classes emerge as shown in Table 8–3. Examples of services in these classes are as follows:

- Class A: constant bit rate such as uncompressed voice plain old telephone service (POTS).
- Class B: variable bit rate video and audio, connection-oriented synchronous traffic, such as that used in video conferencing.
- Class C: connection-oriented data transfer, variable bit rate, asynchronous traffic, such as X.25.
- Class D: Connectionless data transfer, asynchronous traffic such as Internet protocol (IP).

Structure of the ATM Adaptation Layer   The ATM adaptation layer is divided into two major parts, as shown in Figure 8–9. The upper part of the ATM adaptation layer is called the *convergence sublayer*. Its job is to provide the interface to the application. It consists of a subpart that is common to all applications (for a given AAL protocol) and an application-specific subpart. The functions of each of these parts are protocol dependent but can include message framing and error detection.

convergence sublayer

In addition, the convergence sublayer at the source is responsible for accepting bit streams or arbitrary length messages from the applications and breaking

Table 8–3
Service classification for ATM adaptation layer (AAL).

Service Parameter	Class A	Class B	Class C	Class D
Real-time relation	Required	Required	Not required	Not required
Bit rate	Constant	Variable	Variable	Variable
Connection mode	Connection-oriented	Connection-oriented	Connection-oriented	Connectionless
ATM adaptation layer (AAL) type	AAL-1	AAL-2	AAL-3/4 or AAL-5	AAL-3/4 or AAL-5
Examples	DS1, E1, telephone circuit emulation	Packet video, audio	X.25	Internet protocol (IP)

them into units of 48 bytes for transmission. Some applications may use some of the 48-byte payload for their own headers. At the destination, this sublayer reassembles the cells into the original messages. In general, message boundaries are preserved when present. In other words, if the source sends four 512-byte messages, they will arrive as four 512-byte messages, not one 2048-byte message.

SAR sublayer

The lower part of the AAL is called the *segmentation and reassembly (SAR) sublayer*. It can add headers and trailers to the data unit given to it by the convergence sublayer for cell payloads. These payloads are then given to the ATM layer for transmission. At the destination, the SAR sublayer reassembles the cells into messages. The SAR sublayer is basically concerned with cells, whereas the convergence sublayer is concerned with messages.

The generic operation of the convergence and SAR sublayers is shown in Figure 8–10. When a message comes to the AAL from the application, the convergence sublayer may give it a header and/or a trailer. The message is then broken up into 44- to 48-byte units, which are passed to the SAR sublayer. The SAR sublayer may add its own header to each piece and pass them down to the ATM layer for transmission as independent cells. Note that Figure 8–10 shows the most general case, because some of the AAL protocols have null headers and/or trailers. In other words, there are cases where neither CS nor SAR headers are inserted.

The SAR sublayer also has additional functions for some (but not all) service classes. In particular, it sometimes handles error detection and multiplexing. The

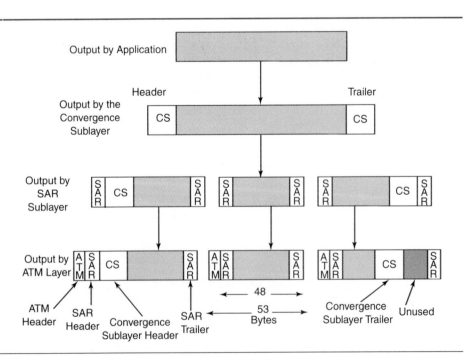

Figure 8–10
The headers and trailers that can be added to a message in an ATM network.

SAR sublayer is present for all service classes but the work it does depends on the specific protocol.

Communication between the application (higher layer) and the AAL layer uses the standard open system interconnection (OSI) *request* and *indication* primitives that were discussed in Chapter 7. A request primitive is issued from the application to the AAL layer and an indication primitive is passed up from the AAL to the application. If you recall, in the OSI model, protocol primitives passing from layer N to layer $N - 1$ are called *requests,* and those passing from layer N to layer $N + 1$ are called *indications.* Communication between the AAL sublayers uses different primitives.

To handle the different classes of service shown in Table 8–3, ITU defined four types of AAL protocols, AAL-1 through AAL-4, respectively. However, ITU later discovered that the technical requirements for classes C and D were so similar that AAL-3 and AAL-4 were combined into AAL-3/4. Then the computer industry, which had been asleep at the switch, realized that none of the protocols were any good. It solved this problem by defining another protocol, AAL-5. In the following sections we will examine all four of these protocols.

AAL-1   AAL-1 is the protocol used for transmitting class A traffic—that is, real-time constant bit rate. Its primary application is to adapt ATM cell transmission to typical T1/E1 and SONET circuits. Typically, AAL-1 is used for voice communications—plain old telephone service (POTS). Bits are fed in by the application at a constant rate and must be delivered at the far end at the same constant rate, with a minimum of delay and overhead. The input is a stream of bits with no message boundaries. For this traffic, error-detecting protocols such as stop-and-wait are not used because of delays that are introduced by timeouts and because retransmissions are unacceptable in voice communications. However, missing cells are reported to the application, which must take its own action (if any) to recover from them.

AAL-1 robs one or two bytes from the payload and uses them as a one-byte header or a one-byte header and a pointer, depending on the cell generated. In other words, the input stream (or messages) is divided into 46- or 47-byte units with one or two bytes from the 48-byte ATM payload being consumed for AAL-1 headers. At the other end, AAL-1 extracts these data units and reconstructs the original input. The segmentation and reassembly (SAR) sublayer is responsible for handling the headers. The convergence sublayer does not have any protocol headers of its own.

In contrast, the AAL-1 SAR sublayer does have a protocol. The format of its cells is given in Figure 8–11. Both formats begin with a one-byte header containing a 3-bit cell sequence number (SN), to detect missing or misinserted cells. This field is followed by a 3-bit sequence number protection (SNP, i.e., checksum) over the sequence number to allow correction of single errors and detection of double errors in the sequence field. It uses a cyclic redundancy check with the polynomial $x^3 + x + 1$. An even parity bit calculated for the header byte further reduces the likelihood of a bad sequence number sneaking in unnoticed. For example, to transmit digitized voice arriving at a rate of 1 byte

Figure 8–11
The AAL-1 cell
format.

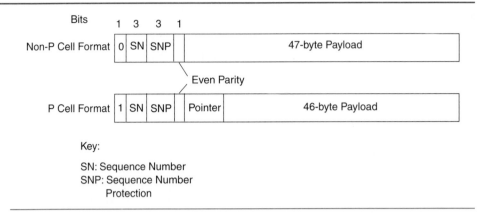

every 125 μsec, filling a cell with 47 bytes means collecting samples for 5.875 msec. If this delay before transmission is unacceptable, partial cells can be sent. In this case, the number of actual data bytes per cell is the same for all cells and is agreed upon in advance.

The *P* cells are used when message boundaries must be preserved. The pointer field is used to give the offset of the start of the next message. Only cells with an even number may be *P* cells. So the pointer is in the range of 0 to 92 to put it within the payload of either its own cell or the one following it. Note that this scheme allows messages to be an arbitrary number of bytes long, so messages can run continuously and need not align on cell boundaries. The high-order bit of the pointer field is reserved for future use.

Alarm indication in this adaptation layer is via a check of the 1s density. When the 1s density of the received cell stream becomes significantly different from the density used for the particular pulse code modulation (PCM) line coding scheme in use, it determines that the system has lost the signal and alarm notification is given.

AAL-2 AAL-2 is designed for simple, connection-oriented, real-time data streams without error detection, except for missing and misinserted cells. For pure uncompressed audio or video, or any other data stream in which having a few bit errors once in a while is not a problem, AAL-1 is adequate.

AAL-2 is intended to handle variable bit rates. The bit rate of compressed audio or video can vary strongly in time. For example, many compression schemes transmit a full video frame periodically and then send frames that only contain the differences between subsequent frames, then a last full frame of a sequence of several frames. When the camera is stationary and nothing is moving, the difference frames are small, but when the camera is painting rapidly, they are large. Message boundaries must also be preserved so that the start of the next full frame can be recognized, even in the presence of lost cells or bad data. For these reasons, a protocol that is more sophisticated than the AAL-1 is needed. AAL-2 has been designed for this purpose.

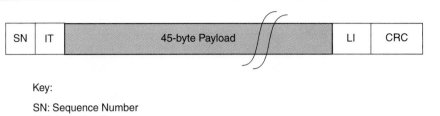

Figure 8–12
The AAL-2 cell format.

Key:

SN: Sequence Number
IT: Information Type
LI: Length Indicators
CRC: Cyclic Redundancy Check

As in AAL-1, the CS sublayer does not have a protocol, but the SAR sublayer does. The SAR cell format is shown in Figure 8–12. It has a 1-octet header and a 2-octet trailer. This leaves 45 data bytes per cell. The sequence number (SN) field is used for numbering cells to detect missing or misinserted cells. The information type (IT) field is used to indicate that this cell is the start or the end of a message. The length indicator (LI) field tells how big the payload is in bytes (it may be less that 45 bytes). Finally, the CRC field is a checksum calculated for the entire cell so errors can be detected. AAL-2 is still evolving. Many standards issues need to be resolved before it can be fully implemented.

AAL-3/4 Initially, ITU defined two AAL specifications, AAL-3 and AAL-4. AAL-3 is for connection-oriented, variable-bit-rate data service and AAL-4 is for connectionless service. ITU then discovered that there was no real need to have two protocols, so they were combined into a single protocol, AAL-3/4.

AAL-3/4 operates in two modes: stream or message. In the message mode, each call from the application to AAL-3/4 inserts one message into the network. The message is delivered such that message boundaries are preserved. In stream mode, the boundaries are not present (AAL-3/4 attempts to deliver a continuous bit stream). We have discussed bit stream operations in previous sections. The message mode service can be used for framed data transfer (e.g., high-level data link control, HDLC frame), and the streaming mode service is suitable for transferring low-speed data with low delay requirements (as seen in AAL-1 for POTS applications). The discussion below will concentrate on the message mode of operations.

Reliable and unreliable (i.e., no guarantee) transports are available in both modes. When the reliable service is used, errored or missing AAL sublayer data units are retransmitted; therefore, flow control is provided as a mandatory feature. On the other hand, if the unreliable service is in use, lost or errored AAL sublayer data units are not corrected by retransmission. The delivery of corrupted data units to the user may be provided as an optional feature, and the user may decide what to do with them. In principle, flow control can be supplied to point-to-point ATM layer connections. No flow control will be provided for point-to-multipoint ATM layer connections.

Figure 8–13
Multiplexing of several sessions onto one virtual circuit.

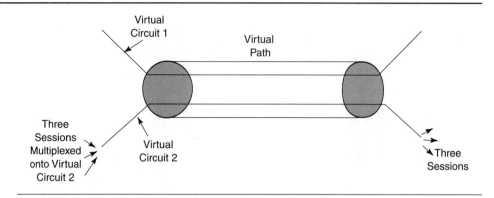

A feature of AAL-3/4 not present in any of the other protocols is multiplexing. This aspect of AAL-3/4 allows multiple sessions (e.g., remote logins) from a single host to travel along the same virtual circuit and be separated at the destinations, as illustrated in Figure 8–13. The reason this feature is desirable is that carriers often charge for each connection setup and for each second that a connection is open. If a pair of hosts have several sessions open simultaneously, giving each session its own virtual circuit will be more expensive than multiplexing all of them onto the same virtual circuit. If one virtual circuit has sufficient bandwidth to handle the job, there is no need for more than one. All sessions using a single virtual circuit get the same quality of service, since this is negotiated per virtual circuit. This issue is the real reason that there were originally separate AAL-3 and AAL-4 formats: the Americans wanted multiplexing and the Europeans did not. So each group developed its own standard. Eventually, the Europeans decided that saving 10 bits in the header was not worth the price of having the United States and Europe not being able to communicate.

AAL-3/4 has protocols for both the convergence sublayer and the segmentation and reassembly (SAR) sublayer, unlike AAL-1 and AAL-2, which have protocols only for the SAR sublayer. Messages as large as 65,535 bytes come into the convergence sublayer from the application. The convergence sublayer pads the message with 4 bytes, then it attaches a header and trailer, as shown in Figure 8–14.

The common part indicator (CPI) field gives the message type and the counting unit for the *BA* size and the *length* field. The *Btag* and *Etag* fields are used to frame messages. The two bytes must be the same and are incremented by one on every new message sent. This mechanism checks for lost or misinserted cells. The BA size field is used for buffer allocation. It tells the receiver how much buffer space to allocate for the message in advance of its arrival. The *length* field gives the payload length again. In AAL-3/4 message mode of operation, the *length* value must equal the *BA* value, but in the stream mode it may differ.

sublayers of protocol, coupled with the surprisingly short checksum (10 bits), led to the creation of a new protocol. It was called the *simple efficient adaptation layer (SEAL)*, suggesting a less complex protocol compared to AAL-3/4. ATM forums accepted SEAL and gave it the designation AAL-5.

AAL-5 handles variable bit rate sources without a real-time relation between the source and destination. It provides services similar to AAL-3/4 and is mainly used for data applications. It offers reduced overhead and provides services identical to those of AAL-3/4: message mode service, streaming mode service, and reliable and unreliable services, as discussed in the previous section. However, one essential difference is that AAL-5 does not support multiplexing functions, so there is no multiplexing ID (MID) field.

In message mode, the application can pass a datagram from 1 to 65,535 bytes to the AAL layer and have it delivered to the destination, either on a guaranteed or a best-effort basis. Upon arrival in the convergence sublayer, a message is padded out and a trailer is added as shown in Figure 8–17. The amount of padding (0 to 47 bytes) is chosen to make the entire message, including the padding and trailer, a multiple of 48 bytes (the payload in an ATM cell). AAL-5 does not have a convergence sublayer header, just an 8-byte trailer.

The user-to-user (UU) field is not used by the AAL layer itself; instead, it is available for a higher layer for its own purposes, for example, sequencing or multiplexing. The higher layer in question may be the service-specific subpart of the convergence sublayer. The *length* field tells how long the true payload is, in bytes, not counting the padding. A value of 0 is used to abort the current message in midstream. The CRC field is the standard 32-bit checksum calculated over the entire message, including padding and the trailer (with the CRC field set to 0). One 8-bit field in the trailer is reserved for future use.

The message is transmitted by passing it to the SAR sublayer, which does not add any headers or trailers. Instead the segmentation and reassembly (SAR) sublayer breaks the message into 48-byte units and passes each of these to the ATM layer for transmission. It also tells the ATM layer to set a bit in the payload type identifier (PTI) field on the last cell, so the message boundaries are preserved. A case can be made that it is an incorrect mixing of protocol layers,

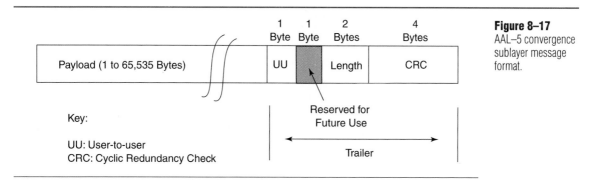

Figure 8–17
AAL–5 convergence sublayer message format.

Key:

UU: User-to-user
CRC: Cyclic Redundancy Check

because the AAL layer should not be using bits in the ATM layer's header. Doing so violates the most basic principle of protocol design and engineering, and suggests that the layering should have perhaps been done differently.

The principle advantage of AAL-5 over AAL-3/4 is that it is a lot more efficient. Whereas AAL–3/4 adds only 4 bytes per message, AAL–5 adds 4 bytes per cell, reducing the payload to 44 octets, a loss of 8 percent on long messages. AAL–5 has a slightly larger trailer per message (8 bytes) but has no overhead in each cell. The lack of a sequence number is compensated for by the longer checksum, which can detect lost, missing, or misinserted cells without using sequence numbers.

Within the Internet community, it is expected that the normal way of interfacing to ATM networks will be to transport Internet protocol (IP) packets using the AAL-5 payload field.

Comparison of AAL Protocols   By now you are probably scratching your head and asking why all these types of protocols are needed for the ATM adaptation layer. Well, you are not alone. AAL protocols seem unnecessarily similar to each other and poorly thought out. The value of having distinct convergence and SAR sublayers is also questionable, especially since AAL-5 does not have anything in the SAR sublayer. A slightly enhanced ATM layer header could have adequately provided for sequencing, multiplexing, and framing. We are obligated, though, to present these protocols as they are given, so we have no choice but to work with them at the present time.

Some of the differences between the various AAL protocols are summarized in Table 8–4. These differences relate to efficiency, error handling, multiplexing, and the relation between the AAL sublayers.

Table 8–4
Some differences among the various AAL protocols.

Criteria	AAL-1	AAL-2	AAL-3/4	AL-5
Service class	A	B	C/D	C/D
Multiplexing	No	No	Yes	No
Message delimiting	None	None	Btag/Etag	Bit in PTI of ATM cell header
Advance buffer allocation	No	No	Yes	No
User bytes available	0	0	0	1
CS padding	0	0	32-bit word	0-47 bytes
CS protocol overhead (bytes)	0	0	8	8
CS checksum	None	None	None	32 bits
SAR payload bytes	46-47	45	44	48
SAR protocol overhead (bytes)	1-2	3	4	0
SAR checksum	None	None	10 bits	None

The overall impression discerned from Table 8–4 is that AAL protocols have many variants with minor differences and a job half done. The original classes, A, B , C, and D, have been effectively abandoned. AAL-1 does not seem necessary. AAL-2 is broken, AAL 3 and AAL 4 never saw the light of day, and AAL 3/4 is inefficient. The future lies with AAL-5, but even here there is room for improvement. The AAL-5 message should have a sequence number and a bit to distinguish data from control messages so it could have been used as a reliable transport protocol. Unused space in the trailer was available to the designers of the protocol, but for one reason or another, they overlooked this issue. As it stands, for reliable transport, the additional overhead of a transport layer is required on top of the AAL-5 protocol, when it could have been avoided. If designing an AAL was a class project assigned to you, we probably would give it to you with instructions to fix it and return it when it was finished.

8.9   MANAGEMENT AND CONTROL

A very important feature of ATM networks is that they can make a number of management and control decisions to discriminate among connections and to provide the variety of *quality of service (QoS)* that different applications need. QoS The decisions are divided into three groups. When a request is made for a connection with a particular QoS, the network must determine whether to accept or reject the request, depending on the resources available. (Recall that QoS involves parameters that include delay, cell loss, and source traffic rate.) If the resources are insufficient to meet the request, the network may negotiate with the user for a connection with a different QoS.

Once the connection is admitted, the network must assign a route or path to the virtual channel that carries the connection. It must inform the switches and other network elements along the path that this virtual channel must allocate certain resources so that the agreed-upon QoS is met.

Lastly, the network must monitor the data transfer to ensure that the source also conforms to the QoS specification and to drop its cells as appropriate. The network may also ask the source to slow down its transmissions. In addition, the network carries a number of information flows to monitor its operation and to detect and identify the location of congested or failed devices.

The broadband ISDN (B-ISDN) standard does not specify how these decisions are to be carried out. The ATM forum has proposed specific frame formats that the network should use to carry its monitoring information and to interact with users. The following is a review of those proposals.

The network uses operation and maintenance information flow for the following functions:

- Fault management.
- Traffic and congestion control.
- Network status monitoring and configuration.
- User/network signaling.

Figure 8–18
Layer arrangement of
ATM network func-
tions, including the
operation and man-
agement functions.

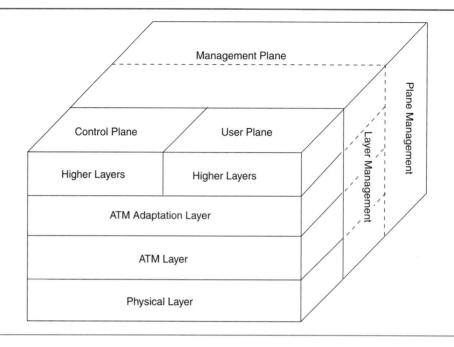

These functions, similar to the other network functions, are organized into layers. Figure 8–18 shows the layer arrangement of all ATM network functions, including those for operation and management. The layers in the *user plane* comprise the functions required for the transmission of user information. For instance, for an Internet protocol over ATM, these layers could be Telnet/ TCP/IP/AAL-5. The layers in the *control plane* are the functions needed to set up a virtual circuit connection. These functions, which include the signaling protocols, are needed only for switched virtual connections and are absent in a network that implements only permanent virtual connections. In a permanent virtual connection, the path or route assigned to a source and destination and the virtual connection identifier (VCI) for that route are fixed. The *layer management plane* contains management functions specific to individual layers. Finally, *plane management* consists of the functions that supervise the operation of the whole network.

8.9.1   Fault Management

Consider a virtual-circuit connection over an ATM network and assume that the connection is implemented by a synchronous optical network (SONET). We know from Chapter 2 that SONET establishes transmission paths for the ATM layer over optical fibers. The transmitters in SONET are all synchronized to the same master clock, which enables time-division multiplexing of different bit streams. The multiplexing is done byte by byte.

The physical layer (SONET) is decomposed into three sublayers: section, line, and path, as discussed in Chapter 2. The section layer transmits bits between any two devices where light is converted back into electric signals or conversely. For instance, there is a section between two successive regenerators or between a regenerator and a multiplexer. The line layer transports bits between multiplexers where SONET signals are added to or subtracted from transmission. Finally, the path layer transports user information. Thus, a path goes across a number of lines (or links) that are switched by the SONET demultiplexers and multiplexers, and a line consists of a number of sections. Each layer inserts and strips its own overhead information used to monitor the transmission functions for which the layer is responsible (see Figure 8–19).

Each of the three sublayers uses overhead bytes in the SONET frames to supervise its operations. The overhead bytes are said to carry a *flow* of operation and maintenance information. The flow carried by the section overhead bytes is called F1. The flows carried by the line and path overhead bytes are called F2 and F3 respectively. The virtual-circuit connection is carried by a virtual-path connection. Accordingly, the network uses a flow of cells to supervise the virtual-path connection and a flow of cells to supervise the virtual-circuit connection. These two flows are called F4 and F5, respectively.

The format of F4 and F5 depends on whether the cells monitor the segment across the user-network interface or the end-to-end connection (see Figure 8–20). The cell formats are shown in Figure 8–21. Note that the F5 cells have the same VPI/VCI as the user cells of the connection they monitor. The F5 cells are distinguished from the user cells by the payload type (PT) field in the ATM cell header. Similarly, the F4 cells have the same VPI as the user cells and are distinguished by their VCI.

The main function of the operation and management (OAM) cells is to detect and manage faults. Fault management OAM cells have the leading 4 bits of the cell payload set to 0001. The next 4 bits, the function type (FT) field, include the type of function performed by the cell: alarm indication signal (AIS), signaled by FT = 0000; far end receiver failure (FERF), signaled by FT = 0001;

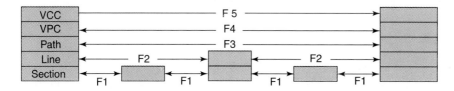

Figure 8–19
Operation and maintenance flow for a virtual-circuit connection over SONET.

Key:

VCC: Virtual Circuit Connection
VPC: Virtual Path Connection
F: Flow

Figure 8–20
A segment indicates a connection across the user-network interface; an end-to-end connection is between the source and destination user equipment.

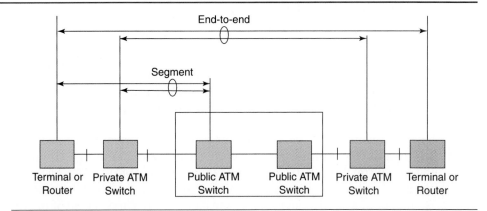

and loopback cell, signaled by FT = 1000. The AIS cells are sent along the virtual-path connection (VPC) or the virtual-circuit connection (VCC) by a network device that detects an error condition along the connection. Those cells are then sent to the destination of the connection. When the equipment at the end of that connection receives the AIS, it sends back FERF cells to the other

Figure 8–21
Format of ATM OAM cells.

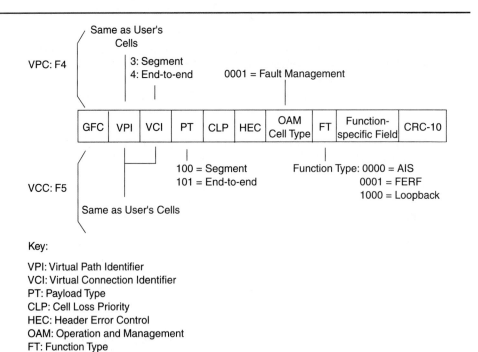

Key:

VPI: Virtual Path Identifier
VCI: Virtual Connection Identifier
PT: Payload Type
CLP: Cell Loss Priority
HEC: Header Error Control
OAM: Operation and Management
FT: Function Type
FERF: Far End Receiver Failure

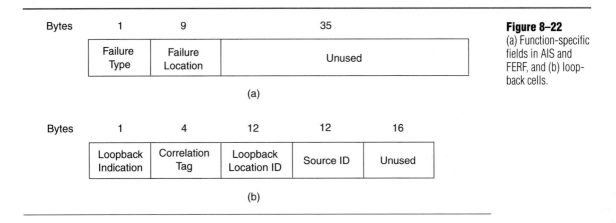

Bytes   1   9   35

Failure Type	Failure Location	Unused

(a)

Bytes   1   4   12   12   16

Loopback Indication	Correlation Tag	Loopback Location ID	Source ID	Unused

(b)

Figure 8–22
(a) Function-specific fields in AIS and FERF, and (b) loopback cells.

end of the connection. As shown in Figure 8–22, the AIS and FERF cells specify the type of failure as well as the failure location.

A loopback cell contains a field that specifies whether the cell should be looped back, a correlation tag, a loopback location identification, and a source identification. These loopback cells are used as shown in Figure 8–23.

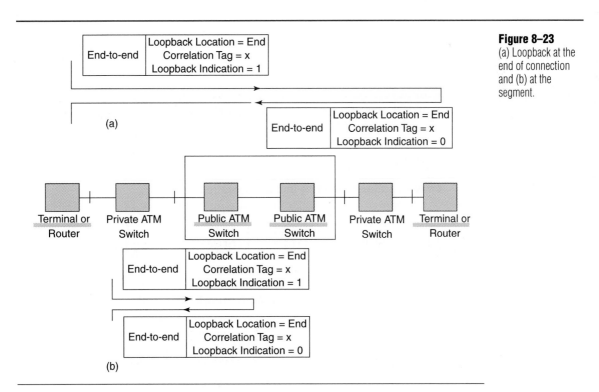

Figure 8–23
(a) Loopback at the end of connection and (b) at the segment.

The device that requests a loopback (called the *source*) inserts a loopback cell and selects a value for the correlation tag. The device can specify where the loopback should take place. The device sets the loopback indication field of the cell to 1 to indicate that the cell must be looped back. When the device where the loopback must occur receives the cell, it sets its loopback indication field to 0 and sends the cell back to the source. The source compares the correlation tag of the cell it receives with the value it selected. The correlation tag prevents a device from getting confused by other loopback cells.

8.9.2   Traffic and Congestion Control

The objectives of traffic and congestion control are to guarantee the contracted quality of service to virtual connections. ATM networks provide a means to allocate network resources and to separate traffic flow according to service characteristics, using a function called *network resource management*. A function called *connection admission control* is used by an ATM network to take a set of actions during the call setup phase or during the call negotiation phase in order to establish whether a virtual channel (VC) or virtual-path connection request can be accepted or rejected. In this case an ATM network can decide whether a request for reallocation can be accommodated. Routing is part of connection admission and control actions.

Feedback controls are another set of actions taken by the network and by the user to regulate the traffic submitted on an ATM connection according to the state of network elements. An ATM network must protect against malicious or unintentional misbehavior, which can affect the quality of service (QoS) of other already established connections. For this purpose, an ATM network uses a set of actions called *usage/network parameters control* to monitor and control traffic in terms of traffic offered and validity of the ATM connection at the user access and network access, respectively. Lastly, the *priority control* set of actions is used to selectively discard cells with low priority, if necessary, to protect as far as possible the network performance for cells with higher priority. Figure 8–24 illustrates those concepts of traffic and congestion control.

8.9.3   Network Status Monitoring and Configuration

A protocol is being defined to facilitate the monitoring and management of an ATM network. This protocol is a version of the simple network management protocol (SNMP), adapted for ATM networks. The 1996 version of this protocol is called the *intermediate local management interface*. The objective of the intermediate local management interface is to supervise the connection across the user-network interfaces. This is illustrated in Figure 8–25.

Figure 8–25 shows a private ATM network connected by a private ATM switch to a public ATM network. Each connection across a user-network interface (UNI) is supervised by two UNI management entities—one for each of the ATM devices. Two such management entities are said to be *adjacent*, and the intermediate local management interface specifies the structure of the management information base that contains the attributes of the connection supervised by the adjacent entities.

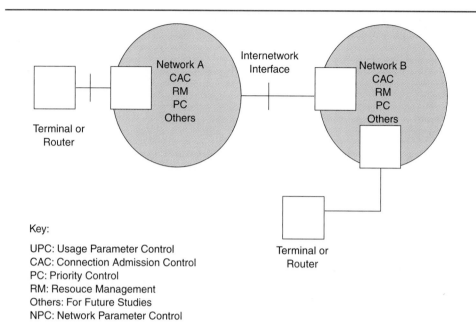

Figure 8–24
Reference configuration for traffic and congestion control.

Key:

UPC: Usage Parameter Control
CAC: Connection Admission Control
PC: Priority Control
RM: Resouce Management
Others: For Future Studies
NPC: Network Parameter Control

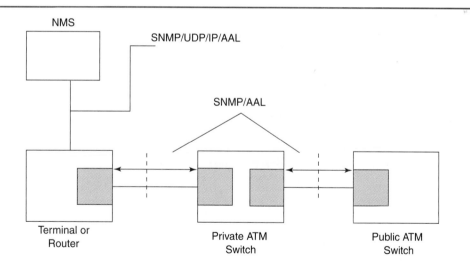

Figure 8–25
The intermediate local management interface (ILMI) protocol is designed to supervise the connection across user-network interfaces.

Key:

SNMP: Simple Network Management Protocol
AAL: ATM Adaptation Layer
IP: Internet Protocol
UDP: User Datagram Protocol
NMS: Network Management System

☐ User-network Interface Management Entity

8.9.4 Signaling

The basic signaling functions between the network and a user are as follows:

- The user requests a switched virtual connection.
- The network indicates whether the request is accepted or not.
- The network indicates error conditions with a connection.

On the one hand ATM-based networks should be available as early as possible, whereas on the other hand a variety of services (very simple as well as highly sophisticated) should be integrated in such a network. Therefore, a phased approach is required for the introduction of ATM-based networks supporting switched services. ITU-T was aware of this fact and developed a timetable of B-ISDN networks and service aspects. This concept comprises three steps that are called release 1, 2, and 3. The main characteristics influencing signaling are summarized in Table 8–5.

In release 1, simple switched services with constant bit rates will be provided and internetworking with the existing 64 kbit/s ISDN is foreseen. This is not to say that variable bit rate services cannot be supported, because they can. However, the bit rate allocated to such service will be the peak rate. Two signaling configurations can be used: point-to-point and point-to-multipoint signaling.

Table 8–5
Timetable for
B-ISDN signaling.

Release 1	Release 2	Release 3

Constant bit rate
Connection-oriented service
  with end-to-end timing
	Variable bit rate
	Connection-oriented service
	Quality-of-service indication by
	  the user
		Multimedia and distributive
		  service
		Quality-of-service negotiation

Point-to-point
Connections (uni- and
  bidirectional, symmetric and
  asymmetric)
Single connection,
  simultaneous establishment
Indication of peak bandwidth
	Point-to-multipoint connection
	Multiconnection, delayed
	  establishment
	Use of cell loss priority
	Negotiation and renegotiation
	  of bandwidth
		Broadcast connections

Peak rate allocation	Bandwidth allocation based on traffic characteristics	

Interworking with 64 kbit/s
  ISDN
Point-to-point or point-to-
  multipoint signaling access
Limited set of supplementary
  services		

Meta-signaling	Supplementary services	

More sophisticated services with variable bit rates, assuming that it is possible to take advantage of statistical multiplexing, point-to-multipoint, and multiconnections, will be supported in release 2. With release 2 *call* and *connection control* will be separated; that is, connections can be set up and released during a call. Finally, release 3 provides the full range of services, including multimedia and distributive services.

The subsequent sections will examine the concept of signaling in ATM networks irrespective of the timetable set up by the ITU-T. We will explore ATM signaling capabilities in a generic sense such that its application may apply to all services outlined for each of the releases.

ATM Call Setup Example   As shown in Figure 8–26, an endpoint—here a piece of video equipment with a terminal adapter—establishes a signaling path with the ATM switch. This path can be either permanent and set up in advance or

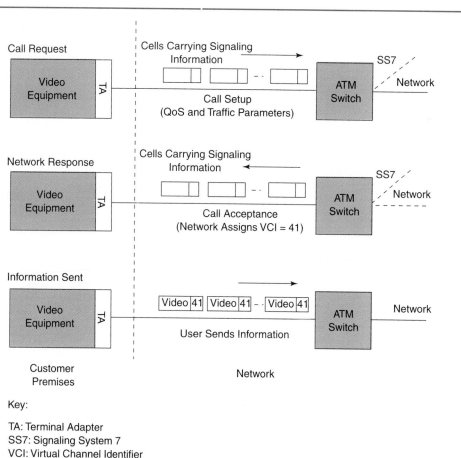

Figure 8–26
ATM call setup example.

Key:

TA: Terminal Adapter
SS7: Signaling System 7
VCI: Virtual Channel Identifier

established during the signaling phase using what is called *meta-signaling protocol*. Once the signaling path is established and the endpoint wishes to establish an ATM call, it populates ATM cells with the appropriate quality-of-service parameters and fields indicating the type of cells and transmits those cells to the ATM switch.

The ATM switch examines the cells to make sure that everything is in order, and if so, takes the proper action to inform the network of the call request. If the network accepts the call, it signals the switch of its decision. The ATM switch then sends a stream of ATM signaling cells back to the endpoint informing it that the connection is accepted. The ATM switch also informs the endpoint of the VCI assigned for the call. At this point the endpoint can transport user cells over the newly established VCC between it and the ATM switch.

In this example we need to keep in mind the following points:

- The network allocates VCI and VPI.
- Allocation of required resources (e.g., throughput) in the network is negotiated between network service providers and users over a separate signaling virtual channel.
- A signaling virtual channel is established and released by using a special meta-signaling channel, which is transported over a preassigned VCI/VPI, defined as a user-network interface.

In general, ATM signaling capabilities must include the ability to establish, maintain, and release ATM virtual channel connections (VCCs) and virtual path connections (VPCs). ATM must also provide for negotiation and renegotiation of traffic characteristics of a connection so that the appropriate resources are allocated. The signaling capabilities must allow for the ability to add and remove calls to existing connections. It should also support asymmetric connections. An asymmetric connection occurs when the bit rate in one direction is different from that of the other direction. An example of asymmetric connection is video on demand. In this case the downstream connection, from the network to the user, should have a large enough bandwidth to accommodate the video being sent to the user. On the other hand, the connection from the user to the network requires a limited bandwidth to handle control functions that allow the user to view the video and so forth. Finally, ATM signaling capability must allow interworking between ATM and non-ATM networks/devices.

Meta-Signaling   A meta-signaling channel manages the signaling of virtual channels only within its own virtual path pair. If the virtual path identifier (VPI) = 0, the meta-signaling virtual channel is always present and has a standardized VCI value. The meta-signaling virtual channel is activated at the virtual path establishment. The signaling virtual channel (SVC) is assigned and removed when necessary. A specific VCI value for meta-signaling is reserved per virtual path at the user-network interface (UNI).

The user negotiates the SVC, signaling virtual bandwidth parameters value. The meta-signaling virtual channel (MSVC) bandwidth has a default value. The bandwidth can be changed by mutual agreement between a network operator and user.

In order to establish and release point-to-point and selective broadcast signaling virtual channel connections, meta-signaling procedures are provided. For each direction, meta-signaling is carried out in a permanent virtual channel connection having a standard VCI value. The channel is called the meta-signaling virtual channel. The meta-signaling protocol is terminated in the ATM layer management entity. The meta-signaling function is required to:

- Manage the allocation of capacity to signaling channels.
- Establish, release, and check the status of signaling channels.
- Provide a means to associate a signaling endpoint with a service profile if service profiles are supported.
- Provide a means to distinguish between simultaneous requests.

Meta-signaling should be supportable on any virtual path (VP); however, meta-signaling can only control signaling VCs within its virtual paths (VPs).

8.10 INTERNETWORKING WITH ATM

An ATM network can be used to carry internetworking traffic. For instance, an ATM network can be used to interconnect various LANs or Internet protocol (IP) subnetworks. Such internetworking can take place at the data link layer (bridging) or at the network layer (routing). The data link layer and networking layer are discussed in Chapter 7. Bridges and routers are discussed in Chapter 9.

Two tasks are required for internetworking over ATM. The first is encapsulation of the protocol data units and the second is the routing or bridging of these protocol data units. Recall from our discussion in Chapter 7 that frames are the protocol data units for the data link layer and packets are the protocol data units for the networking layer. The routing consists of route selection and the switching of the packets. Routing also requires a routing algorithm and address resolution. The address resolution maps an address (such as X.25, IP) into an ATM address. The routing algorithm calculates the routes through the network. A non-ATM network may have its own routing algorithm that is different from the routing algorithms used in ATM networks.

8.10.1 Encapsulation over AAL-5

Two methods are specified for Internet traffic encapsulation depending on whether the ATM virtual channel connection (VCC) carries a single or multiple protocols. In either case, the protocol data units are carried by the payload of the convergence sublayer of AAL-5. That is, a protocol data unit of up to 64 Kbytes is first padded to an exact multiple of 48 bytes. To these bytes, the convergence sublayer adds an 8-byte trailer that specifies the length of the protocol data unit and a CRC, as discussed in the AAL–5 section.

If the VCC carries a single protocol, this protocol is identified implicitly by the virtual path identifier (VPI) and the virtual circuit identifier (VCI). Otherwise, the convergence sublayer protocol data unit starts with a header that specifies the protocol (e.g., routed IP or bridged Ethernet).

media-access
control

If the protocol is bridged, then the *media-access control (MAC)* address of the destination must be specified in the convergence sublayer protocol data unit. Note that the ATM interface must perform the usual functions of a bridge by looking into the MAC address of the encapsulated protocol data unit. We explain this function in the next section.

8.10.2   LAN Emulation with ATM

LAN emulation is a glue that enables ordinary LAN software to operate over ATM and also to interconnect LANs and ATM. This emulation enables the connection of Ethernets or of token rings through an ATM network (not a mixture).

Figure 8–27 shows the LAN emulation in the protocol suite. A packet destined to B invokes the LAN emulation layer that contacts the server to find the ATM address of the bridge. The packet is segmented and sent to the bridge, whose LAN emulation software reconstructs the packet before sending it to the LAN. The reverse transfer, from B to A, is similar.

Figure 8–27
A LAN emulation layer is inserted between the network layer and the AAL layer in ATM nodes.

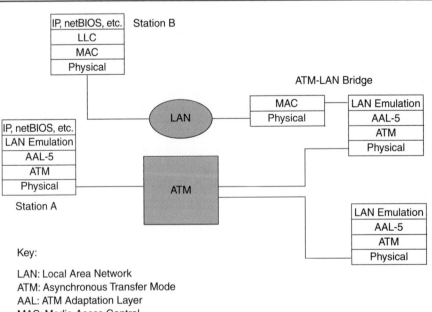

Key:

LAN: Local Area Network
ATM: Asynchronous Transfer Mode
AAL: ATM Adaptation Layer
MAC: Media Acess Control

8.11 SUMMARY

In this chapter, we discussed the need for broadband ISDN and the motivation for ATM. We also examined the differences between the synchronous transfer mode (STM) and the asynchronous transfer mode (ATM). We discussed the concepts of virtual circuits and virtual paths, and the protocol layers as they are mapped into the OSI reference model. In general, ATM is a simple transport technique, but yet it is far-reaching; ATM equipment will be slowly integrated into the network infrastructures in a seamless transition.

REVIEW QUESTIONS AND PROBLEMS

1. What are the rates that B-ISDN is designed for?

2. Describe the advantages and disadvantages of the synchronous transfer mode (STM) and the asynchronous transfer mode (ATM).

3. Describe the structure of an ATM cell.

4. What are the advantages and disadvantages of having a fixed-size ATM cell?

5. Describe the relationship between the virtual channel and the virtual path.

6. Describe the layers used in ATM protocol stack compared to that of the OSI reference model.

7. Why do we need the ATM adaptation layer (AAL)?

8. Describe each of the AAL types and outline the advantages and disadvantages of each.

9. What is meant by meta-signaling?

10. Describe the functions of meta-signaling at the user-network access interface.

11. Describe how internetworking may work when ATM networks are involved, specifically, the interconnections of LANs.

9

Local Area Networks

9.1 INTRODUCTION

LAN is an acronym for *local area network.* Understanding the meaning of *local area* and *network* will start us on our way to appreciating LANs. First of all a LAN is one network type. So, what is a network? A *network* is a set of interconnected computers and/or devices. Examples of computers that can be connected to a LAN are personal computers such as Apple Macintosh or IBM-compatible PCs. Other computers such as mainframes, minicomputers, midrange systems, and workstations can also connect to LANs. The fact that there are multiple interconnected computers or devices in a network contrasts with a single standard system.

For example, many computers such as mainframe and midrange systems do support attachment of multiple terminals and printers. These are considered multiuser systems rather than LANs. One reason is that personal computers, terminals, and workstations are sharing the same single system as shown in Figure 9–1.

Networks consist of two or more interconnected computer systems and/or devices. Each system functions as an independent entity. Some type of transmission medium is used to physically interconnect the computers and/or devices.

What makes a network a local area network? The most obvious answer is in the name. *Local area networks* consist of interconnected computers and

LAN

Figure 9–1
(a) A networked arrangement and a single multiuser system, and (b) network configuration where a number of computers are interconnected with a type of physical medium.

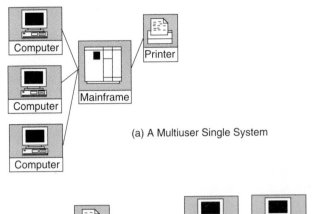

(a) A Multiuser Single System

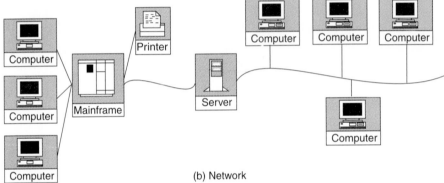

(b) Network

devices located relatively close to one another, that is, within a local area. The local area is typically within an office, a building, or a group of buildings. The distance among the computers and devices is only one characteristic that makes a network a local area network. Even this characteristic is undergoing change. Due to new technologies, LANs now often span much longer distances, up to many miles in some cases.

What features differentiate a LAN from a *wide area network,* known as a *WAN?* Generally, wide area networks are geographically larger and cover much longer distances than LANs. WANs are often used to connect many more computers than LANs.

With newer technologies, however, LANs can extend for many miles and interconnect with other LANs, providing connectivity of many computers covering a much wider area. In general, LANs are typically thought of as limited in distance and area, usually within a company's premises.

Although LANs are owned and operated by a single company in most cases, and LAN facilities (i.e., wiring, systems, and devices) are on the company's premises, a WAN is often provided by common carriers such as the telephone companies. Information flow on a LAN is typically within the confines

WAN

of an organization, but information flow on a WAN can flow through public telecommunications facilities. However, many companies with significant wide-area data transmission requirements provide their own private WANs by leasing telephone circuits or purchasing microwave or satellite transmission equipment.

The transmission techniques for exchanging information are different for LANs than for WANs. As we will see in this chapter, LANs make use of media access techniques that are different than the physical interfaces to WAN. In some cases, LANs use different protocols than do WANs, although protocols are usable in both LANs and WANs.

WANs are often used to connect LANs that are geographically dispersed. For example, a business with offices around the country may have a WAN that connects the individual LANs in each office. This interconnection of LANs over WANs is very common in corporations today.

The bottom line is that the distinction between LANs and WANs is less clear than it used to be. New networking and switching technologies are blurring the lines between the two types of networks. In this chapter, we will examine the many components that a LAN is made up of.

9.2   MOTIVATION FOR USING A LAN

There are two primary reasons for connecting systems and devices in a LAN: to share different resources such as printers or files; and to improve communications between users in a workgroup, office, department, or company. The desire to share disks among multiple computers was the initial driving force behind the growth and popularity of LANs. At the time, disk drives were relatively expensive, often too expensive to purchase for all PC users. The solution was to share one (or a few) disk, which reduced the overall cost of disk storage. Disk drives have become less expensive, but the benefits of centralized disk storage have not diminished. Disk sharing also provides access to commonly used programs. One copy of a program can be stored on the shared disk and accessed by many users on the LAN. This reduces total disk space requirements since users do not need their own copies of the program. Shared disks also provide access to commonly needed data. Many users can reference the same information and databases at the same time via a LAN. With only one shared copy of the information, users can be assured that it is always accurate and up to date. Another benefit of disk sharing is that it allows centralized backup of files. It eliminates the need for individual users on the LAN to back up their files to a floppy disk or tape. The files stored on individual computers on the LAN can be backed up (copied) to a shared disk. Centralized backup can protect the entire workgroup from the disastrous effect of a disk crash.

Probably the most widely shared resource in LANs today is printers. The reason for this is the same as the original motivation for disk sharing. That is, printers were, and still are in many cases, relatively expensive devices. If sharing were not possible, each individual system would need its own printer. This proposition would be expensive and would typically result in that printer being vastly underutilized.

Setting up a LAN is not without additional expense, though. Cables to connect the computers and printers need to be purchased, as do adapters that connect the systems to the LAN cable. Special software is required to permit the printers and other peripherals to be shared. These added expenses are usually lower, however, than the cost of printers and large disks for every system. Another benefit of printer sharing is that users can be provided with access to a wide variety of printers, such as laser printers, high-speed printers, or color printers. Information can be printed on any of the printers connected to the LAN instead of being restricted to printing on the printer that is physically connected to the computer in use.

The ability to connect devices such as printers and disks to LANs, allowing them to be shared among the other systems on the LAN, results in potential cost saving in another way as well. It may be possible to use lower-cost systems for individual users. Rather than each user requiring a fully configured system with printers, multiple disk drives, fax card, etc., the systems can be lower-end with no (or fewer) peripheral devices. These systems cost less than fully configured ones. Multiply the cost saving for each system by the number of systems connected to the LAN and the savings can be substantial. Sharing relatively expensive peripherals remains one of the major benefits of LANs.

In addition to saving on hardware, a LAN can also result in saving money on software. A separate copy of a program may not always need to be purchased for each system on the LAN. Users may be able to share the program that resides on a shared disk (known as a *file server*) on the LAN. The potential saving depends on the pricing structure of the software. Some vendors offer LAN versions of their programs that are capable of supporting multiple users. Other LAN licensing arrangements are based on the number of users. In any case, the total software cost is typically less than it would be if each user had to purchase an individual copy of the software. Figure 9–2 shows the sharing of software on a LAN.

A LAN can also offer improved security. Users get access to programs and data only through servers on the LAN. Security features such as passwords are built into the servers to restrict users' access to only the information and resources they require.

In addition, the system attached to the LAN could be a diskless workstation. Such a system does not have a floppy disk drive. Therefore, no software or data resides on the system, and none can be copied to a floppy disk and taken away. This setup can also limit the introduction of viruses or loading unauthorized programs.

Another benefit of LANs is improving communication among users of the interconnected systems. Two types of popular applications used on LANs for this purpose are electronic mail and workgroup applications. With LAN-based

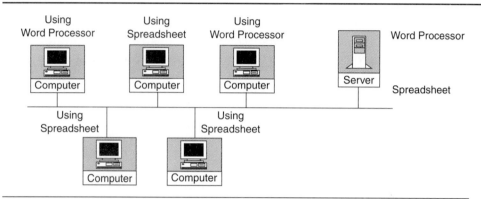

Figure 9–2
In addition to saving cost on hardware, a LAN also results in software savings. (The computers are executing software that resides on server.)

electronic mail, users on the LAN can exchange messages with each other. This function can serve as an alternative to paper-based memos or voice messages. Electronic mail (or E-mail as it is widely known) is typically easy to use. Messages are stored in user mailboxes on disk and later retrieved when users come on line. Messages are saved as long as required so they can be reread or archived as needed. Logs of messages can also be kept. This system eliminates many of the problems such as lost messages that are prevalent when using handwritten messages. Documents, programs, and data files can also be exchanged among users of the LAN using electronic mail. They are sent from one user to another as an attachment.

Workgroup applications allow multiple users to cooperate in performing various tasks. Information is passed to different LAN users who can perform whatever processing is required. Functions such as calendar management and document manipulation can all result in improving productivity, as well as the speed and quality of communication among users. Now that we have seen some of the main reasons for LANs, let's look at the major components found in most LANs.

9.3   COMPONENTS OF LOCAL AREA NETWORKS

Earlier we indicated that LANs consist of interconnected computers and devices. In addition to these, a few other components are important for a LAN to operate. These components include adapters to allow the computers and devices to plug into the LAN and the cables used to physically connect them. Let's briefly describe the role and usage of the various components in the LAN environment.

LANs often consist of interconnected computers. Many different kinds of computers can be connected to a LAN. These include personal computers, UNIX workstations, multiuser systems, and many LANs even include large mainframe computers. The computers connected to a LAN permit the users to execute programs residing on them. The users interact with the software running on these systems to make use of the LAN. Examples of users are secretaries, accounting clerks, and database administrators. To perform work for the users, computers execute application programs, such as electronic mail, file transfer, document processing, or accounting systems. These programs run on computers connected to the LAN and provide users with the resource-sharing functions for which the LAN exists.

Certain computers on a LAN contain files and databases. The hard disks supported on these systems contain the files and databases built by applications and accessed by users. The computers containing these shared resources are called *servers.* The users and applications running on other systems access these servers over the LAN, and are called *clients,* hence the name *client server computing* (see Figure 9–3).

Server computers also support the attachment of printers and allow them to be shared by other systems. These computers are called *print servers.* This computer accepts print requests from other systems on the LAN, temporarily writes the data to be printed onto its disk storage, then sends the data to the printer. The print server manages the print streams from several systems concurrently. Other peripherals can also be attached to servers and shared over the LAN. Plotters, faxes, and CD-ROM drives, when attached to the server computers, become accessible to the other systems over the LAN. Devices such as printers can also be directly connected to a LAN and shared by all users. A directly attached printer acts as a print server. Other systems on the LAN can direct print streams to the printer.

Computers and devices need special LAN adapters to directly connect to the LAN. A *LAN adapter* is a hardware board (e.g., a PC adapter card) that is either integrated with the equipment or purchased separately and inserted into an expansion slot in the PC chassis. LAN adapters are commonly called *network interface adapters (NIAs)* or *network interface cards (NICs).* Their role is to provide

servers
clients

LAN adapter

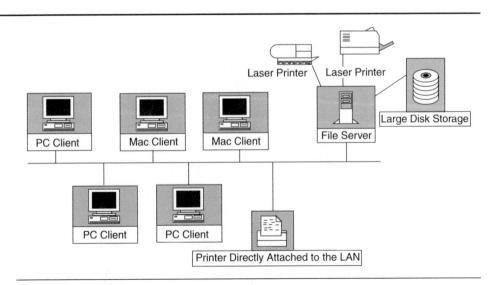

Figure 9–3
LAN client-server configuration.

Laser Printer Laser Printer

Large Disk Storage

PC Client   Mac Client   Mac Client   File Server

PC Client   PC Client

Printer Directly Attached to the LAN

the physical attachment of a device or system to the LAN transmission medium. The adapter provides the interface to the PC or device on one side and the network on the other side. The adapter provides a socket into which the LAN medium, such as a cable, can plug. LAN adapters handle the signals that flow between the connected equipment on the LAN. The adapter has to translate the signals used within the PC or printer into signals used on the LAN cable. The connection of a network interface card is shown in Figure 9–4.

Each computer and device needs to get connected in some manner. This connection is made through the particular transmission medium used in LANs. Cables provide the physical links between the systems and devices. Wireless LANs function without physical cables providing the connections. Instead, wireless LANs use infrared or radio frequencies that broadcast over the air to exchange information between systems. In most LANs, however, information flows over the physical cables. LANs can be constructed using different kinds of cables or wires that have different characteristics, and therefore, make use of different signaling techniques. In the next section we will describe the basics of LANs, including the various types of LAN cabling in use, transmission techniques, methods for controlling access to the LAN, and various ways that the LAN can be configured.

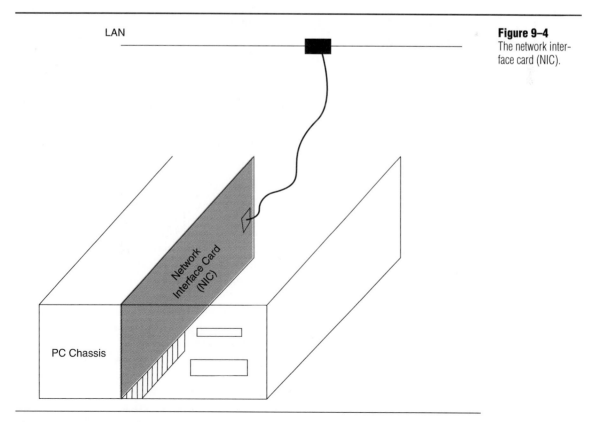

Figure 9–4
The network interface card (NIC).

9.4 BASICS OF LOCAL AREA NETWORKS

IEEE

Standards exist for some of the topics that we will discuss in this section. In particular, a set of LAN standards has been defined by the *Institute of Electrical and Electronics Engineers (IEEE)*. The IEEE is an international association of professional engineers that defines electrical and communications-related standards. The IEEE 802 series of standards is focused on LAN interconnection and operation. We will refer to these standards at various places in this chapter. The LAN standards are related to another set of standards defined by the International Standards Organization (ISO), discussed in Chapter 7. The ISO is an international standards body whose members represent the national standards bodies of various countries. The ISO's charter includes setting internationally accepted computer and communications-related standards. The ISO has defined the open system interconnection (OSI) reference model that serves as the framework for communications standards. As seen in Chapter 7, the OSI reference model consists of seven layers of functions covering a range of communications issues from low-level physical interfaces of equipment to high-level application functions. Each layer of the model defines a particular set of communications functions needed for systems to interoperate with one another. The IEEE LAN standards are a specific set of standards that conforms to the lower layers of the OSI reference model. At various places in the remainder of the chapter, there will be references to the layers of the OSI reference model.

physical medium

One of the most obvious requirements for a LAN is a transmission path between the systems and devices on the LAN, which are interconnected by some type of *physical medium*, typically cables or wires. The medium provides the transmission path between stations on the LAN. Data are carried in the form of signals flowing over the physical transmission medium. The most popular types of LAN physical media are coaxial cable, twisted-pair wire, and fiber optic cable.

Coaxial cable is the type of cable used for cable television (CATV). If you subscribe to cable TV, you have this type of cable coming into your home, connecting the TV to the company providing your cable service. A similar type of coaxial cable can be used to connect personal computers and printers on a LAN. Of course, when used for LAN interconnection, the coaxial cable is used to transmit computer-related information rather than TV signals. Coaxial cable consists of a single-center core wire (an electrical conductor), as seen in Chapter 1, surrounded by insulation that is then surrounded by a wire mesh or sheath; all of this is typically covered by another protective jacket made of rubber or plastic.

There are a variety of coaxial cables. The original type of coaxial cable used in Ethernet LANs is known as *thick coax*. This type of coaxial cable was specified for use in the original Ethernet specification, published jointly by DEC, Intel, and Xerox in 1979. This cable is approximately a half-inch thick and, as a result, is relatively inflexible and difficult to install.

Thin coax came into widespread use because it is less expensive than thick coax and easier to work with. It is approximately a quarter-inch (0.63 cm) thick and very similar in size and weight to the coax used for cable TV. On the downside, this coax supports shorter distances than the thick coax.

The main advantages of coaxial cable are that it is less susceptible to inter-ference than twisted-pair wire, and it supports relatively high rates of data transmission over greater distances. The main disadvantage is that it is more expensive than other media, and in the case of thick coax, its size and inflexi-bility make it more difficult to install. Typical data rates over coaxial cable LANs are 10 Mbps, with distances ranging from hundreds to thousands of feet.

Twisted-pair cable is another type of media popular for LAN connections. As the name indicates, and as seen in Chapter 1, twisted-pair cable consists of pairs of copper wires where each pair of wires is twisted around one another in a manner that enhances its ability to carry electrical signals. The standard telephone wiring in the walls of your home or office is an example of twisted-pair cable.

Twisted-pair wiring comes in two types—unshielded or shielded. *Unshielded twisted-pair (UTP)* is the standard telephone cable. Each wire is insulated and twisted together with another wire to form a pair. Standard office wiring is typically two to four pairs per bundle. Each bundle is encased in a non-metallic (usually) plastic or PVC sheath. The main advantages of UTP wiring are that it is inexpensive, flexible, and easy to install. UTP is readily available since it is a standard telephone cable. Many commercial buildings developed or remodeled in the last few years have been wired with UTP cables that can be used for LANs. unshielded twisted-pair

One problem with UTP is that it is more susceptible to electrical interfer-ence, that is, noise from electrical sources such as other wires, power cables, and machines. *Crosstalk,* in which signals from one wire interfere with signals of another wire, is an example. Such noise and interference results in errors during data transmission. Electrical interference is not a major problem in of-fices, but it can be a problem in factories where electrical machinery is in use. Another problem is that signals lose their strength as they are transmitted over UTP. The loss of signal strength is called *attenuation;* UTP has high attenuation.

These two problems limit the distance that signals can travel over UTP without being regenerated, as well as the transmission rates that can be used. Typically, UTP is useful only over short distances. In the past, only relatively low-speed transmissions were possible over UTP, but newer techniques are now in use that will support UTP speed in the 100 megabits per second range.

Another type of twisted-pair wire is a *shielded twisted-pair (STP).* Unlike UTP, each STP wire pair is covered by a metal foil shield. The shield provides protection from noise, thus eliminating some of the problems associated with susceptibility to noise that plague UTP. This shield makes STP a little more suit-able for transmitting data as opposed to voice, for which UTP was designed. The shield also helps keep signals from emanating out of the wire. This design consideration can be important in certain environments. For example, govern-ment regulations limit the amount of electrical interference a LAN can gen-erate. (A negative aspect of STP, though, is that the shield increases the cost of the wire so it is typically more expensive than UTP.) shielded twisted-pair

In summary, the main advantages of twisted-pair cables (both shielded and unshielded) are that they are widely used, inexpensive, and easy to install. The major disadvantages are that they are more susceptible to interference than other media, and have relatively high attenuation, limiting the distance that signals can travel before requiring regeneration.

Fiber optic cable is different from coaxial and twisted-pair cabling. Instead of being made of copper, fiber optic cable is made of glass. A fiber cable contains one or more glass or plastic fibers surrounded by insulation known as *cladding*, as discussed in Chapter 1. The fibers are very thin and transparent. Instead of electrical signals, fiber optic cable is used to transmit light signals. A light source such as a light-emitting diode (LED) is used to generate pulses of light that are carried along the fiber. A receptor such as a photodiode receives the pulses. The light pulses are eventually converted to electrical signals within a system or device connected to the LAN. Using light rather than electrical signals gives fiber optic cable many of its characteristics. Fiber optic cables support high transmission rates, up to the 100 Mbps range. As a result of the higher frequency of light, fiber optic cables offer greater bandwidth (or more information-carrying capacity) than coaxial cable or twisted-pair wire. In addition, fiber optic cables, because they use light rather than electricity, are immune to electrical interference. They are also very thin and flexible. This means that they are lighter and take up less space than coax or twisted-pair cabling. Thus, a fiber optic cable is easy to install and ideal for bundling many fibers together to create a cable that carries a very large amount of traffic.

Fiber optic media has less attenuation than copper wiring. This means that light signals can travel farther across fiber optic cables before they must be regenerated. The result is that fiber optic cables can be longer before repeaters are required. Another outgrowth of fiber optic cable's use of light is that it is very difficult to tap into without detection. Any tap into the cable interrupts the flow of light and is easily detected. This feature provides greater security than an electrical cable into which unnoticed taps can be made.

The major disadvantage of fiber optic cables is their higher cost compared to other cable types. In addition, network adapters for fiber optic LAN connections are more expensive than other types of LAN adapters. Fiber optic technology is also relatively new and not as widely used in LANs yet. As fiber optic cables and LAN components become more prevalent, their cost will decrease.

Most commonly, fiber optic cables are used for high-speed backbone LANs that interconnect several lower-speed LANs. Standards for fiber optic LANs include the fiber-distributed data interface (FDDI) standard, discussed later in this chapter. Other fiber optic standards and enhancement to the FDDI standards are being developed.

9.5   TRANSMISSION TECHNIQUES

Once the physical transmission medium (cable) is in place, information can flow between stations on a LAN. How is this information actually carried across the cables or wires? Various transmission techniques can be used to send signals, representing information, across the transmission medium. Transmission techniques differ in a number of ways, including whether signals are modified or not before transmission and how signals from multiple sources share the transmission medium. The two basic transmission techniques used in LANs are baseband and broadband. Let's see what is different about them.

9.5.1 Baseband Transmission

Using *baseband transmission,* signals are placed directly onto the transmission medium. By directly, we mean that the signal is not modified (i.e., not modulated) by a carrier signal. The signal is placed directly onto the transmission medium as voltage pulses, as shown in Figure 9–5. This type of signaling is digital in nature, with changes in voltages representing 0s or 1s. A stream of such digital pulses represents the information being transmitted.

baseband transmission

The signal takes up the entire bandwidth of the transmission medium. Therefore, at any point in time, the transmission medium is dedicated to a single signal. However, this does not mean that the transmission medium can be used only by a single application. Signals from multiple sources can be transmitted via the technique known as *time-division multiplexing* or TDM as seen in Chapter 2. With TDM, different time slices are allocated to different applications. For example, data from application A can be sent during time slice 1 and data from application B can be sent during time slice 2. In this manner, the transmission medium is shared by multiple users.

Since signals are placed directly onto the transmission medium, baseband transmission is relatively inexpensive. No special equipment is required to modify the signals. However, one problem with this technique is that the signals lose strength as they are carried over longer distances and, therefore, must be regenerated. Typically, then, baseband transmission is somewhat limited in the distances over which it can be effectively used. Figure 9–5 demonstrates TDM in baseband transmission.

9.5.2 Broadband Transmission

In *broadband transmission,* signals are modulated to different frequency ranges. This form of analog signaling is similar to how radio and TV transmission

broadband transmission

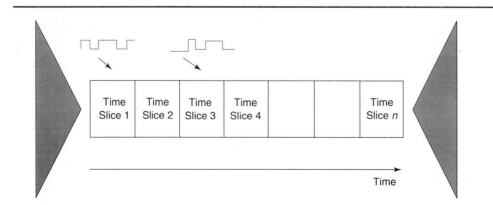

Figure 9–5
Baseband transmission with time-division multiplexing: Signals are placed directly onto the transmission medium.

works. In fact, broadband transmission is typically provided using radio frequency modems. The technique of frequency-division multiplexing (FDM) is used to allow multiple signals to be transmitted concurrently, thus allowing the transmission medium to be shared by multiple applications. As its name indicates, FDM results in different signals being transmitted at different frequency ranges. This technique is as opposed to TDM, in which signals were divided by time slots rather than frequency ranges. Unlike TDM in which the transmission is dedicated to one signal at any point in time, FDM allows multiple signals to be transmitted at the same time. Each signal occupies a different frequency range. The different frequency ranges into which the transmission is divided are called *logical channels*. Therefore, multiple channels are available using broadband transmission. This technique is used in standard cable TV. Each logical FDM channel could also be shared by multiple applications by using TDM within the channel. Figure 9–6 demonstrates FDM in broadband transmission.

Broadband provides greater bandwidth than baseband and is, therefore, able to carry more information and support greater distances than baseband.

Figure 9–6
Broadband transmission with frequency-division multiplexing: Signals are modulated prior to being put onto the medium.

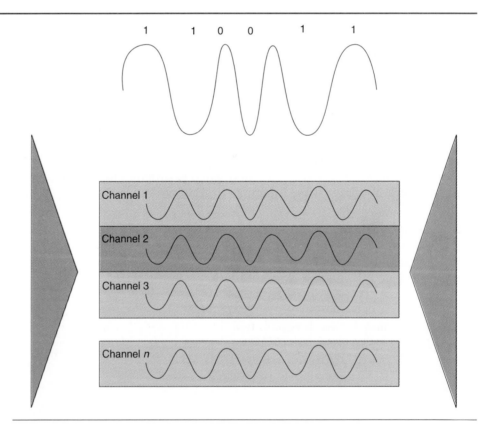

Broadband transmission is well suited to carry voice, data, and video at the same time. One downside is that broadband is typically more expensive than baseband transmission. Broadband is also more difficult to configure and costly to modify, so it is best suited to large installations.

9.6   MEDIA ACCESS METHODS

All devices attached to a LAN share the transmission medium. All devices can send information over the LAN. But, what happens if multiple devices attempt to send data onto the LAN at the same time? If this happens, there is the potential that the signals from multiple devices would interfere with one another, garbling the information. Therefore, there needs to be some method of controlling when devices are allowed to send. In other words, there has to be control over when and how devices access the LAN transmission medium. A *media-access control (MAC)* method determines how multiple devices share the transmission medium. It ensures that all devices have access to the network. There are different methods for providing such control. The most widely used media-access methods are CSMA/CD and token passing, which we examine next.

MAC

9.6.1   CSMA/CD

With *carrier sense multiple access with collision detection (CSMA/CD)*, devices must first listen to the network before sending. What they are listening for is the presence of another signal on the LAN transmission medium. In other words, devices must sense the presence of a carrier signal, hence the *carrier signal* part of the name. The presence of a carrier signal indicates that information is currently being transmitted. Therefore, the device that wants to transmit waits before sending its information. If a carrier signal is not detected, it is an indication that the LAN is free. That is, an active signal is not currently being sent from another device. Therefore, the device can attempt to send.

CSMA/CD

However, things don't end here. What if another device had been listening because it also had something to send? Both devices might detect that no carrier is present and attempt to send their information onto the LAN at the same time, resulting in signals colliding. The *multiple access* part of the name refers to this ability for multiple devices to access the transmission medium. To handle this situation, once a device detects that the transmission medium is free to accept a signal and sends its information, it must continue to listen to what's happening on the transmission medium. If no device had attempted to send at the same time, the signal would be transmitted without any problem. If another device had sent information at the same time, though, the listening devices would detect a collision. This is where the *collision detection* part of the name comes from.

When such a collision is detected, the devices stop transmitting and wait for some period of time before attempting to transmit again. The time delay is different for each device and is based on a random number algorithm. In other words, the period of time a device waits is randomly selected. The result is that devices will retry their retransmission at different times that are unlikely to result in another collision.

One problem with the CSMA/CD access method is that performance degradation can occur in LANs with large numbers of users and heavy traffic loads, because more collisions occur as usage increases. This situation can cause timeouts due to waiting for the medium to be free, and result in retransmissions, thereby slowing application response times.

CSMA/CD is the access method used in Ethernet and 802.3 LANs. It was developed by Dr. Robert Metcalfe at MIT and was first employed in the ALOHANET, a radio-based network linking universities in Hawaii. CSMA/CD became an international standard when IEEE 802.3 was approved. Figure 9–7 summarizes the CSMA/CD access method.

At low LAN traffic levels, CSMA/CD is actually an extremely effective technique. Little housekeeping is required to keep the network running, so the majority of the network's bandwidth is devoted to useful traffic. When a CSMA/CD

Figure 9–7
Summary of the CSMA/CD media access technique.

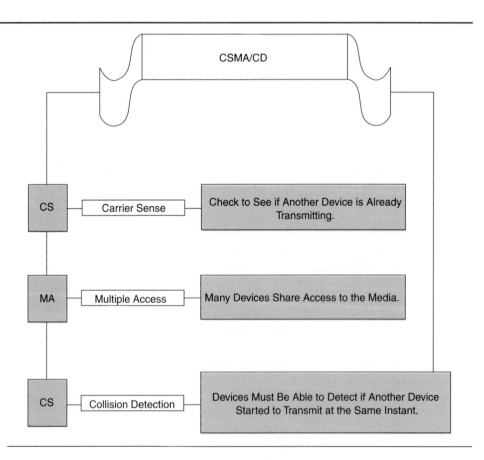

network is loaded too heavily, however, it is possible for collisions to become so frequent that little traffic can find its way onto the network. Under such circumstances, a device might find it impossible to transmit.

A CSMA/CD LAN tends to function very smoothly until some critical threshold is exceeded. At some point, traffic increases result in much more frequent collisions and network performance degrades, often catastrophically. You should make it a habit to monitor your LAN to determine whether usage levels are increasing to a potentially troublesome level. A good rule of thumb is that CSMA/CD LANs (Ethernet) work fine with a constant traffic load of 30 percent of capacity and traffic bursts of about 60 percent of capacity. If your LAN is exceeding these levels, you should take steps to divide your Ethernet into additional segments to reduce the number of stations that are sharing a given segment's bandwidth. We will see later in this chapter the role of LAN bridges in interconnecting LAN subnets (segments).

CSMA/CD is a probabilistic access control method, meaning that a given device probably will have a chance to transmit at roughly the time it needs to, but that the opportunity to transmit is not guaranteed. On most LANs, probabilistic access control works fine. In some applications, however, it might be necessary to be more certain that devices can transmit. Suppose that the network is controlling a manufacturing process, and that things might happen at precise times. The possibility that a device might not gain access to the network at a critical time is unacceptable. This state of affairs is one reason why IBM invented the token ring network.

9.6.2 Token Passing

Token passing is the media access technique used in token ring LANs, as shown in Figure 9–8. A token ring operates as a logical ring. The transmitter of each device is connected to the receiver of the next device in the ring. This enables the devices to pass messages around the ring.

token passing

A token is a special type of data frame that circulates around the ring. A device can transmit only when it is in possession of the token, and it must wait until the token arrives to transmit data. After a data frame is transmitted, the device releases the token to the network so that other devices can transmit.

The apparent simplicity of the token-passing method hides some complexities that had to be accounted for by the network designers, as these examples show:

- Like any network data, tokens can be lost. How is a lost token detected and how is a new token created?
- What steps should a station take if it stops receiving data?
- What if a station that was to receive a frame goes off the network?
- How is the network supposed to identify frames that have circulated the network too many times?

To handle these and other potential problems, one device on a token ring is designated as a *ring error monitor (REM).* The REM can regenerate lost tokens

REM

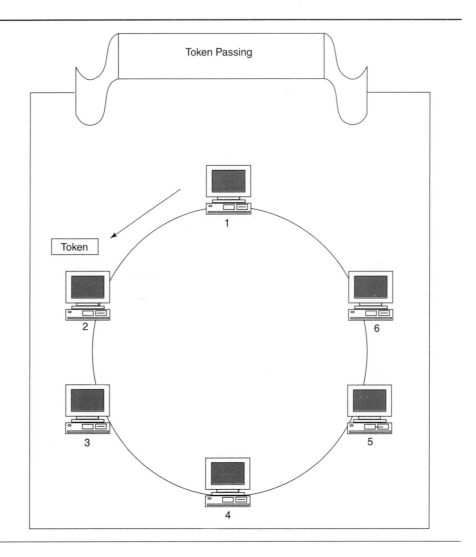

Figure 9–8
Token passing.

and remove bad frames from the network. Additionally, each device is capable of signaling certain problems by transmitting a beacon signal that notifies network management of the problem. The error detection and diagnostic tools available on a token ring are quite extensive. By contrast, no such tools are built into Ethernet (CSMA/CD) networks.

For these and other reasons, development of the token ring concept was a complex process. The mechanism that controls a token ring network is much more involved than the mechanism required for Ethernet. That is one reason that the token ring hardware costs more than an Ethernet.

Token ring is preferred over Ethernet when a deterministic access method is required. Ethernet is probabilistic and does not guarantee device opportunities to transmit. On a token ring, however, every device is guaranteed a chance to transmit each time the token circulates the ring.

Token passing may suffer less performance degradation than CSMA/CD in very large LANs, because contention for the transmission medium is more orderly than with CSMA/CD, eliminating collisions, timeouts, and subsequent retries. Token passing allows stations to transmit whenever a free token is available, but they may have to wait a while to get a token.

A potential problem is that stations that get a token can dominate the LAN. For example, a station may continue sending a large amount of information without putting the token back on the LAN to allow other stations to transmit. Implementations attempt to minimize this occurrence by placing a limit on the amount of time a single station can send before passing on the token.

9.7   LAN TOPOLOGIES

The actual physical layout of a LAN is the arrangement in which devices are interconnected. This layout is called the *topology* of the LAN. LANs can be configured in a variety of ways. The most widely used topologies are bus, star, and ring topologies. Let's discuss the differences among these topologies.

topology

9.7.1   Bus Topology

In a *bus topology*, shown in Figure 9–9, all stations (i.e., systems and devices) are directly connected to the same transmission medium, usually a physical cable. In terms of wiring, a bus topology is simple, and as a result, often inexpensive. Bus topology is very common on LANs, and was the type of topology originally

bus topology

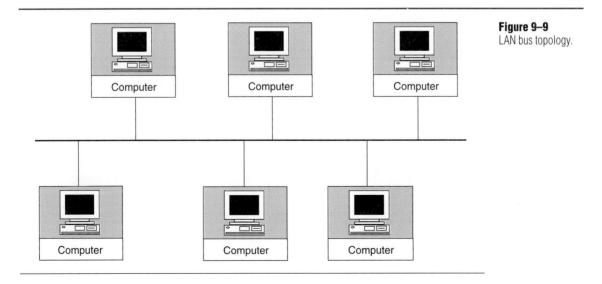

Figure 9–9
LAN bus topology.

specified for Ethernet LANs. Information on bus-based LANs is broadcast to all connected stations. Transmissions go in both directions along the bus or cable. All stations, thus, receive all transmissions. Each station has a unique address, which is assigned to it. A station is able to recognize information directed to it by the station address included in the data *frames* that carry information on the bus.

9.7.2 Star Topology

star topology

In a *star topology*, as seen in Figure 9–10, each station is connected to a central piece of equipment commonly called a *hub*. All communications go through the *hub*. Note that in this topology, the individual stations are not directly connected to one another. They are all connected indirectly through the central hub. The hub amplifies and retransmits signals received on one cable to other cables, thus providing connectivity between stations. One problem with a star is that if the central hub fails, the network is inoperable. To get around this problem, redundancy features are often built into hubs so that they are very reliable. Also, hubs can be configured so that a bypass is possible in the event that a component fails.

Figure 9–10
LAN star topology.

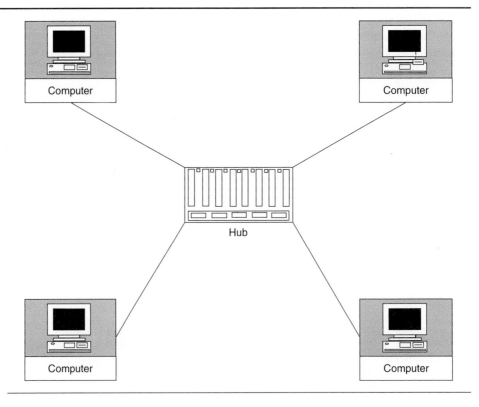

9.7.3 Ring Topology

In a *ring topology*, stations are directly connected to other stations to form what looks like a ring. Unlike a star configuration, adjacent stations on the ring are directly cabled to one another; there is no central hub. Information flows in one direction around a ring. Each station receives all signals from the adjacent station, regenerates them, and retransmits frames to the next station on the ring. Although a ring topology, shown in Figure 9–11, eliminates the problem of depending on a central switch that might fail and take down the entire network, it depends on each individual station on the ring. If a station fails, or the link between stations fails, the ring can become inoperable. There are solutions to this problem, such as redundant links and ways to bypass failed stations.

ring topology

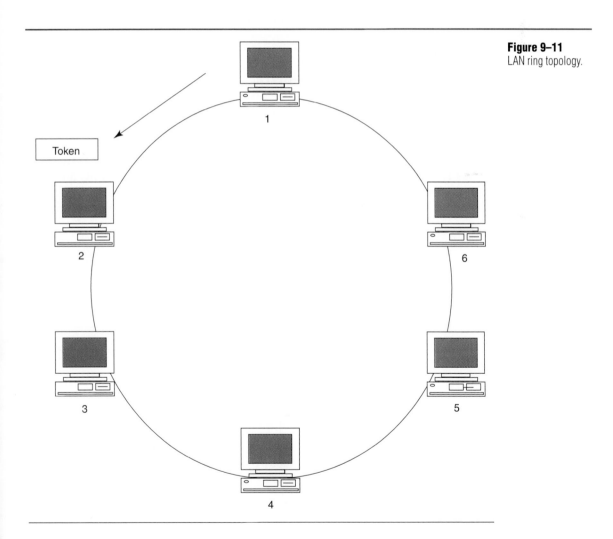

Figure 9–11
LAN ring topology.

As with other LAN topologies, each station on a ring has a unique address. As stations monitor information flowing around the ring, they look at the destination address in the frames to determine if a frame is intended for them. If so, the frame is pulled off the ring. If not, the frame is retransmitted to the next station. Token passing schemes are often used on ring-based LANs. As an example, IBM's token ring LAN makes use of a ring topology on which a token-passing media-access control method is used.

9.8 IMPLEMENTATIONS OF LANS

Now that we have reviewed the fundamentals of LANs, we can look at some specific implementations and see which transmission media, transmission techniques, topologies, and media-access methods are used in some of the most popular types of LANs today.

9.8.1 Ethernet

Ethernet

Ethernet, developed jointly by Xerox, Digital, and Intel, is one of the most popular types of LANs in use today. The original Ethernet specification called for coaxial cable as the transmission medium. Today, Ethernet LANs make use of other types of cabling such as twisted-pair wire. Ethernet uses a bus topology as shown in Figure 9–12.

An Ethernet LAN uses a CSMA/CD contention protocol for controlling access to the bus (cable). This type of access as well as the cabling and signaling specifications are also defined as part of the IEEE 802.3 standard. Ethernet and IEEE 802.3 specifications are very similar, but they are not identical.

Figure 9–12
An Ethernet LAN
using bus topology.

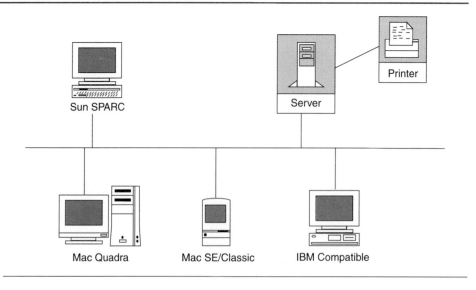

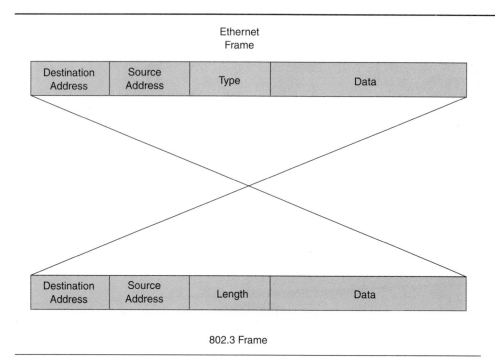

Figure 9–13
Ethernet frame compared to IEEE 802.3 frame.

However, they are compatible and devices adhering to either specification can interoperate on the same LAN.

One difference is that the Ethernet *frame* headers include a type field that indicates the higher layer protocol in use. For example, the type field could indicate that either TCP/IP or XNS protocols were used across the Ethernet LAN. Instead of a type field, the header of an 802.3 frame includes a length field indicating the length of the data contained in the information portion of the frame. This point is illustrated in Figure 9–13.

Ethernet was designed to support transmission rates of 10 megabits per second (Mbps). This rate is supported in Ethernet LANs today, however, there is work going on to support Ethernet-like LANs at higher speeds (fast Ethernet at 100 Mbps).

The main advantage of Ethernet is its relatively low cost. Inexpensive Ethernet adapters are available for most PCs, and Ethernet interfaces are supported in a wide range of computer equipment. The multivendor support of Ethernet makes it a popular choice.

Performance problems can occur in bigger Ethernet LANs having large numbers of users and heavy traffic demands, as indicated earlier in this chapter. This situation is due primarily to the characteristics of the CSMA/CD technique where all stations contend with the use of the transmission medium. This can result in collisions, forcing stations to wait before retrying their transmission. An upper limit of 10 Mbps also becomes a problem when transmitting information requiring high bandwidth such as video images. It also makes Ethernet less attractive as a *backbone* LAN to interconnect other LANs. However, multiple Ethernet LANs can be easily interconnected by devices called *bridges* (discussed later in this chapter), forming large logical LANs.

backbone

Ethernet Cabling   Until recently, all Ethernets were cabled with thick or thin coaxial cable. Customers demanded an unshielded twisted-pair (UTP) configuration, however, and the IEEE responded with the 10BASE-T standard, which defines a 10 Mbps Ethernet LAN using twisted-pair cabling. In this section we present three of the most common Ethernet cabling standards.

Thick Ethernet (10BASE5)   The original Ethernet cabling used a thick, half-inch coaxial cable referred to as *thick Ethernet* or *thicknet*. As with most varieties of Ethernet, thick Ethernet operates in baseband mode with a bandwidth of 10 Mbps. The maximum cable length of a segment is 500 meters. The IEEE label 10BASE5 summarizes these cable characteristics: 10-Mbps bandwidth, baseband operation, and 500 meter segments.

thick Ethernet

Figure 9–14 illustrates the components of thick Ethernet. Workstations do not connect directly to the thick coaxial cable. Instead, they attach by way of a *multistation access unit (MAU)*. An MAU can be installed on the cable by cutting the cable and installing N-type connectors. The more common practice is to use an MAU that clamps onto the cable. Pins in the clamp-on MAU penetrate the cable to make contact with the shielded center conductor. MAUs that attach in this way are called *vampire taps*. For best performance, MAUs should attach to the thick coaxial cable at intervals that are even multiples of 2.5 meters.

MAU

Both MAUs and device network interfaces are equipped with a special 15-pin connector called an *attachment unit interface (AUI) connector*. A multiconductor AUI cable of up to 50 meters can be used to connect the device to the MAU.

Figure 9–14
Components of a thick Ethernet.

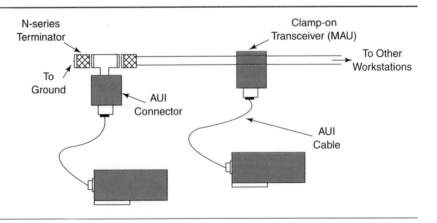

Figure 9–14 shows a common feature of networks based on coaxial cable. The ends of an Ethernet bus must be equipped with termination connectors that incorporate a 50-ohm resistor that matches the characteristic impedance of the cable. The terminators absorb the signals that reach them and prevent the signals from being reflected back into the cable as noise. One of the terminators must be connected to an electrical ground. Without the ground connection, the cable's shield might not do an effective job of preventing against electromagnetic interference (EMI). Only one termination should be grounded. Grounding both connectors may produce electrical problems in the cable. A common source of an electrical ground is the screw in the center of an electrical wall plate; however, you should have your outlet tested to ensure that a good ground is present.

Two sets of terminology are used with thick Ethernet. The IEEE 802.3 standard refers to the connectors as *attachment unit interface connectors*. Practitioners of the older Ethernet II standard, however, refer to these connectors as *Digital Intel Xerox (DIX) connectors*. Secondly, although the IEEE 802.3 standard refers to the cable interface as a *medium attachment unit*, it is more commonly called a *transceiver* when used in an Ethernet II environment. You will often find these terms used interchangeably.

Apart from naming conventions, cabling is identical for Ethernet II and IEEE Ethernet 802.3. In fact, messages from both standards can coexist on the same cable segment.

An Ethernet *segment* consists of the coaxial cable between two terminators. The maximum length of a thick Ethernet segment is 500 meters, and each segment can support a maximum of 100 connections. When it is necessary to construct larger networks, segments can be connected using repeaters, as shown in Figure 9–15. Repeaters serve several purposes: they electrically isolate the coaxial segments, they amplify signals that pass between segments, and they reshape waveforms to correct distortions that arise as signals travel through cables.

segment

A repeater segment is called a *link segment*. At most, an Ethernet can include three coaxial cable segments and two link segments. Repeaters connect to coax segments using AUI cables, which are limited to 50 meters. The maximum length of a link segment, therefore, is 100 meters. If longer link segments are required, fiber optic cable can be used to create a fiber optic repeater link segment that can be up to 1 kilometer.

Thick Ethernet is seldom used in new installations, because the cable is expensive and bulky. Network components are also expensive and difficult to install.

Thin Ethernet (10BASE2) The IEEE 10BASE2 specification describes a CSMA/CD network with these features: 10-Mbps bandwidth, baseband operation, and segment length of approximately 200 meters (the precise limit is 185 meters). The coaxial cable used with 10BASE2 is considerably thinner than thicknet coax. Consequently, this Ethernet cabling system is often called *thin Ethernet*, or *thinnet*.

thin Ethernet

Figure 9–15
Extending a thick
Ethernet with a link
segment.

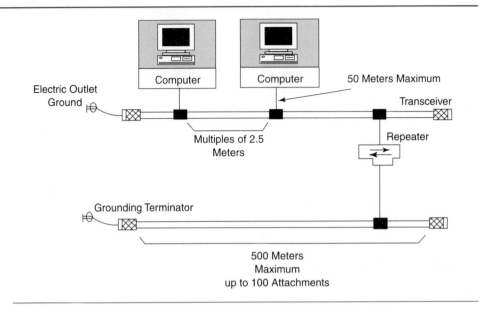

Figure 9–15
Extending a thick
Ethernet with a link
segment.

Thin Ethernet connectors are shown in Figure 9–16. The connector used in a bayonet connector (BNC) attaches and locks by twisting into place. Network interface cards (NICs) attach to the network by means of T-connectors, which attach directly to the NIC. Although it is possible to use MAUs and AUI cables with thin Ethernet, this is seldom done due to the high cost of the components.

As shown in Figure 9–16, thinnet coaxial cable must have a terminator at each end. One but not both of the terminators should be grounded. A thin Ethernet

Figure 9–16
Components of a
thin Ethernet.

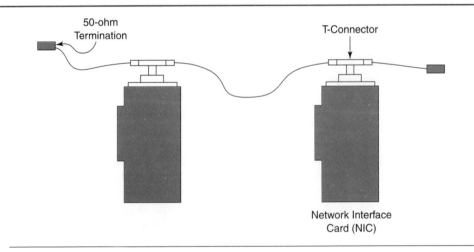

segment should not exceed 185 meters or include more than 30 attached devices. For best performance, devices should be spaced along the cable at even multiples of one-half meter.

BNC connectors are easy to install with a few simple tools, or you can buy premade cables. Manufactured cables have two advantages: the connectors are securely installed, and cable lengths conform to the rule of one-half meter multiples. Thin Ethernet is easy to install in most situations.

Like thick Ethernet, thinnet networks can be extended using repeater links. A thin Ethernet that includes two coax segments is shown in Figure 9–17. Each work station connection involves three separate BNC connectors; these are the cause of most problems with thin Ethernet. It is quite common for users to accidentally dislodge connectors or to pull the cable out of a connector—perhaps when moving the PC. Any break in the coaxial segment can disrupt the network communications for the entire segment.

Although thin Ethernet is gradually being replaced by a preference for UTP networks, it remains an excellent choice, especially for small networks. The cost of a connection for thin Ethernet is the lowest of any network cabling system—easily less than $100 per device. No hubs, which can raise the cost of a network connection by $20 to more than $100 per device, are required. Thinnet cannot be surpassed as an inexpensive medium for networking small organizations in a fairly restricted area.

One problem with thin Ethernet networks is that they can be difficult to reconfigure. Adding a device requires you to add a T-connector to the coax, which cannot be done without cutting the cable. As a result, you can add users only when the network is shut down.

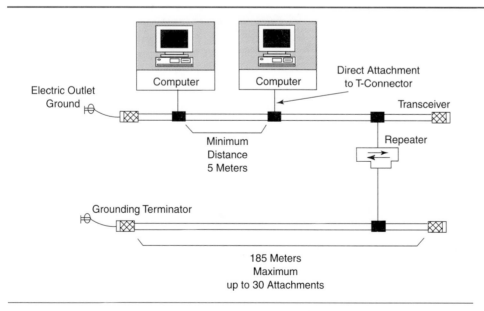

Figure 9–17
A thin Ethernet that includes a repeater link segment.

For a large network or networks that must be reconfigured frequently, it is a good idea to choose a star-wired network that includes intelligent hubs. These networks are easy to reconfigure and manage.

Twisted-Pair Ethernet (10BASE-T)   The IEEE has developed a *twisted-pair Ethernet* standard called *10BASE-T*, which describes a 10-Mbps Ethernet that uses unshielded twisted-pair cable. Unlike coax Ethernet, 10BASE-T uses a star-wiring topology based on hubs. The length of the cable that connects a device to the hub is limited to 100 meters. A simple 10BASE-T network is shown in Figure 9–18.

The cables used with 10BASE-T include two twisted-pairs. These cables must be configured so that the transmitter of the device at one cable end connects to the receiver of the device at the other end. Figure 9–19 shows that this configuration can be accomplished in two ways:

- By crossing over the pairs of wires in the cable itself.
- By reversing the connection at one of the devices. The device in which connections are reversed is marked with an X by the manufacturer. This is a common approach with 10BASE-T hubs.

10BASE-T networks require at least category 3 UTP cable connected with RJ-45 connectors. Never be tempted to use installed telephone wiring without determining whether it meets category 3 specifications, and never use telephone-grade patch cables for connecting devices to hubs. Pairs of wires are not twisted in the flat-satin cables that are normally used to attach telephones

Figure 9–18
Connecting workstations to a 10BASE-T hub.

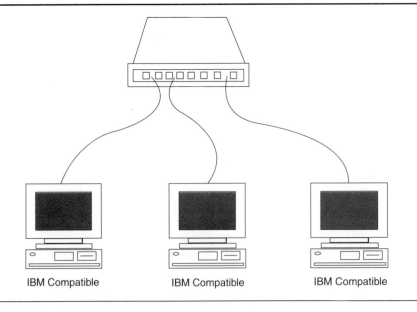

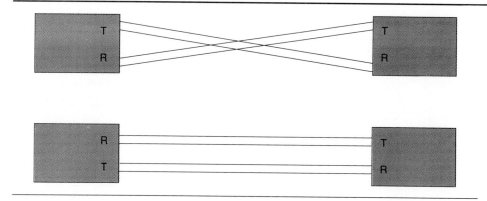

Figure 9–19
Methods of crossing pairs in 10BASE-T connections.

to the wall. Also, flat-satin cables are designed for voice-grade service and cannot support LAN data rates.

It is important to note that the physical topology of a 10BASE-T network does not alter the logical topology of Ethernet; a 10BASE-T network remains a logical bus. All devices on the network share the 10-Mbps bandwidth of the network, and only one device is permitted to access the network at a given time.

The primary reason for choosing 10BASE-T should not be the goal of saving money with unshielded twisted-pair (UTP) cables; rather, it should focus on the advantage a star network offers in terms of reconfiguration and management.

802.3 Ethernet versus Ethernet II   Although IEEE 802.3 Ethernet and Ethernet II networks can operate with the same cabling components, one significant difference exists between the two standards, as mentioned earlier. This difference is illustrated by Figure 9–20, which compares the frame format defined for the two standards.

As shown in Figure 9–20, data frames are organized into fields. In most cases, the fields used in 802.3 Ethernet and Ethernet II have the same formats, although the names are often different. The following list summarizes the similarities and differences between the two frame formats:

• An Ethernet II frame starts with an 8-octet preamble, which has the same format as the preamble and the start frame delimiter of the IEEE 802.3 frame.

• Ethernet II uses 6-octet source and destination address fields. 802.3 Ethernet can operate with 2- or 6-octets, although 6 octets are most commonly used. Manufacturers are granted blocks of addresses by an Ethernet address registry, and a unique 6-octet address is burned into each Ethernet network interface card (NIC) during manufacturing. This hardware address typically is used in the address field.

• Ethernet II employs a type field, using values that are defined by IEEE. The 802.3 specification replaces the type field with a length field that describes

Figure 9–20
Frame formats for
802.3 Ethernet and
Ethernet II.

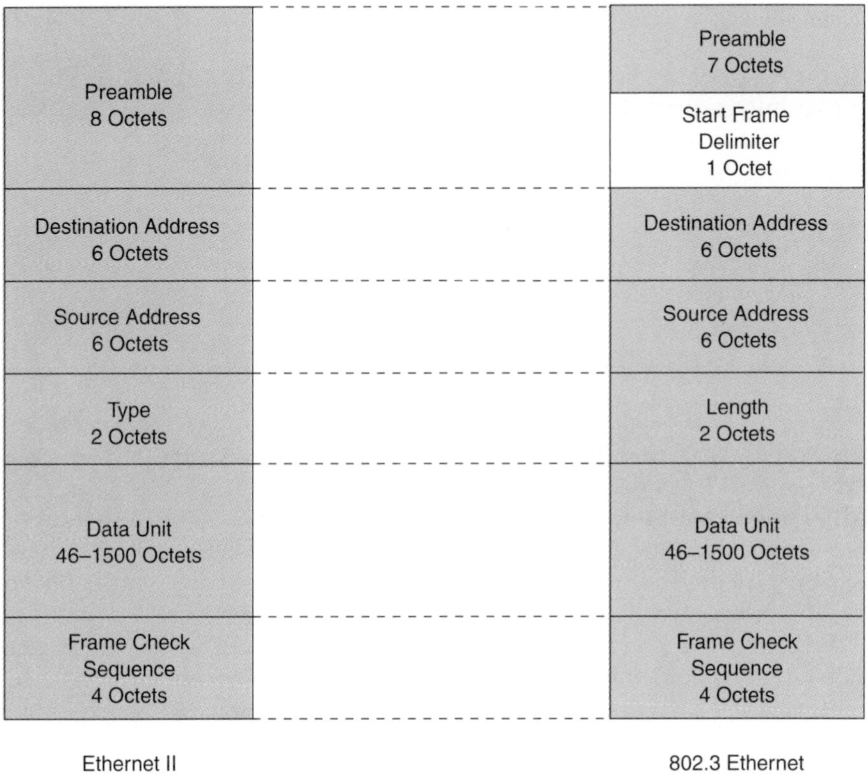

Ethernet II		802.3 Ethernet
Preamble 8 Octets		Preamble 7 Octets
		Start Frame Delimiter 1 Octet
Destination Address 6 Octets		Destination Address 6 Octets
Source Address 6 Octets		Source Address 6 Octets
Type 2 Octets		Length 2 Octets
Data Unit 46–1500 Octets		Data Unit 46–1500 Octets
Frame Check Sequence 4 Octets		Frame Check Sequence 4 Octets

the length in bytes of the LLC data field. Because the LLC data field has a maximum length of 1500 octets, it is easy to distinguish an 802.3 frame from an Ethernet II frame by examining the value of the type/length field. If the value is 1501 or greater, the frame is an Ethernet II.

• The data fields have different names but are otherwise identical. The 802.3 standard calls it the LLC data field because it contains data that is passed down from the logical link control (LLC) sublayer. Data fields have a minimum length of 46 octets and are padded with extra data if they fall short of the minimum. The maximum length of an Ethernet data field is 1500 octets.

• The frame check sequence field contains a value that is used to detect transmission errors. Before it transmits a frame, a device calculates a mathematical value called the cyclic redundancy check (CRC), discussed in Chapters 2 and 7. The receiving device calculates the CRC for the frame it receives; if the calculated CRC matches the value in the CRC field of the received frame, then it is assumed that the frame was transmitted correctly.

High-Speed Ethernet A significant marketplace change has recently taken place. Fast Ethernet has become very popular, driven by dozens of vendors and hundreds of products. Fast Ethernet offers 100-Mbps speed technologies similar to 10BASE-T. Three years ago, the technology was experimental and expensive, but now 100-Mbps Ethernet is available at attractive prices. Fast Ethernet hubs are available for less than $200 per port.

Organizations will often establish a mix of 10-Mbps and 100-Mbps Ethernet. Slower Ethernet is used for workstations with conventional needs. Fast Ethernet is used for workstations that need greater LAN bandwidth, and also for establishing high-speed backbones that interconnect servers. Two technologies are competing for your attention in the 100-Mbps Ethernet area: 100BASE-T, the IEEE standard derived from 802.3 Ethernet; and 100VG-AnyLAN, which uses completely different technology.

Do you need 100-Mbps LAN? First, you need a large number of high-performance workstations equipped with PC network cards. An ISA bus connection can't use the bandwidth available in 10-Mbps Ethernet, and a 100-Mbps connection would be a wasted expense. Second, you need applications that cannot be resolved using a well-designed 10-Mbps network in which traffic is well managed. Video conferencing is a good example of an application that needs a lot of bandwidth.

100BASE-T 100BASE-T was standardized by the IEEE 802.3 committee and is part of the established IEEE Ethernet standards. Three variations of 100BASE-T are defined:

- 100BASE-TX uses two pairs of category 5 unshielded twisted-pair (UTP) cable.
- 100BASE-T4 uses four pairs of category 3, 4, or 5 UTP.
- 100BASE-FX uses multimode optical fiber.

Ethernets at speeds of 10-and 100-Mbps can coexist on the same network. Ethernet switches that support ports with both speeds are available. These 10/100-Mbps Ethernet switches can be used to organize the network into different speed segments or as tools for migrating the network to 100-Mbps speeds.

100VG-AnyLAN Although 100VG-AnyLAN is not an Ethernet technology, it is aimed directly at the 100BASE-T market. The technology for 100VG-AnyLAN was developed and promoted by Hewlett-Packard and was standardized by IEEE 802.12. At this writing, Hewlett-Packard is the only major hub manufacturer supplying 100VG-AnyLAN products. The technology is supported by several smaller vendors.

The access method used is called *domain-based priority access,* which eliminates collisions. This enables 100VG-AnyLAN networks to exceed the size limits imposed on 100BASE-T and greatly reduces the need for repeaters.

Although 100VG-AnyLAN operates on completely different principles than Ethernet, it supports Ethernet frame formats and can be integrated with standard Ethernet networks. 100VG-AnyLAN also supports token-ring frames, and can be used as a high-speed LAN that integrates Ethernet and token ring network segments. 100VG-AnyLAN may, therefore, be a better choice than 100BASE-T in environments that include both Ethernet and token-ring LANs.

Gigabit Ethernet   In late 1995, the IEEE 802.3 began investigating means of transferring Ethernet packets at speeds in the gigabit per second range. The idea behind the gigabit Ethernet is similar to those of the 10- and 100-Mbps Ethernets. Although it defines new medium and transmission specifications, gigabit Ethernet retains the CSMA/CD media-access method and the Ethernet frame format of its 10- and 100-Mbps predecessors. It is compatible with 100BASE-T and 10BASE-T, thus preserving a smooth migration path. As more organizations move to 100BASE-T, increasing the traffic load on backbone networks, demand for gigabit Ethernet has intensified.

The medium and transmission specification for gigabit Ethernet calls for the use of optical fiber over relatively short distances, although unshielded and shielded twisted pairs are also allowed.

9.8.2 Token Ring

Another popular type of LAN is token ring, developed by IBM. Although Ethernet LANs have a broader appeal due to multivendor support and lower cost, both LAN types have a strong following. Token ring LANs are especially popular in corporations that have IBM mainframes and equipment, since token ring is an IBM-strategic LAN.

Token ring LANs are configured in a star-wired topology as shown in Figure 9–21. Note that in this arrangement, the systems do not form a physical ring directly with one another. All stations are connected to hub devices, called *multistation access units (MAUs)*, much like in a star topology. Internally within the MAU, the stations are linked into a logical ring.

This configuration has several advantages. For one, the wiring is much easier because all stations are connected to one or more centralized MAUs, which would typically be located in a wiring closet. This makes it easy to plug stations into and out of the LAN without disrupting its operation. It also makes it easier to bypass stations in the case of failures. A station going down or being turned off does not take the entire ring down.

The token ring LAN uses a token-passing media-access method that is specified in the IEEE 802.5 standard. Token ring LANs operate at 4 or 16 Mbps over a variety of media, including unshielded and shielded twisted-pair cables.

The IBM Cabling System   Unlike the IEEE 802.3 standard, the 802.5 token-ring standard does not specify a cabling technology. Therefore, it is common practice for vendors to base their equipment design on the IBM cabling system. IBM has specified a wide variety of cables for its token ring network. The most common cable types follow:

- *Type 1.* Includes two data-grade twisted pairs consisting of solid copper conductors. Both pairs are enclosed by a single shield.
- *Type 2.* Includes two solid-conductor, data-grade twisted pairs enclosed by the same shield. In addition, the cable incorporates two voice-grade cables.

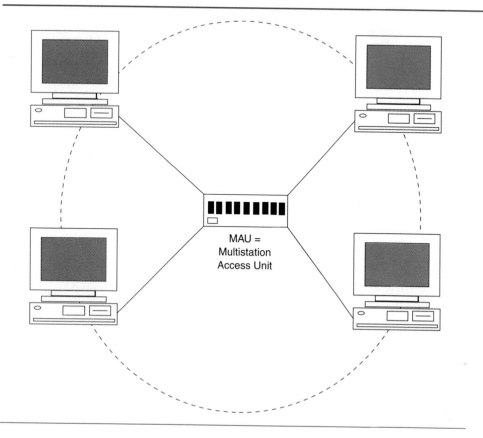

Figure 9–21
Token ring LAN.

MAU =
Multistation
Access Unit

- *Type 3.* An unshielded twisted-pair cable similar to category 3 UTP. IBM specifications originally limited use of type 3 cable to 4-Mbps token ring.
- *Type 6.* Includes two data-grade twisted pairs constructed with strand wires. The strand wires make type 6 cables more flexible and, therefore, better suited for cables that are frequently moved. Strand cables, however, cannot carry signals as far as cables with solid conductors.

Cable types 1, 2, and 6 are fairly thick and use a special IBM data connector (see Figure 9–22). The connector is fairly bulky but has several interesting features. When disconnected, it shorts pairs of conductors together, enabling you to disconnect devices from the network without introducing a break into the ring.

IBM's token ring cabling uses a type of hub called a *multistation access unit* (*MAU* or *MSAU*), as described previously. The original IBM MAU was called 8228. It was a passive hub that used relays (electromechanical switches) to control

Figure 9–22
An IBM data
connector.

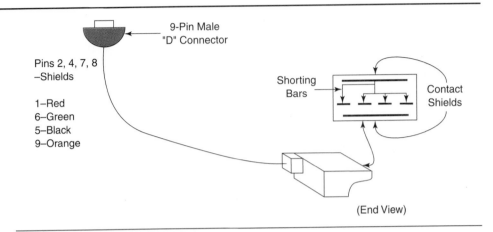

network signal flow. Although 8228 is still available, more recent MAU designs invariably use active electronic switches, replacing the relays. Active hubs are desirable because they enable hubs to correct signal distortion, to detect errors, and to be managed from a network management console.

Figure 9–23 shows how MAUs are used to build a network. The 8228 has eight ports, but modern MAUs often have 16 or more ports to which devices can be connected. Each MSAU has two special ports called ring in (RI) and ring out (RO). A ring network is created by connecting the ring out port of each MSAU to the ring in port of the next MSAU in the ring.

Token networks have a ring logical topology, but they are wired as physical stars (as indicated earlier). Each device on the network is connected to an MSAU by additional cable. Figure 9–24 shows how the relays in an MSAU operate to configure the network as a logical ring. Relays in the MSAU are normally closed, permitting signals to flow around the ring. When a device connects to the network, the relay for that station opens, forcing signals to flow out to the device. The device returns the signal to the MSAU, and the signal continues around the logical ring.

STP Token Ring   Originally, IBM sanctioned two speeds for token ring operation. IBM specified 4-Mbps data rates for use with shielded twisted pair (STP) and unshielded twisted pair (UTP) cabling. STP cable was required for 16-Mbps data rates. Type 1 and type 2 cables are commonly used for running cables to workstations. These cables are called *lobe cables* in IBM token-speak. In general, lobe cables are limited to 100 meters.

Up to 260 devices and up to thirty-three 8228 MAUs can be attached to a ring. Repeaters that count as ring devices can be used to extend the physical size of a ring. The actual number of devices a ring can accommodate depends on the ring size, the number of MAUs, and the length of the longest cable. Figuring out the capacities of a token ring is a fairly involved process. One IBM

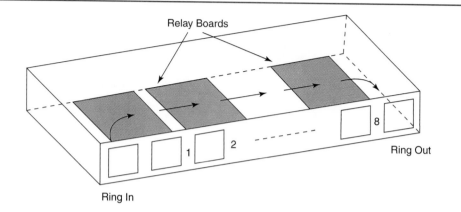

Figure 9–23
How devices and
MSAUs are con-
nected to build a
token ring network.

(a) 8228 Multistation Access Unit
Uses Internal Relays to Break
the Network Ring and Insert
an Operating Workstation

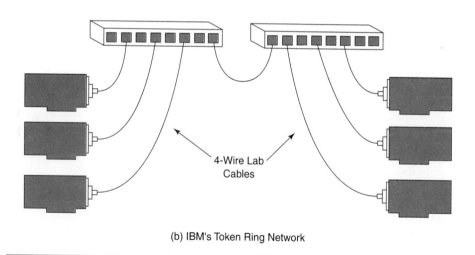

(b) IBM's Token Ring Network

token ring network planning guide devotes 22 pages to the subject of ring capacities and dimensions.

Most networks are now cabled with active MAUs. IBM active token ring hubs are called *controlled access units (CAUs)*. Active hubs and heavy use of UTP cable have rendered many of the old guidelines obsolete. The manufacturer's specifications of your network's token-ring components will be a better guide to the capacities you can expect on your network.

Figure 9–24
Operation of an 8228
MAU.

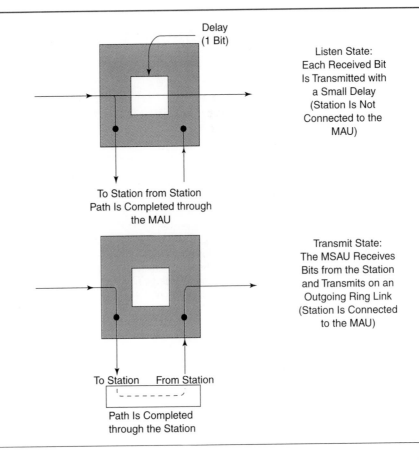

Delay
(1 Bit)

Listen State:
Each Received Bit
Is Transmitted with
a Small Delay
(Station Is Not
Connected to the
MAU)

To Station from Station
Path Is Completed through
the MAU

Transmit State:
The MSAU Receives
Bits from the Station
and Transmits on an
Outgoing Ring Link
(Station Is Connected
to the MAU)

To Station   From Station

Path Is Completed
through the Station

Originally IBM sanctioned UTP cable only for 4-Mbps data rates. In the early 1980s, numerous IBM engineers expressed the opinion that 16-Mbps data rates over UTP cable would never be practical.

UTP Token Ring   Other vendors, however, did not listen to that theory and did not wait for IBM to develop a 16-Mbps UTP specification. Vendors such as Proteon delivered MAUs that could support 16-Mbps token ring networks with UTP a couple of years before IBM announced comparable products. In fact, IBM, the inventor of token ring, was among the last manufacturers to develop a 16-Mbps token ring network that satisfied customer demands for UTP. UTP now has wide support for use at both 4-Mbps and 16-Mbps data rates. RJ-45 connectors are used with UTP cables.

In order to control noise on the network, token-ring devices must be connected to the UTP cable by a media filter. On older token-ring NICs, the media filter was often an extra item. Most currently manufactured NICs include the media filter on the network board and provide an RJ-45 connector for a network connection.

Because most token-ring NICs and MAUs now support 4-Mbps and 16-Mbps data rates, there is no cost advantage to opting for 4-Mbps operation. The 16-Mbps token ring has been widely available since the early 1990s, and can now be regarded as a mature technology.

9.8.3 LocalTalk

LocalTalk was developed by Apple Computer, which built LocalTalk adapters into all Apple Macintosh computers, LaserWriters, and other equipment. This makes LocalTalk LANs inexpensive as well as very easy to install. All you need is cables to connect the equipment that already has built-in LocalTalk adapters.

LocalTalk

LocalTalk was designed for small LANs and, therefore, has some limitations. For example, LocalTalk operates at 230.4 Kbps. This is sufficient for file and printer sharing, but cannot handle the demands of a large number of systems or the bandwidth requirements of graphic, image, video, or multimedia applications. LocalTalk best supports a small number of users, so it is utilized mainly in small businesses or departments in larger companies.

LocalTalk LANs typically use shielded twisted-pair cables and employ the carrier sense multiple access with collision detection (CSMA/CD) media-access method similar to what was described earlier. LocalTalk is an Apple-proprietary LAN that has no corresponding IEEE LAN standards.

9.8.4 FDDI

Because they are based on fiber optic technology—that is, light signaling rather than electrical signaling—*fiber-distributed data interface (FDDI)* LANs can operate at very high speeds. This makes them capable of supporting a large number of users and applications that require very high bandwidth (e.g., video, multimedia, and image-based applications). The original FDDI specification called for operation at 100 Mbps, now a 200-Mbps version has also been defined. FDDI uses two fiber optic rings, as shown in Figure 9–25.

FDDI

The most common use of FDDI LAN technology is to create corporate high-speed backbone LANs to interconnect departmental Ethernet or token ring LANs. FDDI LANs can also span long distances, another characteristic of fiber optic technology. Figure 9–26 illustrates the connectivity of FDDI as a backbone network.

The major disadvantage of fiber-distributed data interface (FDDI) LANs is that they are still relatively expensive compared to other types of LANs. Cost should drop, however, as FDDI LANs become more widespread.

One feature to note is the manner in which FDDI networks can incorporate two rings that are configured to route signals in opposite directions. Devices that attach to both rings are called *dual-attached stations.* Devices that attach to only one ring are called *single-attached stations.*

Because of the way in which FDDI rings are configured, the network can recover from a cable break between any two dual-attached stations. Figure 9–27a shows what happens when a link is broken. The two stations next to the break rework their connections so the two connector-rotating rings are reconfigured into a single ring.

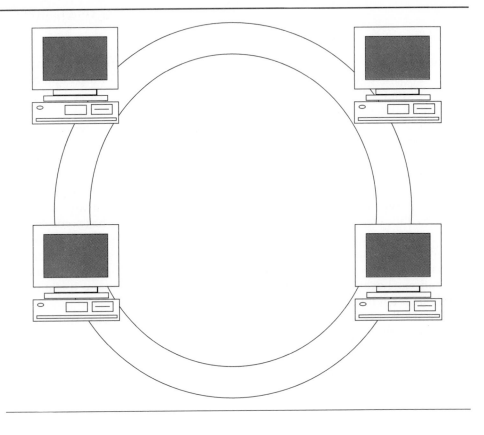

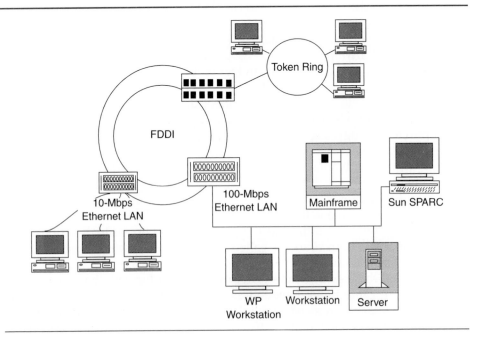

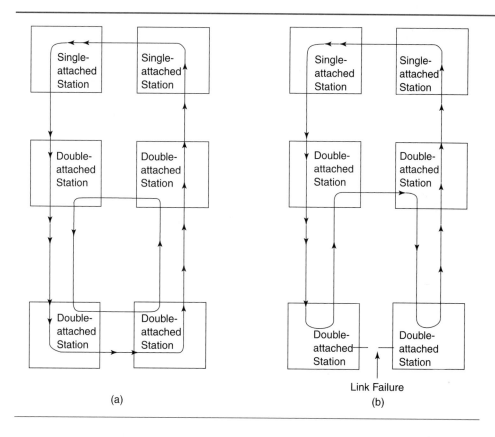

Figure 9–27
Operation of an FDDI
network: (a) Normal
operation with
single- and dual-
attached stations,
and (b) dual-
attached FDDI
stations correcting
for a failed link.

FDDI is not well suited to workstations because it does not expect devices to be turned on and off. Also, only the fastest PCs can keep pace with FDDI networks due to the limitations of most PC I/O bus designs. FDDI does make an excellent high-speed backbone, however. Also, because dual-attached FDDI rings can extend to 100 kilometers in diameter, they can be adapted for use as a campus or metropolitan area network.

FDDI is superior to fast Ethernet (100-Mbps) in much the same way that token ring is superior to standard Ethernet. If your organization needs a deterministic high-speed network that is fault-tolerant and highly manageable, FDDI is worth considering.

9.9 LAN NETWORK OPERATING SYSTEMS

It is one thing to physically connect systems and devices to create a LAN, but it is another to make them all work together. The hardware components, such as network interface cards and cables, provide the transmission infrastructure for sending and receiving signals, but a lot more is needed. For example, there needs to be a means for making sense of transmitted signals. How is data to be interpreted and how are commands to be handled? The ability to detect errors and retransmit information is required. Ways to control the information between programs communicating across the LAN must be provided. Finally, some useful work actually needs to be carried out, such as exchanging messages, transferring files, or updating databases.

What provides these kinds of functions? Software does. The protocols used for exchanging information between users on the LAN are implemented in software. In addition, several other functions, known as network services, are provided in LANs and implemented in software.

network operating system

Many LAN functions that we will discuss are provided by specialized software called the *network operating system (NOS).* As indicated in its name, a network operating system extends the function of the operating system (OS) used to internally manage a computer and its resources, such as disks and printers. The NOS extends OS functions to a network environment and supports any type of LAN hardware. Typical NOS functions include:

- Print services.
- File and database services.
- Remote access services.
- Messaging services.
- Network management services.
- Communications services.

Print services allow information to be passed from stations on the LAN to printers attached elsewhere on the LAN. Information can then be directed to any printer in addition to one that may be locally attached to the computer. It also allows a printer to be started by multiple stations on the LAN. Print services usually operate as a two-step process. First the information is sent over the LAN from the computer to a *print server*. There it is temporarily saved on disk storage. When scheduling of the print job is complete, the information is directed to the desired printer. This function is called *spooling.* The use of spooling allows stations to send data to printers without worrying about whether the printer is busy, or mixing up the data with data from another source. The print server takes care of keeping the printer streams separate and printing data when the printer is available. Figure 9–28 illustrates the use of a print server.

File and database services provide access to files and databases stored on other systems on the LAN. The systems where the files or databases reside and that provide such services are called *file* or *database servers.* Typically, the file or database server includes relatively large amounts of disk storage that can be shared by other systems on the LAN. This capability eliminates the need for each system to have large storage capacity. It can also avoid duplication of files and databases. In addition to storing the files and databases, these servers also monitor access to the data. This task is accomplished by use of login names and passwords. Access to information on the server can be limited by the system administrator to only those who require it.

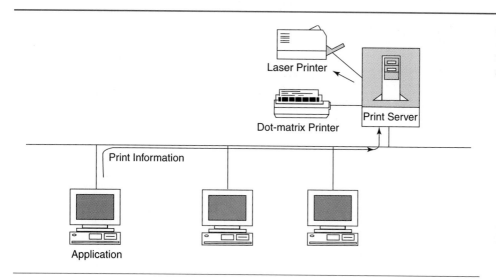

Figure 9–28
Printing services of
the network oper-
ating system.

Network operating systems also provide services allowing programs on one system to access programs and resources on other systems. This remote access capability allows users and applications running on the system to interact with applications that reside on other systems.

Messaging services are often provided in a network operating system. These services allow users on different systems to exchange messages. Users can send messages to one or more other users and can receive messages from multiple sources. The NOS typically maintains mailboxes of received messages that users can check when they sign on to their system and retrieve their messages.

Network management is an important issue in LANs, and some network operating systems provide network management support. Statistics covering operating characteristics such as media errors, collisions, and retransmissions are compiled and can be reported. Additionally, file and database servers might also keep statistics to report disk media errors, file access information, or security violations. See Figure 9–29.

When reported and analyzed, network management information can be used to identify failing components, analyze usage statistics and predict future capacity requirements, and spot security breaches. As a result, network management has become an area of in-house interest as LANs become larger and more critical to the operation of corporate computing.

Popular network management protocols such as the simple network management protocol (SNMP) or OSI-based common management information protocol (CMIP) are supported for this purpose.

Network operating systems may also provide communications services, permitting communications outside of the LAN. The communication may be with other LANs (LAN to LAN), with microcomputer or mainframe systems,

Figure 9–29
Network management services.

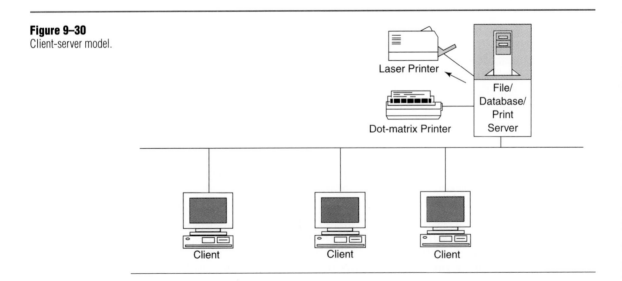

Server

Computer Running
Network Manager
Software

or across wide area networks (LAN to WAN). This is done through support for widely used protocols such as TCP/IP, SNA, X.25, and others.

Most network operating systems are structured on the client-server model, as shown in Figure 9–30. One or more systems act as servers to other systems on the LAN. Servers include print, file, database, and communication servers. The systems that request these types of services are called *clients*, and servers communicate across the LAN using a particular set of protocols. The protocols vary depending on the type of system, NOS, server, and application.

Figure 9–30
Client-server model.

Laser Printer

File/
Database/
Print
Server

Dot-matrix Printer

Client   Client   Client

Many different network operating systems exist on a variety of platforms. The most widely used PC-based network operating systems are Novell Net-Ware, Microsoft LAN Manager, IBM OS/2 LAN Server, and Banyan VINES. Choosing a network operating system involves a number of variables, including the types of systems in use, the size and geographical scope of the LAN, its expected growth, types of application usage, speed and bandwidth requirements, and others.

Some network operating systems are designed for small LANs and are known as *peer-to-peer networking systems.* They are most appropriate for small workgroups or office environments where the need is simply to share printers or disks. These systems would not be appropriate where more extensive networking facilities are required to support many users or provide connectivity to a wide range of other networks and systems.

Other network operating systems, called *client-server systems,* support more robust requirements. These systems are appropriate for large LANs, but they may be overkill for smaller LANs. They are not only more expensive than peer-to-peer networks, but they also require more powerful platforms and dedicated servers, which also adds to the cost of the LAN.

NetWare, provided by Novell, is a widely used LAN network operating Netware system. It is function-rich, runs on a range of computers, and supports several different operating environments, including DOS, Windows, OS/2, UNIX, and the Macintosh. It was originally developed by Novell in 1983 as the server operating system for their S-Net, a star-based network with a proprietary server, and then ported to the Intel 8088 processor so it could be used on PC XTs. Since then, NetWare has been rewritten to be media-independent, so that it runs over all popular network hardware, including Ethernet, token ring, Arcnet, and others. It has also been enhanced to take full advantage of the 386, 486, and Pentium processors.

NetWare is available to run on IBM's system/370/390 mainframes and AS/400 systems. A version of NetWare also runs on DEC VAX systems and many manufacturers' UNIX-based systems. Its support on a wide range of platforms makes NetWare a popular choice in a corporate environment. With non-PC systems it is relatively new, and NetWare is still widely used in 80×86-based PCs. Because of its relative maturity as a product, NetWare provides a wide range of services:

- File and print services.
- Database services.
- Communication services.
- Messaging services.
- Network management services.

NetWare services are packaged as add-on software products called *value-added processes (VAPs)* and *NetWare loadable modules (NLMs).* They provide a modular approach to assembling the software functionality of a server. For example, client-server applications run as NetWare loadable modules. Backup and restore functions are implemented in NetWare v3.11 by NetWare Backup NLM.

NetWare supports a number of application programming interfaces (APIs), including NetBIOS, DOS, and OS/2 Named Pipes. Programmers use these APIs to build sophisticated, multiuser, network-based applications. NetWare also supports a wide range of networking protocols such as TCP/IP, SNA, and others.

The Microsoft LAN Manager was developed by Microsoft in conjunction with 3Com and first appeared in 1988. IBM also adapted the LAN Manager as the basis of its OS/2 LAN server offering. Microsoft has now taken most of the LAN Manager functionality and rewritten it to run on a Windows/NT server.

VINES, developed by Banyan Systems, uses the UNIX operating system on the server. It is designed along the lines of the OSI reference model. VINES is best known for its extensive support of large, geographically distributed networks incorporating many servers. Due to this extensive internetwork support, VINES provides seamless access to network resources, whether they are located on the same LAN or another LAN and accessed across a backbone LAN or a WAN.

VINES' sophisticated naming service, known as Street Talk, keeps track of many users and their resources on multiple LANs. These LANs are usually interconnected via LAN or WAN backbones. This ability allows a user to log on to the network anywhere and retrieve mail or access files, even if that information is physically located in another city or country. VINES supports DOS, Windows, OS/2, and Macintosh requesters, in addition to UNIX requesters.

9.10   LAN PROTOCOLS

LAN protocols

Many different protocols can be used to manage the exchange of information among users connected on a LAN. The *LAN protocols* define the rules of interaction allowed among users. For example, rules are defined to indicate when a user is allowed to send information, how errors are reported, and how much information can be sent at any one time. Without these rules, LAN operation would be chaotic. The rules are implemented using specific commands and control messages that indicate the types of things described earlier. If such control were not used, messages could get lost or become out of sequence, users may not be prepared to receive information, errors could go undetected, and the network could be clogged with traffic.

Different protocols have their own unique ways of providing these kinds of functions and services. The most popular protocols used in LAN environments are:

- NetBIOS.
- TCP/IP.
- XNS.
- SPX/IPX.

Let's take a look at each of them now.

9.10.1 NetBIOS

NetBIOS provides services similar to the basic input output system (BIOS) services provided by DOS. Whereas DOS's BIOS offers services for accessing local devices such as disks and printers, NetBIOS provides a similar set of services for accessing devices connected to a LAN. NetBIOS became the *de facto* standard for DOS-based networks when IBM incorporated it into its PC network in 1984. NetBIOS made use of a proprietary set of protocols, and the protocols' support was implemented on some of the PC LAN adapters. Because of its early availability with DOS-based PCs and its adoption by IBM, NetBIOS achieved widespread use. Figure 9–31 shows how NetBIOS fits within the DOS operating system.

NetBIOS became popular as other network vendors emulated its functions, allowing applications using NetBIOS services to work on other LANs. NetBIOS is now used as an application program interface (API) between programs running over a LAN. It makes the underlying LAN transparent to applications. A NetBIOS interface provides access to a set of basic services, which includes general control services, name support services, datagram support services, and session support services.

The general control services include miscellaneous types of networking services, such as status-of-network hardware adapters and canceling previously issued commands.

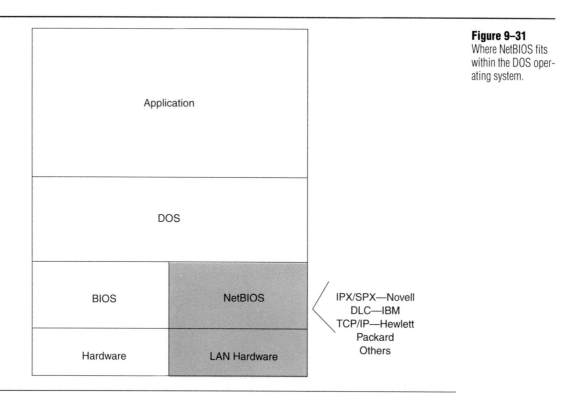

Figure 9–31
Where NetBIOS fits within the DOS operating system.

Name support services are used to assign and delete names to and from local name tables. NetBIOS names are used to identify logical entities such as programs or processes within a program. These names must be registered before any data can be exchanged. Programs then communicate with one another over the network by referring to each other by name. The name support services include the following:

- Add names.
- Remove names.
- Show list of names.

The NetBIOS datagram services provide transfer services that are *unreliable*. As you may recall, unreliable does not literally mean that; it means there may be errors, but we don't check for them. The datagrams are sent on a best-effort basis with a guarantee that they will be successfully received. Individual data packets are not sequences and, therefore, are not tracked or acknowledged. Datagram services include:

- Send packet.
- Receive packet.

NetBIOS session support services provide a reliable, connection-oriented data transfer service. With connection-oriented operation, the sender and receiver first set up a logical connection, sometimes called a *session*, between themselves before they exchange information. These connections are established by using the names registered with the name support services mentioned earlier. The packets of information exchanged across a session are assigned sequence numbers by NetBIOS. The sequence numbers are used to track and acknowledge packet receipt as the packets flow across the LAN. The NetBIOS services ensure that packets are received without errors, in the same order that they are sent, and that duplicate packets are automatically discarded. Session services include:

- Establish a connection (call another station).
- Send data.
- Receive data.
- Terminate a connection (hang up).

All this checking may seem like overkill; however, it is necessary when the two communicating programs are running on computers located on different LANs, and the information must flow over a WAN. In the telephone network that carries much of the WAN traffic, data errors are not uncommon, and single packets may be lost or damaged. It is also possible in some WANs for the packets to arrive out of sequence. The session support services both detect missing information and request its retransmission, as well as reorder any out-of-sequence packets.

Figure 9–32 shows the concept of the application program interface with NetBIOS. It demonstrates that network services can be used through both intercepted DOS calls and direct network calls. NetBIOS is a significant advancement over the board-level programs written to control and manage network adapter cards. Programmers no longer have to worry about the type of network adapter card being used, the size of packets allowed, or proper reception of messages; nor do they have to worry about the hexadecimal address of each station. These are some of the extra services offered by NetBIOS. There are disadvantages of course—extra processing, network traffic increases, and quite often, additional memory use.

NetBIOS vaguely fits into the OSI model as a horizontal line between the session and presentation layers. It is an interface specification, not a description of how protocols work or how packets are formed. On more recent IBM networks, NetBIOS is implemented by NETBEUI protocol. On UNIX networks, NetBIOS, if used, rides on top of TCP/IP. In NetWare, it rides on IPX. In short, NetBIOS is not the key to connecting diverse computers and networks.

9.10.2 TCP/IP

Transmission control protocol/Internet protocol (TCP/IP) refers to a suite of protocols originally developed for the Department of Defense to interconnect different government and university research networks. Thus, TCP/IP was designed

TCP/IP

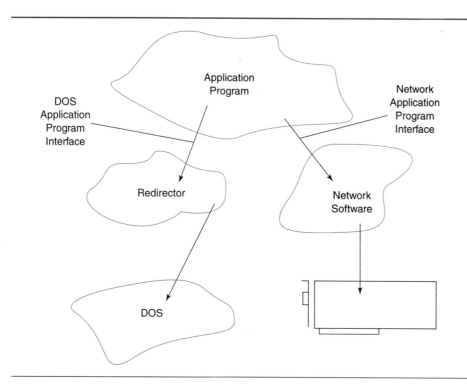

Figure 9–32
Network services can be used through both intercepted DOS calls and direct network calls.

Figure 9–33
Relation of TCP/IP
to the OSI reference
model.

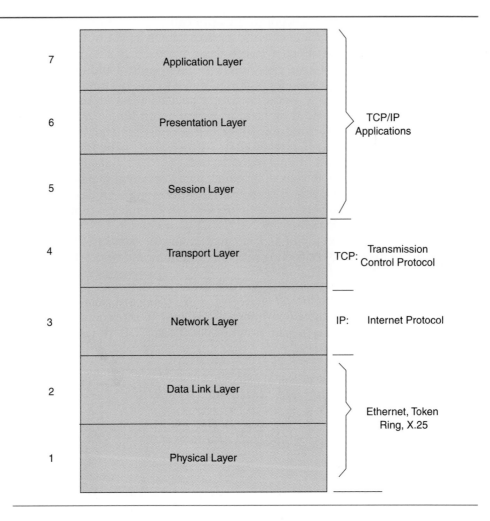

for internetworking. Today it is the most widely used set of protocols for tying together multiple networks.

TCP/IP is ideally suited for connecting multiple physical LANs, but it can also be used within a single LAN. The TCP/IP protocol suite includes a range of protocols that supports the exchange of information among users connected via either LANs or WANs. Figure 9–33 shows the relationship between TCP/IP and the open systems interconnection (OSI) reference model.

TCP and IP are two of the protocols used in the TCP/IP environment. The term, *TCP/IP*, typically refers to the entire suite of protocols and applications shown in Figure 9–34.

In terms of using TCP/IP protocols in LAN environments, a few points should be noted. As shown in Figure 9–35, the TCP/IP protocol suite does not address the lower networking layers. These lower layers are network dependent

APPLICATION	PROTOCOL
Mail	Simple Mail Transfer Protocol (SMTP)
Remote Terminal or Display	Telnet, X-Window
File Transfer	File Transfer Protocol
Directory	Domain Name Service (DNS)
File Service	Network File System (NFS)
Program-to-Program Communication	Remote Procedure Call (RPC)
Security	Kerberos

Figure 9–34
TCP/IP applications.

and are determined by the particular hardware being used. That is, each individual network across which TCP/IP protocols are used provides its own set of lower-layer protocols. TCP/IP, therefore, can be used between or among different kinds of LANs and WANs in conjunction with different lower-layer protocols. For LANs, this means that TCP/IP protocols can be used on top of Ethernet, token ring, or other LAN hardware. We will say more about TCP/IP in Chapter 10.

9.10.3   XNS

Xerox network systems (XNS) is a suite of protocols introduced in the early 1980s, soon after the introduction of Xerox's Ethernet support. Although they are proprietary, Xerox has licensed the protocols to other vendors, and they are used in a wide variety of network equipment and applications.

As an historical note, these protocols were developed at Xerox's Palo Alto Research Center (PARC) in California. This is the R&D center acknowledged to be the birthplace of Ethernet, as well as the idea that later became Apple's Macintosh computer.

Figure 9–35
TCP/IP architecture.

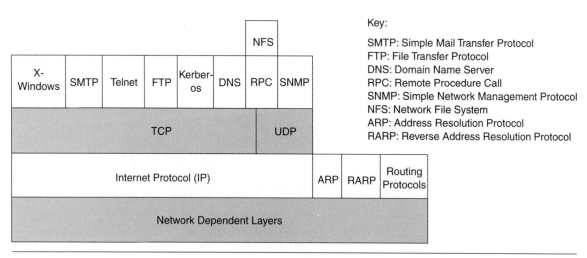

Key:

SMTP: Simple Mail Transfer Protocol
FTP: File Transfer Protocol
DNS: Domain Name Server
RPC: Remote Procedure Call
SNMP: Simple Network Management Protocol
NFS: Network File System
ARP: Address Resolution Protocol
RARP: Reverse Address Resolution Protocol

The XNS suite covers five different levels of protocols. The levels are similar to the layers of the OSI reference model, which describes a seven-layer networking model intended to serve as a guide for developing standards to allow communications between dissimilar systems. Although the number of layers differs, XNS corresponds closely to the OSI model from a functional point of view. Figure 9–36 shows the relationships between the OSI layers and the XNS protocol suite.

The lowest XNS level, level 0, includes functions of OSI's physical and data link layers (layer 1–2). XNS level 0 deals with the physical transmission of data over some type of transmission medium, such as coaxial cable or twisted-pair cables in the case of LANs.

XNS protocols can be used in LANs or WANs. For LANs, the CSMA/CD media-access control method is supported at level 0. For WANs, the RS-232-C interface is supported along with X.25 protocols.

Figure 9–36
XNS protocol suite in relation to the OSI reference model.

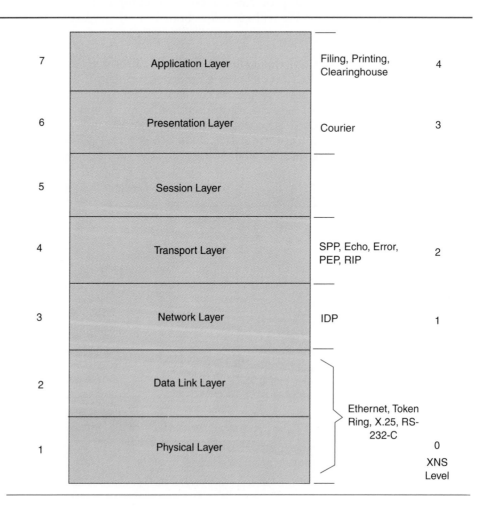

XNS level 1 corresponds to the network layer (layer 3) of the OSI reference model. This level is responsible for routing data through a network or Internet. The one protocol specified by XNS at this level is called the Internetwork datagram protocol (IDP). A number of protocols are included at XNS level 2, which corresponds to the OSI transport layer (layer 4). These protocols are used to give structure to streams of related packets exchanged among users. Functions handled at this level include retransmitting packets received with errors, handling packet sequencing and suppressing duplicate packets, and providing flow control mechanisms. Level 2 protocols include the echo protocol (Echo), error protocol (Error), sequence packet protocol (SPP), packet exchange protocol (PEP), and routing information protocol (RIP).

Level 3 corresponds to the OSI presentation and application layers (layers 6 and 7). There is no XNS level that corresponds to the OSI session layer (layer 5). XNS level 3 defines a set of control protocols that establish conventions for data structuring and process interaction. Protocols that operate at level 3 include Courier (operates at OSI layer 6).

XNS level 4 corresponds to the OSI application layer (layer 7). It is at this level that application programs operate and application-level protocols are handled. Xerox's filing, printing, and network directory applications run at this level.

9.10.4   IPX/SPX

Internet packet exchange (IPX) and sequenced packet exchange (SPX) are key networking protocols supported in Novell's NetWare. Because of the huge installed base of NetWare systems, IPX/SPX is the most widely used LAN protocol. Internet packet exchange (IPX) is based on the XNS IDP protocol, and the packet formats used by IPX are the same as those used with IDP.

IPX is a connectionless, datagram-oriented service that handles routing of data through an Internetwork. Because of its connectionless operation, it does not require the sender and receiver to first set up a logical connection before they exchange information. The individual packets of information exchanged are called *datagrams*. As a datagram service, IPX does not guarantee delivery   datagram
of packets. Each packet is treated as an independent entity and is transported over the network on a best-effort basis. When the destination device or computer is on the same LAN, datagram service is very efficient. From a practical standpoint, packets are rarely lost or damaged during transmission over a single LAN. (Figure 9–37 shows the relationship between IPX/SPX protocols and the OSI reference model.)

Since IPX does not guarantee delivery of packets, higher-layer protocols must provide the services for reliable delivery. Such services include packet retransmission requests, packet acknowledgment, flow control, and error detection. In NetWare, this is provided by SPX—Novell transport layer protocol.

SPX is a connection-oriented transport protocol providing reliable data stream delivery. SPX ensures that logical connections are established between sender and receiver prior to data exchange. Once a connection is established, data can be transferred. SPX guarantees that packets are delivered in the same sequence they were submitted and without error.

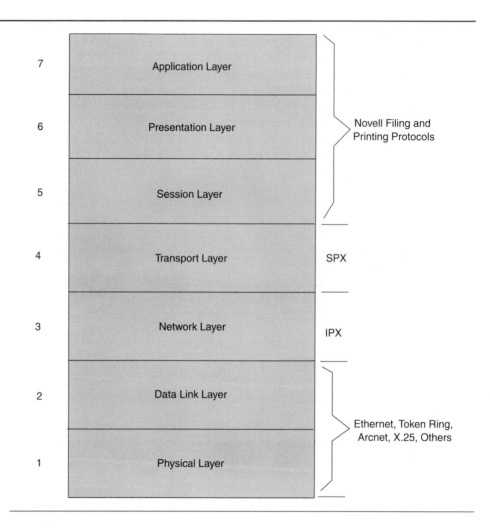

Figure 9–37
Relationship between IPX/SPX and the OSI reference model.

7	Application Layer	
6	Presentation Layer	Novell Filing and Printing Protocols
5	Session Layer	
4	Transport Layer	SPX
3	Network Layer	IPX
2	Data Link Layer	Ethernet, Token Ring, Arcnet, X.25, Others
1	Physical Layer	

The relationship between SPX and IPX is similar to the relationship between TCP and IP, and SPP and IDP of XNS. As with the TCP/IP and XNS protocols, IPX and SPX can be used in both LAN and WAN environments. Because network protocols are layered, many network applications can run independent of the underlying networking protocols used (see Figure 9–38).

9.11 LINKING LANS

There are a number of reasons why it may be desirable to link multiple LANs together. These reasons include extending the range of the LAN and increasing the number of users supported. The reason why single LANs cannot always satisfy these needs is that LANs have limits on the distances, transmission

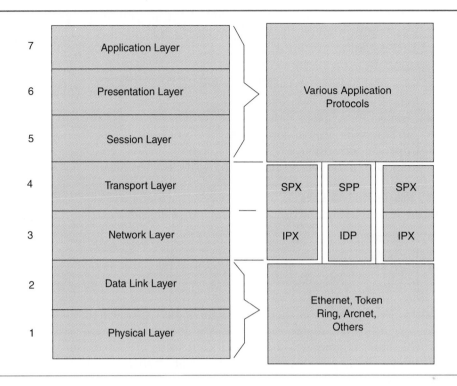

Figure 9–38
OSI Reference
Model: TCP/IP, XNS,
and IPX/SPX.

speed, and number of devices supported. These limitations vary by the type of LAN, transmission medium used, and other factors. For example, a standard thick Ethernet coaxial cable is limited to a length of 500 meters. What if there are users outside of this distance range? A way to connect users at distances greater than the LAN supports is to link multiple LANs, thereby extending the total area covered.

LANs also have limits on the number of devices or computers that can be supported. When more connections are required than the media or topology supports, these other devices can be connected in another LAN and two LANs can be linked together.

Another reason for linking LANs is to improve overall performance. If the traffic is heavy on a single LAN, resulting in poor performance, a solution may be to divide the LAN into multiple independent LANs (or segments) and then link them together. This action increases total throughput potential.

LANs with very different characteristics may also need to be tied together. For example, some LANs may only need to support a small number of users and relatively low transmission speeds. The desire may be to keep the cost down for these LANs. On the other hand, there may also be a need for higher-speed LANs connecting a larger number of users. Linking these different LANs allows the users on each LAN to communicate and share data.

For cost reasons, it may be desirable to have several lower-speed, lower-cost LANs that are linked by a higher-speed backbone LAN. The interconnected LANs can be treated as a single logical LAN providing communications among users on all the LANs.

The reasons to link LANs can be summarized as follows:

- Extend the distance.
- Connect more users.
- Add more devices or computers.
- Improve performance.
- Integrate LANs with different physical chracteristics.

9.11.1   How Do We Link LANs

LANs are linked using one or more of the following devices:

- Repeaters.
- Bridges.
- Routers.
- Brouters.

repeater

Repeaters   A *repeater* is a device that links two physical LAN segments, thereby extending the distances of the LAN. It takes the signals coming from one LAN cable, regenerates them, and then transmits them onto the other LAN cable(s). See Figure 9–39. The repeaters are transparent to the protocols used for exchanging information across the LAN, which operates totally at the physical level (layer 1) of the OSI reference model. Repeaters are most commonly used in Ethernet networks, as we have seen earlier.

Since repeaters perform no processing other than signal regeneration, they are the least expensive devices to link LANs. However, they are limited to connecting LANs with the same physical characteristics. The LANs linked by the repeater must be running the same speed and using the same media-access method.

bridge

Bridges   A *bridge* is the next step up from a repeater on the LAN linking hierarchy. Bridges link LANs at the data link layer (layer 2) of the OSI reference model. Bridges pass signals from one LAN to another, but they perform some additional processing as well. See Figure 9–40.

Figure 9–39
Using repeaters to link LANs.

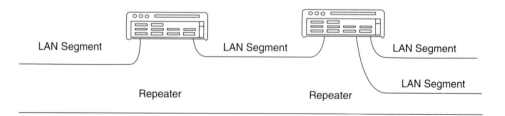

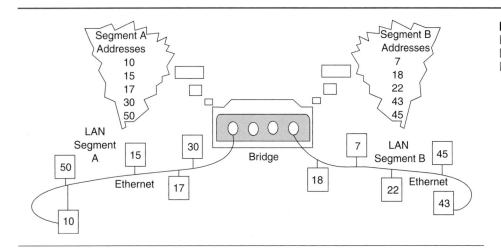

Figure 9–40
LAN bridge links two
LAN segments at
layer 2 level.

A bridge monitors all traffic on both of the LANs that it connects. It builds a list of addresses of devices on each LAN segment. It checks the source and destination addresses in each frame to determine whether that frame should be passed on to the other LAN segment or whether it is destined for a user on the same LAN from which it originated. If the destination address in the frame is for a user who is not on the same LAN as the user sending the frame, the bridge forwards the frame to the other LAN. If the destination address is for a user on the same LAN as the sender, the frame is not forwarded. Basically, the bridge ignores such frames. This function is called *filtering*. The result is that traffic on each LAN segment is reduced, because traffic between two devices on the same segment is not forwarded to the other segment, as with a repeater.

An advantage of bridges over repeaters is that bridges can connect different types of LANs. They can connect LANs that have different data link layers, more specifically different media-access controls. For example, an Ethernet LAN can be connected to a token ring LAN via a bridge (see Figure 9–41). The bridge recognizes both Ethernet and token ring frames. Frames from the Ethernet LAN that are destined for users on the token ring LAN will be reformatted by the bridge to allow transmission on the token ring LAN and vice versa.

Since bridges operate at the data link layer level, they are transparent to higher-layer protocols. They can easily interconnect networks with devices from many manufacturers running different protocols. A problem with bridges is that, because they operate only at the link layer level, and have no routing information, they can create multiple paths to destination networks. An advantage of bridges and routers is that they provide connections across wide area networks. That is, a bridge or router can interconnect LANs across telephone lines or satellite links.

Figure 9–41
LAN bridge connects
Ethernet LAN and
token ring LANS.

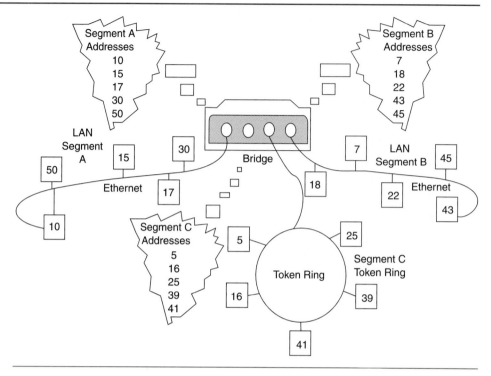

Routers   Some of the limitations of bridges are overcome by routers. A *router* is a device that interconnects multiple LANs at the network layer (layer 3) of the OSI reference model. (You can see that we are again moving up the hierarchy of the layers.) Because it operates at the network layer, a router has access to network addresses and operates with respect to a specific protocol— TCP/IP or IPX/SPX, for example. This system allows routers to act as intermediate forwarding devices. Thus, routers are widely used in internetworking (see Figure 9–42).

Routers can receive information from one link and, based on network addresses, route it over multiple links, even using different lower-level protocols. This ability means that information from one network is stored and then routed along the next step toward its destination network. Bridges, on the other hand, do not use network addresses and are therefore restricted to sending all information across the same data links.

Another advantage of routers is that they can route traffic based on class-of-service requirements. This capability means that they can take into consideration speed and security requirements in determining the appropriate route to take. This information is handled on a logical connection (session) basis. Bridges are not able to route on a session-by-session basis since they only have access to data link information.

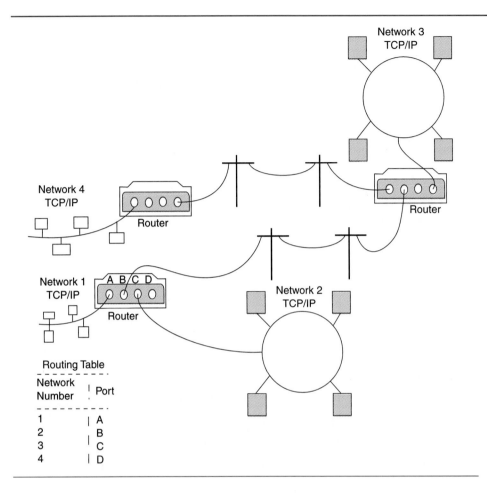

Figure 9–42
Routers connect multiple LAN networks at layer 3 level of the OSI reference model.

A problem with routers is that they are higher-level protocol-dependent. That is, they must know how to handle the specific protocols, such as TCP/IP, that they are routing. The more protocols supported, the more complex the router becomes.

Many routers typically are able to route many of the widely used networking protocols. These include TCP/IP, IPX/SPX, XNS, OSI, and DECnet protocols. Some protocols such as SNA and NetBIOS are regarded as un-routable protocols because they do not include network IDs as part of their addresses. Routers typically support the transport of these unroutable protocols by encapsulating them into other protocols such as TCP/IP.

Brouters   A *brouter* is a term used to describe systems that implement the functions of both a bridge and a router in one compartment. In fact, most routers today can perform the functions of a bridge and a router.

brouter

9.12   LAN STANDARDS

There are official standards for many of the LAN interfaces and protocols described in this chapter. They are specified as part of the IEEE's 802-series of LAN standards, which were developed to make it easier to connect different kinds of systems and devices to LANs. This standardization allows mixing and matching of equipment from different vendors. The standards cover a number of issues, including the physical connections and signaling used, and the protocols used to exchange information with other systems on the LAN. Following is a list of LAN standards:

802.2
802.3

802.4
802.5

- *802.2:* logical link control.
- *802.3:* carrier sense multiple access with collision detection (CSMA/CD).
- *802.4:* token-passing bus.
- *802.5:* token-passing ring.

The 802.2 standard describes a common set of data link protocols called logical link control (LLC) that are usable across different types of LANs. These protocols are used to manage the orderly exchange of information across a LAN.

The 802.3 standard describes a CSMA/CD media-access method similar to that used in Ethernet LANs. This standard includes the 10BASE-2, 10BASE-5, and 10BASE-T specifications, which describe 10-Mbps transmission over coaxial cable and unshielded twisted-pair cables.

The 802.4 standard describes a token-passing media-access method for use on bus-based LANs. LANs used in a manufacturing floor environment are often based on 802.4 as specified in the manufacturing automation protocol (MAP).

The 802.5 standard describes a token-passing media-access method for use on ring-based LANs. IBM token ring LANs conform to the 802.5 standard. Many other manufacturers also make IEEE 802.5-compliant LAN products.

9.13   SUMMARY

The intent of this chapter was to introduce the reader to the basics of local area networks and some of the terminology used in that type of network. We presented the motivations behind using a local area network in one's organization. We examined many of the components that make up a LAN.

LAN operations, topology, and protocols were also presented. Specific implementation of Ethernet and token rings was examined to illustrate the system with an emphasis on operations, cabling, and technical issues. An introduction to high-speed LANs, specifically Ethernet II and FDDI, was also discussed. We also presented an overview of LAN interconnections that consisted of bridges and routers.

REVIEW QUESTIONS AND PROBLEMS

1. What are the components of a LAN?

2. What type of computers and workstations can attach to a LAN?

3. What would the primary reason be for using a LAN?

4. What are the benefits of sharing printers on a LAN?

5. A _____ workstation could be used on LANs to improve security (graphics, PC, diskless, technical).

6. Savings on _____ can result from using a LAN (cabling, software, mice, adapter boards).

7. _____ is an example of an application on a LAN (NetBios, CD-ROM, token passing, electronic mail).

8. Computers on a LAN can function as _____ _____ (adapters, NOSs, servers, UTPs).

9. What do we use to connect a computer directly to a LAN?

10. What is another name for a LAN adapter?

11. The _____ is the cables or wires used to construct a LAN.

12. The _____ provides the transmission path between systems and devices on a LAN.

13. Which is less susceptible to noise: coaxial cables or twisted pairs?

14. Unshielded twisted-pair wire is standard _____ _____ wire.

15. Fiber optic cable is immune to electrical interference because of the use of _____ instead of electrical signaling.

16. Twisted-pair wire has relatively high _____ _____, limiting the distance signals can travel without regeneration (bandwidth, frequency, attenuation, speed).

17. Fiber optic LANs are sometimes used as a _____ _____network (hub, backbone, TCP/IP, WAN).

18. A _____ is the type of physical signaling used on a LAN (transmission technique, transmission medium, token passing, token-passing ring).

19. Placing signals directly onto the transmission medium is called (broadband, transmission, CSMA/CD, baseband).

20. With _____ transmission, signals are modulated to different frequency ranges.

21. What is the technique of dividing signals from sources into different time slots called?

22. Baseband signaling is _____ _____ in nature (expensive, FDM, analog, digital).

23. Broadband signaling is _____ _____ in nature.

24. Broadband transmission is typically done using ____ _____ (radio-frequency modems, TV, bridges, repeaters).

25. A _____ is the manner in which station access to the LAN is controlled.

26. _____ method determines how multiple devices can share the transmission medium.

27. MAC is an acronym for _____ _____.

28. _____ and _____ _____ are two MACs.

29. Using _____, a device must first listen before attempting to transmit on it.

30. CSMA/CD is a technique used in _____ _____ LANs.

31. A _____ is used to control access to the medium using token passing (token, media-access method, transmission technique, token ring).

32. The _____ is the physical layout of a LAN.

33. In a _____ topology, all devices are connected directly to a single cable.

34. Signals are _____ on a bus (lost, broadcast, repeated, regenerated).

35. In a ring, each station _____ signals to the next station (retransmits, acknowledges, intercepts, bypasses).

36. Ethernet uses a _____ access method.

37. Ethernet is specified to operate at _____ over _____ .

38. 10BASE-T networks operate over _____ _____.

39. BASE-T LANs utilize a _____ topology

40. _____ and _____ are two reasons that companies install 10BASE-T LANs.

41. 10BASE-T LANs can connect to _____ _____ LANs (FDDI, Ethernet, token ring, LocalTalk).

42. _____ is the media-access method used in fast Ethernet.

43. _____ is the speed at which fast Ethernet operates.

44. Two major differences between thick and thin Ethernet are _____ and _____.

45. _____ is an IBM-strategic LAN.

46. Token ring LANs are configured in a _____ topology.

47. Token ring LANs operate at 4 or _____ _____.

48. Token ring is based on the IEEE _____ standards.

49. _____-adapters are built into Apple Macintosh computers and laserwriter printers (LocalTalk, token ring, Ethernet, NetWare).

50. FDDI is an acronym that stands for _____ _____.

51. LocalTalk is an _____ proprietary LAN.

52. What is the operation speed of LocalTalk LAN?

53. FDDI is based on _____ technology.

54. FDDI LANs are good for transmitting information with _____ _____ requirements.

55. An FDDI LAN will often be used as a _____ network.

56. What is the network operating system (NOS)?

57. Is it true that most network operating systems support a single type of LAN?

58. How does a NOS support printer sharing?

59. Which NOS is widely used for PC-based LANs?

60. NetWare services are packaged as _____ (NLMs, pipes, APIs, SQLs).

61. _____ is a PC-based network operating system that runs UNIX on the server.

62. Some network operating systems support peer-to-peer networking, but others support _____.

63. Is it true that datagram services of an NOS guarantee delivery of packets?

64. General control, name support, datagram support, and session support are basic services provided by _____

65. NetBIOS is widely used in _____-based LANs.

66. _____ is a de facto standard API for communication between programs on a LAN.

67. _____ was designed to support Internetworking (FDDI, TCP/IP, NetBIOS, SNA).

68. Is it true that the TCP/IP protocols are independent of the lower-layer protocols used in LANs?

69. _____ is the most widely used protocol for tying together Internetworks.

70. _____ is a suite of _____- proprietary protocols (TCP/IP, IBM, XNS, Xerox, NetBIOS, Novell, FDDI, 3Com).

71. _____ is based on XNS Internet datagram protocol (IDP).

72. IPX and SPX are protocols supported in _____.

73. _____ provides reliable data delivery (UDP, IPX, IP, SPX).

74. Because of the huge installed base of _____ network _____ is the most widely used protocol on PC LANs (LAN Manager—Ethernet, token ring—XNS, NetWare—IPX/SPX, IBM—SNA).

10

Internet*working* with TCP/IP

OBJECTIVES

In this chapter we will examine the following:

- Transmission control protocol/Internet protocol (TCP/IP).
- World Wide Web (WWW).
- Other Internet-related topics, including:
  a. The network layer in the Internet.
  b. IP packet formats (IPv4 and IPv6).
  c. Internet control protocols.
  d. Domain name servers.
  e. Network sockets used in UNIX and Windows.

10.1   INTRODUCTION

Thousands of telecommunication networks are in use around the world, and the pace of installation is increasing. Internetworking techniques allow users at different networks to work with each other. The Internet represents a vast number of networks (public and private) that work together to provide the familiar services we use today. We can view the Internet as a collection of subnetworks or autonomous systems that are connected together. There is no real structure, but several backbone networks exist. These networks are constructed from high-bandwidth lines and fast nodes. Attached to the backbones are regional (midlevel) networks, and attached to these regional networks are the LANs at many universities, companies, and Internet service providers. Figure 10–1 shows this quasi-hierarchical organization.

In this chapter, we examine the Internet protocol suite called *transmission control protocol/Internet protocol (TCP/IP).* This name represents a collection of related protocols, ranging from application (ISO layer 7) to data link (ISO layer 2). TCP provides end-to-end communicators with a reliable virtual circuit, which is then transported across the Internet using the Internet protocol (IP). TCP also provides a connectionless unreliable network, where unreliable means that delivery of packets is not guaranteed.

The "old" Internet used the government-funded National Science Foundation Network (NSFnet) Internet backbone consisting of T3 pipes

Figure 10–1
The Internet is an interconnected collection of many networks.

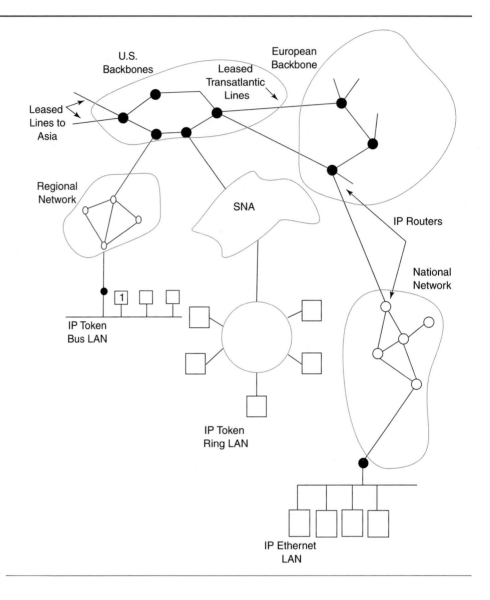

U.S. Backbones

European Backbone

Leased Transatlantic Lines

Leased Lines to Asia

Regional Network

SNA

IP Routers

National Network

IP Token Bus LAN

IP Token Ring LAN

IP Ethernet LAN

to carry transient traffic without any fee. Under this circumstance private network operators such as telephone companies were effectively prevented from becoming Internet service providers (ISPs). The U.S. government realized this and now forbids free use of the NSFnet for transient traffic, meaning the telephone companies and other for-profit organizations could become ISPs. Thus the "new" Internet has IP traffic interchange points called *network access points (NAPs)*. ISPs may use the NAPs or any other interconnection mechanism. Figure 10–2 shows an Internet configuration with a variety of NAPs.

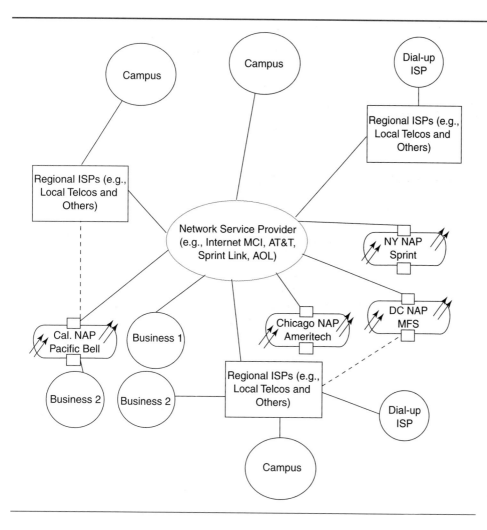

Figure 10–2
Current Internet configuration.

TCP/IP is a five-layer protocol suite that includes the physical, data link, network, and transport layers. The physical layer can be any one of the physical layers we previously examined, ranging from the standard modem, to the T3 and synchronous optical network (SONET) interfaces. The data link layer includes layer protocols such as media-access control (MAC), point-to-point protocol (PPP), serial line Internet protocol (SLIP), the asynchronous transfer mode (ATM), and many more. The network layer (ISO layer 3) consists of the IP, the Internet control message protocol (ICMP), and the Internet group management protocol (IGMP). The transport layer (ISO layer 4) consists mainly of two protocols; transmission control protocol (TCP) and user datagram protocol (UDP). Figure 10–3 shows The TCP/IP protocol suite and Figure 10–4 shows the TCP/IP terminology. Layer 5 involves Telnet, FTP, and SMTP riding over TCP, and NFS, SNMP, and TFTP riding over UDP, as shown in Figure 10–3.

Figure 10–3
TCP/IP protocol
suite.

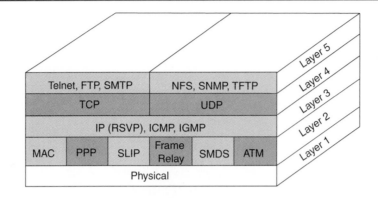

Telnet, FTP, SMTP		NFS, SNMP, TFTP	
TCP		UDP	
IP (RSVP), ICMP, IGMP			
MAC	PPP	SLIP	Frame Relay	SMDS	ATM
Physical			

ATM   = Asynchronous Transfer Mode
SMDS   = Switched Multimegabit Data Service
PPP   = Point-to-point Protocol
SLIP   = Serial Line Interface Protocol
FTP   = File Transfer Protocol
SMTP   = Simple Mail Transfer Protocol
NFS   = Network File System
ICMP   = Internet Control Message Protocol
SNMP   = Simple Network Management Protocol
MAC   = Media-access Control
RSVP   = Reservation Protocol
IGMP   = Internet Group Management Protocol
TCP   = Transmission Control Protocol
UDP   = User Datagram Protocol
IP   = Internet Protocol
TFTP   = Trivial File Transfer Protocol

The culture of the Internet has its roots in the UNIX and C programming language environment but is independent of both. This culture is one of "make it work" and is without strong central authority or direction. Somehow this system works without official standards bodies or proprietary protocols. We can summarize the TCP/IP culture as follows:

- Intertwines with UNIX and C languages but is independent of both.
- Disdains proprietary anything.
- Disdains official standard bodies.
- Consists of "bottom up" experimenters and doers.
- Disdains software license fees.
- Reacts to efforts to institute Internet-wide policies with, "Conspiracy! No! No!"

TCP is an "off net" layer 4 protocol, meaning only the end communicators (computers and devices) execute it; the network IP packet switches know nothing of layer 4. TCP provides a virtual circuit with flow and congestion

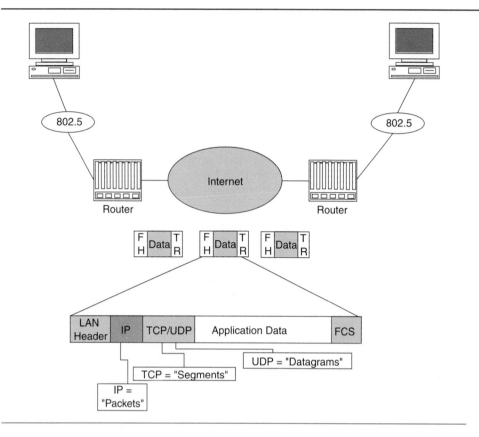

Figure 10–4
TCP/IP terminology.

control; it is a very "smart" protocol. Applications are indicated via *port numbers;* the destination port number addresses a particular application.

An alternate layer 4 protocol is the UDP, which provides the port numbers but little else, meaning the UDP provides an unreliable service to applications.

Layer 3, the network layer, is IP and contains the source and destination addresses. Because the destination address is in the IP header, we say that IP is connectionless; the packet is launched into the IP network where the packet switches (commonly called a *router*) forward the packet out on a link toward the destination. Routers maintain a route table for this purpose.

The IP addresses, examined later in this chapter, historically are for networks of any size. Each address has a network part assigned by a central authority and a host part assigned by a local administrative authority. The latter is further subdivided into a subnet and a host via a bit mask mechanism. The subnet usually corresponds to a physical LAN like Ethernet.

TCP/IP and the Internet are *the* technology for computer telecommunications, and they are becoming the *de facto* standard supported by all vendors. The Internet's popularity has reached beyond those who are computer literate. If you follow the stock market, you will see that Internet-based service companies' stock prices are soaring, even though they are significantly in debt. It is a phenomenon that wise investors and programmers should take advantage of. Think of how many "xyz.com" companies are out there. They all need computer systems, network interconnection devices, and application software to provide their services.

10.2   THE TCP PROTOCOL

In this section we will provide a general overview of the TCP protocol. In the following subsections we will go over the protocol header—field by field, TCP connection management, and TCP transmission policy.

The sending and receiving machines exchange data in the form of segments. A segment consists of a fixed 20-byte header (plus an optional part) followed by 0 or more data bytes. The software executing on the communicating machine decides how big the segments should be. Two limits restrict the segment size. First, each segment, including the TCP header, must fit into the 65,535-byte IP payload. Second, each network has a maximum packet size, and each segment must fit into the network maximum packet size. The network maximum packet size is generally a few thousand bytes and thus defines the upper boundary on the segment size. If a segment passes through a sequence of networks without being fragmented, then hits one whose maximum packet size is smaller than the segment, the router at the boundary fragments the segment into two or more smaller segments.

A segment that is too large for a network to carry can be broken into multiple segments by a router. Each new segment gets its own IP header, so fragmentation by routers increases the total overhead (because each additional segment adds 20 bytes of extra header information in the form of an IP header, discussed in Section 10.3).

The basic protocol used by TCP software is the sliding window protocol. When a sender transmits a segment it also starts a timer. When the segment arrives at the destination, the receiving TCP software sends back a segment (with data if any exists, otherwise without data) bearing an acknowledgment number equal to the next sequence number it expects to receive. If the sender's timer goes off before the acknowledgment is received, the sender transmits the segment again.

Although this protocol sounds simple, there are many ins and outs. For example, since segments can be fragmented, it is possible that part of a transmitted segment arrives and is acknowledged by the receiving TCP software, but the rest is lost. Segments can arrive out of order, so bytes 2075–4095 can arrive but cannot be acknowledged because bytes 1048–2074 have not turned up yet. Segments can also be delayed so long in transit that the sender times out and retransmits them. If a retransmitted segment takes a different route than the original and is fragmented differently, bits and pieces of both the original and the duplicate can arrive sporadically, requiring careful administration to

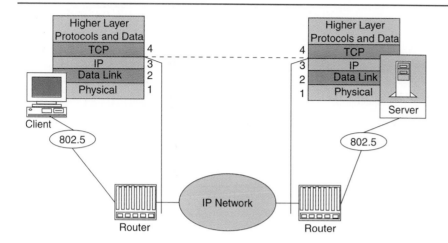

Figure 10–5
Transmission control protocol is implemented on the communicating computers but not the network nodes.

Notes:
* TCP exists only in end equipment (i.e., an IP network does not process the TCP protocol).
* Provides reliable transport (sequencing, acknowledgment, retransmission, flow control).
* Maintains connection between application layers.

achieve a reliable stream. Finally, with so many networks making up the Internet, it is possible that a segment may occasionally hit a congested (or broken) network along the path.

TCP must be prepared to deal with and solve these problems efficiently. A considerable amount of effort has gone into optimizing the performance of TCP software handling TCP streams, even in the face of network problems.

Remember that TCP is "off net" protocol, which means that the TCP software resides on the communicating end-to-end computer. Network nodes neither include TCP implementation nor care about it, since network nodes are only interested in the first three layers of the ISO model. TCP is a layer 4 protocol. Figure 10–5 shows the layers of the protocol suite that need to be implemented in the communicating computers over the Internet.

10.2.1   The TCP Segment Header

In this section we will examine the details of a TCP segment header. Figure 10–6 shows the layout of a TCP segment. Every segment begins with a 20-byte header. The fixed header my be followed by header options. After the options, if any, up to $65,535 - 20 - 20 = 65,495$ data bytes may follow, where the first 20 refers to the IP header and the second to the TCP header. Segments that contain no data are mainly used for acknowledgments and control messages.

Let us now dissect the TCP header field by field. The *source* and *destination port* fields identify the applications making the connection. Each host may decide for itself how to allocate its own ports starting at 256. A port plus its host's

Figure 10–6
The transmission
control protocol
(TCP) header.

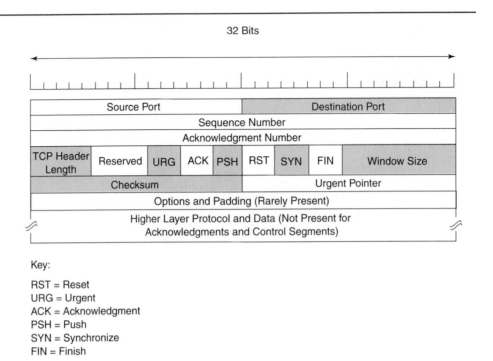

TSAP

IP address forms a 48-bit unique *transport service access point (TSAP).* The source and destination socket numbers together identify the connection. We will talk more about sockets later in this section. The *sequence number* and the *acknowledgment number* fields perform their usual functions. Note that the latter specifies the next segment expected, not the last segment correctly received. Both fields are 32 bits long because every byte of data is numbered in the TCP stream.

The *TCP header length,* also known as the *data offset,* tells how many 32-bit words are contained in the TCP header. This information is needed because the *option* field is of variable length, so the header is, too. Technically, this field indicates the start of the data within a segment, measured in 32-bit words, but that number is the header length in words, so the effect is the same. Following the header length field is a 6-bit field that is not used.

The next six bits are flags. *URG* is set to indicate that the *urgent pointer* is in use. The urgent pointer is used to indicate a byte offset from the current sequence number at which urgent data can be found. This facility is in lieu of interrupt messages. As we mentioned previously, this function is a simple way of allowing the sender to signal the receiver without getting TCP itself involved in the reason for the interrupt. The *ACK* bit is set to 1 to indicate that *acknowledgment number* is valid. If *ACK* is 0, the segment does not contain an acknowledgment so the *Acknowledgment number* field is ignored. The *PSH* bit indicates pushed data. The receiver is hereby requested to deliver the data to the application upon arrival without buffering it until a full buffer has been received (which it might otherwise do for efficiency and performance reasons). The *RST* bit is used to reset a connection that has become inoperable because of a computer crash or some other reason. It is also used to reject an invalid segment or to refuse an attempt to open a connection. In general, if you get a segment with the RST flag on, you have a problem that you need to take care of in order to maintain communication with the far end. The *SYN* bit is used to establish connections. The connection request has $SYN = 1$ and $ACK = 0$ to indicate that the piggyback acknowledgment field is not in use. In essence, the *SYN* bit is used to denote connection request and connection accepted, with the ACK bit used to distinguish between those two possibilities. The *FIN* bit is used to release a connection. It specifies that the sender has no more data to transmit.

Flow control in TCP is handled using a variable-size sliding window. The *window size* field tells how many segments may be sent starting at the segment acknowledged. A *window size* field of 0 is legal and says that segments up to and including *acknowledgment number*–1 have been received, but that the receiver is currently too busy, is in need of a rest, and would like no more data for the moment. Permission to send can be granted later by sending a segment with the same *acknowledgment number* and a nonzero *window size* field. The *checksum* provides extreme reliability. It is calculated over the header and data.

The *options* field was designed to provide a way to add extra facilities not covered by the regular header. The most important option allows each host to specify the maximum TCP payload it is willing to accept. Using large segments is more efficient than using small ones, because the 20-byte header can then be amortized over more data. Small hosts may not be able to handle very large segments. During connection setup, each side can announce its maximum size and see its partner's size. The smaller of the two numbers wins. If the host does not use this option it defaults to a 536-byte payload. All Internet hosts are required to accept TCP segments of $536 + 20 = 556$ bytes.

Another option being widely used is selective repeat instead of go-back N protocol. If the receiver gets one bad segment and then a large number of good ones, the normal TCP protocol will eventually time out and retransmit all the unacknowledeged segments, including all those that were correctly received. The selective repeat request introduced negative acknowledgments (NAKs) to allow the receiver to ask for a specific segment (or segments). After it gets these, it can acknowledge all the buffered data, thus reducing the amount of data retransmission.

10.2.2 TCP Connection Management

TCP uses a three-way handshake to establish connections. To establish a connection, one side, say the server, passively waits for an incoming connection by having its software execute the *listen* and *accept* primitives, either specifying a source or no one in particular. We will talk more about that in the subsection dealing with UNIX sockets.

The other side, say the client, executes a *connect* primitive, specifying the IP address and port to which it wants to connect, the maximum TCP segment size it is willing to accept, and optionally some user data (e.g., a password). The *connect* primitive sends a TCP segment with the *SYN* bit on (1) and *ACK* bit off (0), as discussed earlier, and waits for a response.

When this segment arrives at the destination, the listening software can then either accept or reject the connection. If it accepts, an acknowledgment segment is sent back. The sequence of TCP segments sent in the normal case is shown in Figure 10–7.

Although TCP connections are full duplex, to understand how connections are released it is best to think of them as a pair of simplex connections. Each simplex connection is released independently. To release a connection, either side can send a TCP segment with the *FIN* bit set, which means that it has no more data to transmit. When the *FIN* is acknowledged, that direction is shut down for new data. Data may continue to flow indefinitely in the other direction, however. When both directions have been shut down, the connection is released. Normally, four TCP segments are needed to release a connec-

Figure 10–7
Normal TCP connection establishment.

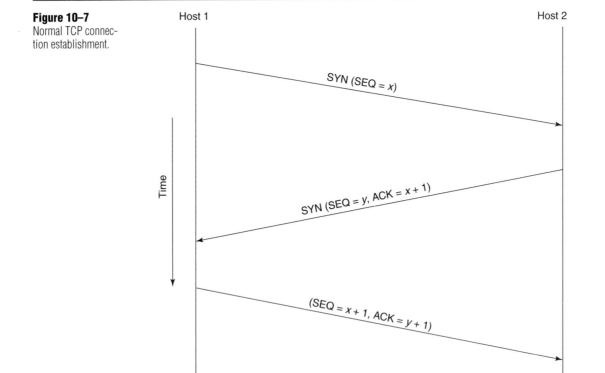

tion—one *FIN* and one *ACK* for each direction. However, it is possible for the first *ACK* and the second *FIN* to be contained in the same segment, reducing the total count to three.

Just as with telephone calls in which both people say goodbye and hang up the phone simultaneously, both ends of a TCP connection may send *FIN* segments at the same time. These segments are each acknowledged in the usual way, and the connection shuts down. There is, in fact, no difference between the two computers releasing sequentially or simultaneously.

If an acknowledgment to the *FIN* segment never arrives, the sender of the *FIN* will eventually time out and release the connection anyway. The other side will eventually notice that nobody is listening to it anymore and time out as well. The timeout solution is not perfect, but since a perfect solution is theoretically impossible, it will have to do. In practice, problems rarely arise.

The steps required to establish and release a connection can be represented in a finite state machine, with the eleven states listed in Table 10–1. In each state, certain events are legal. When a legal event occurs, some action may be taken. If some other event happens, an error is reported.

The finite state machine itself is shown in Figure 10–8. The common case of a client actively connecting to a passive server is shown with heavy lines—solid for the client, dotted for the server. The light-faced lines are unusual event sequences. Each line in Figure 10–7 is marked by an *event/action* pair. Two types of events exist. The user may cause the software to execute *connect, listen, send,* or *close.* The other type of event is that initiated by a segment arrival over the wire (SYN, FIN, ACK, or RST). The action is the sending of a control segment (SYN, FIN, or RST).

The diagram in Figure 10–8 can best be understood by first following the path of a client (the heavy solid line), then later the path of a server (the dashed

State	Description
CLOSED	No connection is active or pending
LISTEN	The server is waiting for an incoming call
SYN RCVD	A connection request has arrived
SYN SENT	The application has started to open a connection
ESTABLISHED	The normal data transfer state
FIN WAIT 1	The application has said it is finished
FIN WAIT 2	The other side has agreed to release the connection
TIMED WAIT	Wait for all packets to die off
CLOSING	Both sides have tried to close simultaneously
CLOSE WAIT	The other side has initiated a release
LAST ACK	Wait for all packets to die off

Table 10–1
The states used in the finite state machine of TCP connnection management.

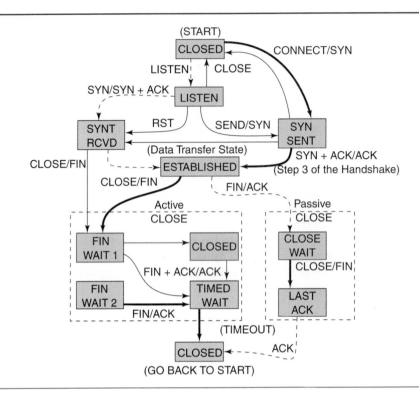

line). When an application on the client machine issues a *connect* request, the local TCP software application (executing on the client machine) creates a connection record, marks it as being in the SYN SENT state, and sends a SYN segment to the server. Note that many connections may be open (or being opened) at the same time on behalf of multiple applications, so state is per connection and is recorded in what is called *the connection record*. When the SYN + ACK arrives, TCP sends the final ACK of the three-way handshake, shown in Figure 10–7, and switches into the *established* state. Data can now be sent and received. When the software application is finished, it executes a *close* command, which causes the local TCP software to send a FIN segment and wait for the corresponding ACK (dashed box marked active close). When the ACK arrives, a transition is made to the FIN WAIT 2 state and one direction of the connection is now closed. When the other side closes, too, a FIN comes in, which is acknowledged. Now both sides are closed, but TCP waits a time equal to the maximum packet lifetime to guarantee that all packets from the connection have died off, in case the acknowledgment was lost. When the timer goes off, TCP deletes the connection record.

Now let us examine connection management from the server's viewpoint. The server initiates a *listen*, and settles down to see who may turn up. Executing a *listen* is asking the software to listen to the TCP/IP ports to see if anyone wants to make a connection (i.e., it looks for TCP segments that request connection establishment). When a SYN comes in, it is acknowledged and the server goes to the SYN RCVD state. When the server's SYN is acknowledged, the three-way handshake is complete and the server goes to the *established* state. Data transfer can now start.

When the client's application software cannot take it anymore, it executes a *close*, which causes a FIN to arrive at the server (dashed box marked passive close). The server is then signaled. When it, too, sends a *close*, a FIN is sent to the client. When the client's acknowledgment shows up, the server releases the connection and deletes the connection record.

10.3   THE NETWORK LAYER IN THE INTERNET

The glue that holds the Internet together is the network layer protocol, or Internet protocol (IP). Unlike most of the older network layer protocols, it was designed from the beginning with internetworking in mind. The network layer's job is to provide a best-efforts way to transport datagrams from their source to their destination without regard to whether or not the machines doing the exchange are on the same network, or whether or not there are networks between them.

Communication in the Internet can be described as follows. The transport layer (layer 4 in the OSI model) on the sending side takes data streams and breaks them into datagrams. In theory, datagrams can be up to 64 Kbytes each; in practice, they are around 2 Kbytes. Each datagram is transmitted through the Internet, possibly being fragmented into smaller units as it travels. When all the pieces get to the final destination machine, they are reassembled by the network layer into the original datagram. This datagram is then handed to the transport layer, which inserts it into the receiving application's input stream.

10.3.1   The Internet Protocol
At this point it is appropriate for us to examine the format of the IP datagrams, which will help us understand the working of the network layer in the Internet. An IP packet consists of a header part and a text part. The header is divided into a 20-byte fixed section and a variable-length optional section. The header format is shown in Figure 10–9. It is transmitted in big endian order: from left to right, with the high-order bit of the *version* field going first. (In Motorola machines such as Sun workstations, big endian transmission is used. In Intel machines such as the Pentium, little endian is used.) On little endian machines, software conversion is required on both transmission and reception, because the header is transmitted and received from right to left.

The *version* field indicates to which version of protocol the packet belongs. Including the version field makes it possible for machines running with different versions of the protocol to communicate. In some cases a machine may run with a version that is months or even years older than another version on another machine.

Figure 10–9
The Internet protocol
(IP) header.

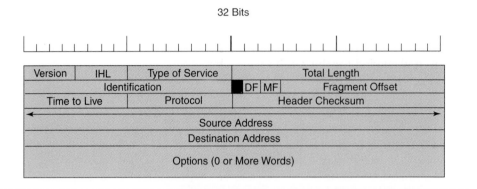

Because the header length is not constant, a field in the header, *IHL*, is provided to tell how long the header is in 32-bit words. The minimum value is 5, which applies when no options are present. The maximum value of this 4-bit field is 15, which limits the header to 60 bytes, and thus the options field to 40 bytes. For some options, such as one that records the route a packet has taken, 40 bytes is far too small, making the option useless.

The *type of service* field allows the host to tell the subnet what kind of service it wants. Various combinations of reliability and speed are possible. For digitized voice, fast delivery beats accurate delivery. For file transfer, error-free transmission is more important than fast transmission. The field itself contains (from left to right) a three-bit *precedence* field, three flags *D, T,* and *R,* and two unused bits. The precedence field is a priority from 0 (normal) to 7 (network control packet). The three flag bits allow the host to specify what is important (i.e., delay, throughput, and reliability). In theory, these fields allow routers to make choices between, for example, a satellite link with high throughput and high delay or a leased line with low throughput and low delay. In practice, current routers ignore the *type of service* field altogether.

The *total length* field includes everything in the datagram (packet)—both header and data. The minimum length is 65,535 bytes. At present the upper limit is tolerable, but with future gigabit networks larger datagrams will be needed.

When a datagram is fragmented, we need a way to identify which fragment belongs to which datagram. The *identification* field is needed to allow the destination host to determine to which datagram a newly arrived fragment belongs. All fragments of a datagram contain the same *identification* value.

The identification field is followed by an unused bit then two 1-bit fields. *DF* stands for *don't fragment*. It tells the routers to not fragment the datagram because the destination is incapable of putting the pieces back together again. For example, when a computer boots, its ROM might ask for a memory image to be sent to it as a single datagram. By sending the datagram with a DF bit, the sender knows it will arrive in one piece, even if this means that the datagram must avoid a small-packet network on the best path and take a less optimum route. All machines are required to accept fragments of 576 bytes or less.

MF stands for more *fragments*. All fragments except the last one have this bit set. It is needed for the receiving machine to know when all fragments of a datagram have arrived.

The *fragment offset* field tells where in the current datagram this fragment belongs. All fragments except the last one in a datagram must be a multiple of 8 bytes, the elementary fragment unit. Since 13 bits are provided, there is a maximum of 8192 fragments per datagram, giving a maximum datagram length of 65,536 bytes, one more than the total length field.

The *time to live* field is a counter used to limit packet lifetimes. It is supposed to count time in seconds, allowing a maximum lifetime of 255 seconds. It must be decremented on each hop and is supposed to be decremented multiple times when queued for a long time in a router. In practice, it just counts hops. When it hits 0, the packet is discarded and a warning packet is sent to the source host. This feature prevents datagrams from wandering around forever, something that otherwise might happen if the routing tables become corrupted.

When the network layer has assembled a complete datagram, it needs to know what to do with it. The *protocol* field tells it which transport process to use. TCP is one possibility but so are UDP and some others. The numbering of protocols is consistent across the entire Internet.

The *header checksum* verifies the header only. The checksum is useful for detecting errors generated by bad memory words inside a router. The algorithm is used to add up all the 16-bit half-words as they arrive, using the 1's complement arithmetic and then taking the 1's complement of the result. For the purpose of this algorithm, the *header checksum* is assumed to be 0 upon arrival. This algorithm is more robust than using a normal add. Note that the *header checksum* must be recomputed at each hop, because at least one field (the *time to live* field) always changes; however, tricks can be used to speed up the computation.

The *source address* and *destination address* fields indicate the network number and host number. We will discuss Internet addresses in the next section.

Fragmentation Example   At this point it is appropriate to consider an example to illustrate the fragmentation and header fields. Assume that a datagram with 636 data bytes arrives at a network with a maximum length restriction of 256 bytes (including the header). Also assume that the IP header length is the minimum possible of 20 bytes. We consider two approaches to fragmentation. In the first approach, each fragment except for the last is the maximum length possible. With 20-byte IP headers, this allows 236 bytes per fragment, but the number in all fragments except for the final fragment must be divided by 8. Hence, this number is adjusted downward to $29 \times 8 = 232$ and fragments with 232, 232, and 172 data bytes are used. The header length is four times the IHL value, the total length is the data field length plus the header length, and the fragment offset is the number of data bytes (from the fragmented datagram) in the fragment preceding the one under consideration. Thus the values for pertinent IP header fields are as follows:

First fragment
IHL = 5
Total Length = 252
Fragment Offset = 0
More = 1

Second fragment
IHL = 5
Total Length = 252
Fragment Offset = 29
More = 1

Third fragment
IHL = 5
Total Length = 192
Fragment Offset = 58
More = 0

For the second approach, all fragments are as nearly equal in length as possible. The average number of data bytes per fragment is $636 \div 3 = 212$. This is not a multiple of 8, so the lengths of the first two must be modified. The most nearly equal lengths are found by increasing one length and decreasing the other to yield $27 \times 8 = 216$, $26 \times 8 = 208$ and 212 bytes. This gives the following

First fragment
IHL = 5
Total Length = 236
Fragment Offset = 0
More = 1

Second fragment
IHL = 5
Total Length = 228
Fragment Offset = 27
More = 1

Third fragment
IHL = 5
Total Length = 232
Fragment Offset = 53
More = 0

Option	Description
Security	Specifies how secret the datagram is
Strict source routing	Gives the complete path to be followed
Loose source routing	Gives a list of routers not to be missed
Record route	Makes each router append its IP address
Timestamp	Makes each router append its address and timestamp

Table 10–2
IP options.

The *options* field was designed to provide an escape to allow subsequent versions of the protocol to include information not present in the original design. It was also designed to permit experimenters to try out new ideas and to avoid allocating header bits to information that is rarely used. The options are of variable length. Each begins with a 1-byte code identifying the option. Some options are followed by a 1-byte option length field, and then 1 or more data bytes. The *options* field is padded out to a multiple of four bytes. Five options are currently defined, as listed in Table 10–2, but not all routers support all of them.

The *security* option tells how secret the information is. In theory, a military router might use this field to specify not to route information through certain countries the military considers to be hostile. In practice, all routers ignore it, so its only practical function is to help spies find the good stuff more easily.

The *strict source routing* option is used to give the complete path from the source to the destination as a sequence of IP addresses. The datagram is required to follow that exact route. This option is most useful for system managers to send emergency packets when routing tables are corrupted, or for making timing measurements. A datagram can be sent with a sequence of IP addresses that leads back to the source (loopback). By measuring how long it takes the packet to come back to the source, an indication of timing can be obtained.

strict source routing

The *loose source routing* option requires the packet to traverse the list of routers specified and in the order specified, but it is allowed to pass through other routers on the way. Normally, this option would only provide a few routers to force a particular path. For example, this option could be used to force a packet from London to Cairo to go west instead of east. This option may specify routers in New York and Washington DC. This option is most useful when political or economic considerations dictate passing through or avoiding certain countries.

loose source routing

The *record route* option is used to tell the routers along the pass to append their IP address to the option field. This allows system maintenance people to track down problems ("Why are packets from London to Paris visiting Tokyo first?").

Finally, the *timestamp* option is similar to the *record route option*, except that in addition to recording its 32-bit IP address, each router also records a 21-bit timestamp. This option, too, is mostly for tracking problems related to routing of packets. By appending timestamps to the IP packet, the parts of the route that take the longest for the packet to traverse can be isolated.

10.3.2 The IP Address

Every router, computer, and device on the Internet has an IP address. The IP address encodes its network number and host number. This combination must be unique; no two devices have the same IP address. All IP addresses are 32 bits long and are used in the *source address* and *destination address* fields of the IP packet. The formats for the IP addresses are shown in Figure 10–10. A device connected to multiple networks must have a different IP address on each network. For example, if a computer in a school campus is connected to two local area networks (one is Ethernet and the other is a token ring), that computer must have different IP addresses for each network.

As shown in Figure 10–10, there are five classes of IP addresses: A, B, C, D, and E. The class A, B, C, and D formats allow different combinations of network host addresses:

- Up to 126 networks with 16 million hosts each.
- 16,382 networks with up to 64,000 hosts.
- 2 million networks (e.g., LANs) with up to 254 hosts each.
- *Multicasting,* in which a datagram is directed to multiple hosts.

multicasting

Addresses beginning with 1110 are reserved for future use. Since tens of thousands of networks are now connected to the Internet, and the number doubles

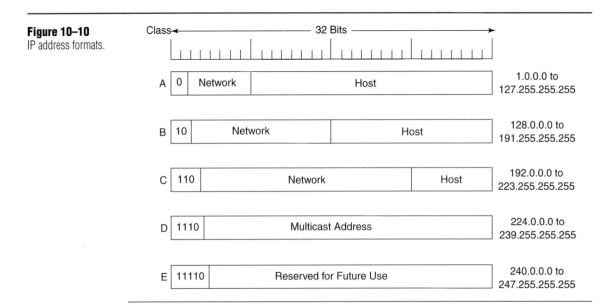

Figure 10–10
IP address formats.

Class			
A	0	Network	Host	1.0.0.0 to 127.255.255.255
B	10	Network	Host	128.0.0.0 to 191.255.255.255
C	110	Network	Host	192.0.0.0 to 223.255.255.255
D	1110	Multicast Address		224.0.0.0 to 239.255.255.255
E	11110	Reserved for Future Use		240.0.0.0 to 247.255.255.255

Binary Representation

1 1 0 0 1 1 1 0	0 0 0 0 0 0 0 0	1 1 0 1 0 0 1 0	0 0 0 0 0 0 1 0

Figure 10–11
IP address example
(three representations).

Decimal Representation

206	0	210	2

*Used for more convenient representation.
*Each byte is represented by a decimal number in the range of 0–255.
*Dotted decimal notation 206.0.210.2.

Symbolic Representation

barnes.khader.com

*Used to make it easy for a human being to remember.
*A domain name server provides the translation of symbolic to actual IP addresses.

each year, it is important to find a mechanism to assign network numbers to avoid confusion. Network numbers are assigned by the *Network Information Center (NIC)*. An example of an IP address is given in Figure 10–11.

NIC

Network addresses, which are 32-bit numbers, are usually written in *dotted decimal notation*. In this format, each of the 4 bytes is written in decimal form, from 0 to 255. For example, the hexadecimal address C0290616 is written as 192.41.6.22. The lowest IP address is 0.0.0.0 and the highest is 255.255.255.255. The values 0 and –1 have a special meaning, as shown in Figure 10–12. The value 0 means this network or this host. The value –1 is used as a broadcast address to mean all hosts on the indicated network.

dotted decimal notation

The IP address 0.0.0.0 is used by hosts when they are being booted but is not used afterward. IP addresses with 0 as the network number refers to the current network. These addresses allow machines to refer to their own networks without knowing the number (but they have to know its class to know how many 0s to include). The address consisting of all 1s allows broadcasting on the local network, typically a LAN. The address with a proper network number and all 1s in the host field allows machines to broadcast packets to distant LANs anywhere in the Internet. Finally, all addresses of the form 127.xx.yy.zz are reserved for *loopback* testing. Packets sent to that address are not put out onto the wire; they are processed locally and treated as incoming packets. This function allows packets to be sent to the local network without the sender knowing the number. Again, this feature is used for debugging the network when tracking a problem.

loopback

Figure 10–12
Special IP ad-
dresses.

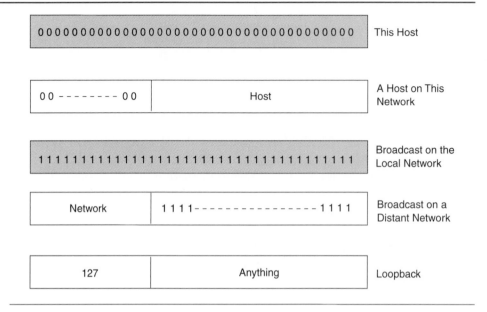

10.4 SUBNETS

As stated in the previous section, all the hosts in a network must have the same network number. This property of the IP addressing can cause problems as networks grow. For example, consider a company with one class C LAN on the Internet. As time goes on, it might acquire more than 254 machines, and thus need a second class C address. Alternatively, it might acquire a second LAN of a different type and want a separate IP address for it (the LANs could be bridged to form a single IP network, but bridges have their own problems). Eventually, it might end up with many LANs, each with its own router and each with its own class C network number.

As the number of distinct local networks grows, managing them can become a serious problem. Every time a new network is installed, the network administrator has to contact the NIC to get a new network number. Then this number must be announced to the world. Furthermore, moving a machine from one LAN to another requires changing its IP address, which in turn may mean modifying its configuration files and also announcing the new IP address worldwide. If some other machine is given the newly released IP address, that machine will get E-mail and other data intended for the original machine until the address has propagated all over the world.

The solution to these problems is to allow a network to be split into several parts for internal use but to still act like a single network to the outside world. In Internet language, these parts are called *subnets*. If our growing company started

subnets

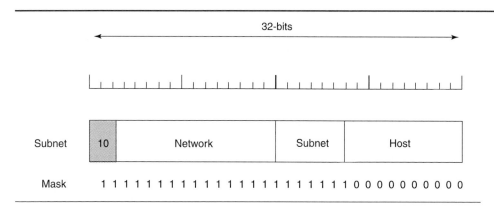

Figure 10–13
One of the ways to subnet a class B network.

up with a class B address instead of a class C address, it could start out by numbering the hosts from 1 to 254. When the second LAN arrived, it could decide, for example, to split the 16-bit host number into a 6-bit subnet number and a 10-bit host number, as shown in Figure 10–13. This split allows 62 LANs (0 and –1 are reserved), each with up to 1022 hosts.

Outside the network, the subnetting is not visible, so allocating a new subnet does not require contacting the network interface center or changing any external databases. In this example, the first subnet might use an IP address starting at 130.50.4.1, the second subnet might start at 130.50.8.1, and so on.

To see how subnets work, it is necessary to understand how IP packets are processed at a router. Recall that routers operate at level 3 of the protocol model. Each router has a table listing some number of (network, 0) IP addresses and some number of (this network, host) IP addresses. The first kind tells how to get to distant networks. The second kind tells how to get to local hosts. Associated with each table is the network interface to use to reach the destination, and certain other information.

When an IP packet arrives, its destination address is looked up in the routing table. If the packet is for a distant network, it is forwarded to the next router on the interface given in the table. If the network is not present in the router's table, the packet is forwarded to a default router with more extensive tables. This algorithm means that each router has to keep track only of other networks and local hosts, not (network, host) pairs, generally reducing the size of the routing table.

When subnetting is introduced, the routing tables are changed, adding entries of the form (this network, subnet, 0) and (this network, the subnet, host). Thus, a router on subnet k knows how to get to all other subnets and also how to get to all the hosts on subnet k. It does not have to know the details about hosts (computers) on other subnets. In fact, all that needs to be changed is to have each router do a Boolean AND with the network's *subnet mask* (see Figure 10–13) to get rid of the host number and look up the resulting address in its tables (after determining which network class it is). For example, a packet addressed to

130.50.15.6 and arriving at a router on subnet 5 is ANDed with the subnet mask of Figure 10–13 to give the address 130.50.12.0. The address is looked up in the routing tables to find out how to get to hosts on subnet 3. The router on subnet 5 is spared the work of keeping track of the data link address of hosts other than those on subnet 5. Subnetting reduces router table space by creating a three-level hierarchy.

10.5   INTERNET CONTROL PROTOCOLS

In addition to IP, which is used for data transfer, the Internet uses several control protocols in the network layer. Internet control message protocol (ICMP), address resolution protocol (ARP), and reverse address resolution protocol (RARP) are three of many Internet control protocols. In this section we will examine those three Internet control protocols.

10.5.1   Internet Control Message Protocol

ICMP

The *Internet control message protocol (ICMP)* is a required companion to IP. Its function is to provide feedback from routers to hosts about problems in the communication environment. It also allows hosts, routers, and gateways to interact with Internet-monitoring centers. Although ICMP is basically at the same level as IP, it uses the IP to translate its messages; it constructs an ICMP message and passes it to IP. The IP encapsulates the ICMP message with an IP header and transmits it to the destination router or host. About a dozen messages are defined for the ICMP. We now examine those that are most important. A list of those messages is given in Table 10–3.

The *destination unreachable* message is used when a router cannot locate the destination or a packet cannot be delivered because a small-packet network is standing in its way.

Table 10–3
Important ICMP message types.

Message Type	Description
Destination unreachable	Packet could not be delivered
Time exceeded	Time to live field hit 0
Parameter problem	Invalid header field
Source quench	Chock packet
Redirect	Teach a router about geography
Echo request	Ask a machine whether it is alive
Echo reply	The machine is saying "Yes, I'm alive"
Timestamp request	Same as echo request, but with timestamp
Timestamp reply	Same as echo reply, but with timestamp

The *time exceeded* message is sent when a packet is dropped due to its counter reaching zero. This is a symptom that packets are looping, that there is enormous congestion, or that the timer value is set too low. Recall that the *time to live* field in the IP header is basically a counter, which is set to a particular value when a packet is first put onto the Internet. The counter value is then decremented one as it passes by each of the routers along the route.

The *parameter problem* message indicates that an illegal value has been detected in a header field. This problem indicates a bug in the sending host's IP software, or possibly in the software of a router in transit.

The *source quench* message was formerly used to throttle hosts that were sending many packets. When a host received this message, it was expected to slow down its data transmission. It is rarely used anymore, though, because when congestion occurs, these packets tend to add to the congestion rather than ease it. Congestion control in the Internet is now done largely in the transport layer.

The *redirect* message is used when a router notices that a packet is being routed the wrong way. The router uses this message to tell the sending host about the probable error.

The *echo request* and *echo reply* messages are used to see if a given destination is reachable and alive. Upon receiving the echo message, the destination is expected to send an echo reply message back. The *timestamp request* and *timestamp reply* messages are similar, except that the arrival time of the message and the departure time of the reply are recorded in the reply. This facility is used to measure network performance.

In addition to these messages, four other messages deal with Internet addressing to allow hosts to discover their network numbers and to handle the case of multiple LANs sharing a single IP address.

10.5.2 Address Resolution Protocol

Although every machine on the Internet has one (or more) IP address, these cannot actually be used for sending packets because the data link layer hardware does not understand Internet addresses. Most hosts are attached to a LAN by an interface board that only understands LAN addresses. For example, every Ethernet board comes equipped with a 48-bit Ethernet address. Manufacturers of Ethernet boards request a block of addresses from a control authority to ensure that no two boards have the same address (to avoid conflicts in case two boards ever appear on the same LAN). The boards send and receive frames based on their 48-bit Ethernet addresses. They know nothing at all about 32-bit IP addresses.

Now, we have a question on our hands. How do IP addresses get mapped onto data link layer addresses, such as Ethernet? To explain how this works, let us use the example of Figure 10–14, in which a small university with several class C (refer to class C IP addresses in Figure 10–10) networks is illustrated. Here we have two Ethernets, one in the Engineering Technology department with IP address 192.31.65.0 and one in the Computer Science department with

Figure 10–14
Three interconnected
class C networks:
Two Ethernets and
an FDDI ring.

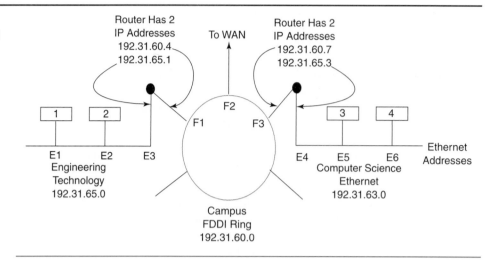

IP address 192.31.63.0. These Ethernets are connected by a campus FDDI ring with IP address 192.31.60.0. Each machine on an Ethernet has a unique Ethernet address, labeled F1 and F3.

Let us start out by seeing how a user on host 1 sends a packet to a user on host 2. Assume that the sender knows the name of the intended receiver, possibly something like khader@eagle.et.univ.edu. The first step is to find the IP address for host 2 known as eagle.et.univ.edu. This lookup is performed by the domain name server, which we study later in this chapter. For the moment we will assume that the domain name server (DNS) returns the IP address for host 2 (192.31.65.5)

The upper-layer software on host 1 now builds a packet with 192.31.65.5 in the *destination address* field and gives it to the IP software to transmit. The IP software can look at the address and see that the destination is on its own network, but it needs a way to find the destination's Ethernet address. One solution is to have a configuration file somewhere in the system that maps IP addresses onto Ethernet addresses. This solution is certainly possible, but for organizations with thousands of machines, keeping these files up to date is an error-prone, time-consuming task.

A better solution is for host 1 to output a broadcast packet onto the Ethernet asking: "Who owns IP address 192.31.65.5?" The broadcast will arrive at every machine on the Ethernet 192.31.65.0, and each one will check its IP address. Host 2 alone will respond with its Ethernet address (E2). In this way host 1 learns that IP address 192.31.65.5 is on the host with Ethernet address E2. The protocol for asking this question and getting the reply is called *address resolution protocol (ARP)*. Almost every machine on the Internet runs it.

The advantage of using ARP over configuration files is its simplicity. The system manager does not have to do much except assign each machine an IP address and decide about subnet masks; ARP does the rest.

ARP

At this point, the IP software on host 1 builds an Ethernet frame addressed to E2, puts the IP packet (addressed to 192.31.65.5) in the payload field, and dumps it into the Ethernet. The Ethernet board of host 2 detects this frame, recognizes it as a frame of itself, scoops it up, and somehow interrupts the Ethernet driver to take the frame. The Ethernet driver extracts the IP packet from the payload and passes it to the IP software, which sees that it is correctly addressed and processes it.

Various optimizations are possible to make ARP more efficient. To start with, once a machine has run ARP, it caches the result in case it needs to contact the same machine shortly. Next time it will find the mapping in its own cache, thus eliminating the need for a second broadcast. In many cases host 2 will need to send back a reply, forcing too, to run ARP to determine the sender's Ethernet address. This ARP broadcast can be avoided by having host 1 include its IP to Ethernet mapping in the ARP packet. When an ARP broadcast arrives at host 2, the pair (192.31.65.7, E1) is entered into host 2's ARP cache for future use. In fact, all machines on Ethernet can enter this mapping into their ARP caches.

Another optimization is to have every machine broadcast its mapping when it boots. This broadcast is generally done in the form of an ARP looking for its own IP address. In this case there should not be a response, but a side effect of the broadcast is to make an entry in everyone's ARP cache. If a response does arrive, two machines have been assigned the same IP address. The machine being booted, which initiated the broadcast, should somehow inform the system manager and abort the booting process. This is because another machine has been assigned the same IP address, which is not allowed.

To allow mappings to change, for example when an Ethernet board malfunctions and is replaced with a healthy one (and thus a new Ethernet address), entries in the ARP cache should time out after a few minutes.

Now let us look at Figure 10–14 one more time, only this time host 1 wants to send a packet to host 4 (192.31.63.8). Using ARP will fail because host 4 will not see the broadcast (this is because routers do not forward Ethernet-level broadcasts). There are two solutions to this problem. First the Engineering Technology router could be configured to respond to ARP requests for network 192.31.63.0 (and possibly other local networks). In this case, host 1 will make an ARP cache entry of (192.31.63.8, E3) and gladly send all traffic for host 4 to the local router. This solution is called *proxy ARP*. The second solution is to have host 1 immediately see that the destination is on a remote network and send all traffic to a default Ethernet address that handles all remote traffic, in this case E3. This solution does not require the ET router to know which remote network it is serving.

Either way, what happens is that host 1 packs the IP packet into the payload field of an Ethernet frame address to E3. When the Engineering Technology router gets the Ethernet frame, it removes the IP packet from the payload field and looks up the IP address in its routing tables. It discovers that packets for network 192.31.63.0 are supposed to go to router 192.31.60.7. If it does not already know the FDDI address of 192.31.60.7, it broadcasts an ARP packet onto the ring and learns that its ring address is F3. It then inserts the packet into the payload field of an FDDI frame addressed to F3 and puts it on the ring.

At the computer science router, the FDDI driver removes the packet from the payload field and gives it to the IP software, which sees that it needs to send the packet to 192.31.63.8. If this IP address is not in its ARP cache, it broadcasts an ARP request on the Computer Science Ethernet and learns that the destination is E6. So it builds an Ethernet frame addressed to E6, puts the packet in the payload field, and sends it over the Ethernet. When the Ethernet frame arrives at host 4, the packet is extracted from the frame and passed to the IP software for processing.

Going from host 1 to a distant network over a WAN works essentially the same way, except that this time the Engineering Technology router's tables tell it to use the WAN router whose FDDI address is F2.

You will notice that many "what if scenarios" must be handled by the ARP to successfully map IP addresses into Ethernet addresses. Also, remember that Ethernet boards only understand Ethernet addresses. That is why we needed an address mapping protocol in the first place. Understanding the ARP protocol can be helpful when tracing problems among interconnected networks.

10.5.3   Reverse Address Resolution Protocol

ARP solves the problem of finding out which Ethernet address corresponds to a given IP address. Sometimes the reverse problem has to be solved, however. Given an Ethernet address, what is the corresponding IP address? In particular, this problem occurs when booting a diskless workstation. (A diskless workstation is a computer that does not have a hard disk, instead it uses the hard disk on a server to use and manipulate files and information.) Such a machine will normally get the binary image of its operating system from a remote file server. But how does it learn its IP address?

RARP

The solution is to use *reverse address resolution protocol (RARP)*. This protocol allows a newly booted workstation to broadcast its Ethernet address and say: "My 48-bit Ethernet address is 14.02.07.18.27. Does anyone out there know my IP address?" The reverse address resolution protocol server sees this request, looks up the Ethernet address in its configuration files, and sends back the corresponding IP address.

Using RARP is better than enabling an IP address in the memory image, because it allows the same image to be used on all of the diskless machines. If the IP address were burned inside the image, each diskless workstation would need its own image.

A disadvantage of RARP is that it uses a destination address of all 1s, limited broadcasting, to reach the RARP server. However, such broadcasts are not forwarded by routers, so an RARP server is needed on each network. To get around this problem, an alternative bootstrap protocol called *BOOTP* has been developed. Unlike RARP, BOOTP uses UDP messages, which are forwarded over routers. It also provides a diskless workstation with additional information, including the IP address of the file server holding the memory image, the IP address of the default router, and the subnet mask to use.

10.6 INTERNET MULTICASTING

Normal IP communication is between a single sender and a single receiver. However, in some cases it is useful for a communication software program to be able to send to a large number of receivers simultaneously. Examples of when this function is useful include updating replicated, distributed data-bases; transmitting stock quotes to multiple brokers; and handling multiparty digital telephone calls.

IP supports multicasting using a class D address, which identifies a group of hosts (computers). Twenty-eight bits are available for identifying groups, so over 250 million groups of computers can exist at the same time. When a soft-ware program sends a packet to a class D address, a best-effort attempt is made to deliver it to all the members of the group addressed, but no guaran-tees are given. Some members may not get the packet.

Two kinds of group addresses are supported: permanent addresses and temporary ones. A permanent group is always there and does not have to be set up. Each permanent group has a permanent group address. Some exam-ples of permanent group addresses are:

224.0.0.1 All systems on a LAN
224.0.0.2 All routers on a LAN

Temporary groups must be created before they can be used. A software program can ask its host computer to join a specific group. It can also ask its host to leave the group. Each host keeps track of which group it belongs to.

Multicasting is implemented by special multicast routers, which may or may not be co-located with the standard routers. About once a minute, each multicast router sends a hardware (i.e., data link layer) multicast to the host on its LAN (address 224.0.0.1) asking it to report back on the group its software program currently belongs to. Each host sends back responses for all the class D addresses it is interested in. These query and response packets use a protocol called Internet Group Management Protocol (IGMP), which is vaguely analo-gous to ICMP. IGMP has only two kinds of packets—query and response—each with a simple fixed format containing some control information, the first word of the payload field, and a class D address in the second word.

10.7 CLASSLESS INTERDOMAIN ROUTING (CIDR)

IP has been in heavy use for quite some time. It has worked extremely well, as proven by the exponential growth of the Internet. Unfortunately, IP is rapidly becoming a victim of its own popularity—it is running out of addresses. This looming catastrophe has sparked a great deal of controversy within the Internet community about what should be done about it. In this section we will present the problem and a number of the proposed solutions.

In the late 1980s, some people predicted that the Internet might grow to 100,000 networks. Most experts at that time ridiculed this prediction as being decades away in the future, if ever. The 100,000th network was connected in 1996. The problem simply stated is that the Internet is rapidly running out of IP addresses. In principle, over 2 billion addresses exist, but the practice of organizing the address space by class (see Figure 10–10), wastes millions of them. In particular, the real culprit is the class B network. For most organizations, a class A network, with 16 million addresses, is too big, and a class C network, with 256 addresses, is too small. A class B network, with 65,536 addresses, is just right.

In reality, a class B address is far too large for most organizations. More than half of all class B networks have less than 50 hosts. A class A address would have done the job, but no doubt every organization that asked for a class B address thought that one day it would outgrow the 8-bit host address field.

One solution that is now being implemented and that will give the Internet a bit of extra breathing room is classless interdomain routing (CIDR). The basic idea behind CIDR is to allocate the remaining class C networks, of which there are almost two million, into variable size blocks of 2048 addresses (eight contiguous class C networks) and not a full class B address. Similarly, a site needing 8000 addresses gets 8192 addresses (32 contiguous class C networks).

In addition to using blocks of contiguous class C networks as units, the allocation rules for the class C address were changed. The world was partitioned into four zones, and each one was given a portion of the class C address space. The allocation was as follows:

Addresses 194.0.0.0 to 195.255.255.255 are for Europe
Addresses 198.0.0.0 to 199.255.255.255 are for North Africa
Addresses 200.0.0.0 to 201.255.255.255 are for the Americas
Addresses 202.0.0.0 to 203.255.255.255 are for Asia and the Pacific

In this way, each region was given about 32 million addresses to allocate, with another 320 million class C addresses from 204.0.0.0 through 223.255.255.255 held in reserve for the future.

Let us now consider an example. Suppose that Rutgers University needs 2048 addresses and is assigned addresses 200.24.0.0 through 200.24.7.255, along with mask 255.255.248.0. Next, the New Jersey Institute of Technology asks for 4096 addresses. Since a block of 4096 addresses must lie on a 4096-byte boundary, they cannot be given addresses starting at 2000.24.8.0. Instead they get 200.24.16.0 through 200.24.31.255, along with mask 255.255.240.0. Now the University of North Carolina asks for 1024 addresses and is assigned addresses 200.24.8.0 through 200.24.11.255 and mask 255.255.252.0.

10.8   DOMAIN NAME SERVER (DNS)

Programs rarely refer to host computers, mailboxes, and other resources by their binary network addresses. Instead of binary numbers, they use ASCII strings, such as nadia@english.ucla.edu. Nevertheless, the network only understands binary addresses. Therefore, some mechanism is needed to convert the ASCII strings to network addresses. In this section we will study how this mapping is accomplished in the Internet.

The domain name server (DNS) is a hierarchical, domain-based naming scheme, which uses distributed databases to essentially implement the mapping of names (mainly host names and E-mail destinations) into network addresses. Very briefly, DNS is used as follows. To map a name onto an IP address, an application program calls a software library procedure called the *resolver*, passing it the name as a parameter. The resolver sends a UDP packet to a local DNS, which then looks up the name and returns the IP address to the resolver, which in turn returns it to the caller (the application program). After it knows the IP address, the application program can then establish a TCP connection with the destination or send it UDP packets.

10.8.1   The DNS Name Space

The postal service requires letters to specify (implicitly or explicitly) the country, state, province, city, and street address of the intended recipient of the mail. By using this kind of structure, there is no confusion between the William Barnes on Main Street, in White Plains, New York, and the William Barnes on Main Street, in Seattle, Washington. The DNS works the same way.

Conceptually, the Internet is divided into several hundred top-level *domains*, where each domain covers many hosts. Each domain is partitioned into subdomains, which are further partitioned. All domains can be represented by a tree, as shown in Figure 10–15. The top-level domains come in two types: generic and countries. The generics are *com, edu, gov, int, mil, net* and *org*. The country domains include one entry for every country, as defined in ISO 3166. Table 10–4 lists the top-level domains and their meanings.

Each domain is named by the path upward from it to the (unnamed) root. The components are separated by periods (pronounced "dot"). Thus, the Intel engineering department might be *eng.intel.com*, rather than a UNIX-style name such as */com/intel/eng*. With this structure, similar to the postal service structure, eng.intel.com does not conflict with a potential use of *eng* in *eng.harvard.edu*, which might be used by the Harvard English department.

Domain names are not case sensitive, so edu and EDU mean the same thing. Component names can be up to 63 characters long, and full path names must not exceed 255 characters.

In practice, nearly all organizations in the United States are under a generic domain, and nearly all organizations outside the United States are under the domain name of their country. Registering under two top-level domains is not prohibited, but doing so might be confusing, so few organizations do it.

Figure 10–15
Part of the Internet domain name space.

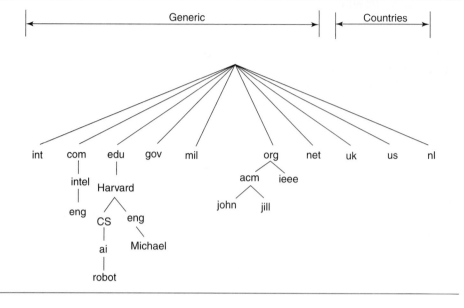

Creating a new domain requires permission of the domain in which it will be included. For example, if a telecom group is started at Harvard University and wants to be known as *telecom.cs.harvard.edu*, it needs permission from whomever manages *cs.harvard.edu*.

Naming does not follow physical networks; instead it follows organizational boundaries. For example, if the Engineering Technology and Computer Science departments are located in the same building and share the same LAN, they can nevertheless have distinct domains. Similarly, even if Engineering Technology is physically split between the Alumni Hall and the Information Technology Building, all hosts in both buildings will normally belong to the same domain.

Table 10–4
Top-level domain names.

Domain	Meaning
com	Commercial organization
edu	Educational institution
gov	Government
int	International organization
mil	Military
net	Networking organization
org	Nonprofit organization
uk, au, nz, etc.	Country abbreviations

10.8.2 Name Servers

Conceptually speaking, a single name server could contain the entire DNS database and respond to all inquiries about it. In practice, this server would be overloaded beyond usefulness. Furthermore, if it ever went down, the entire Internet would be crippled.

The DNS name space is divided into nonoverlapping *zones*. One possible way to divide the name space shown in Figure 10–15 is illustrated in Figure 10–16. Each zone contains some part of the tree and also contains name servers holding the record information about the zone. Normally, a zone will have one primary name server, which gets its information from a file on its disk, and one or more secondary name servers, which gets its information from the primary name server. To improve reliability, some servers for a zone can be outside the zone.

How the zone boundaries are placed within a zone is up to that zone's administrator. This decision is made in large part based on how many name servers are desired, and where. For example, in Figure 10–16, Harvard has a server for harvard.edu that handles eng.harvard.edu but not cs.harvard.edu, which is a separate zone with its own name server. Such a decision might be made when a department such as English does not wish to run its own name server, but a department such as Computer Science does. Consequently, cs.yale.edu is a separate zone but eng.yale.edu is not.

When a resolver has a query about a domain name, it passes the query to one of the local name servers. If the domain being sought falls under the jurisdiction

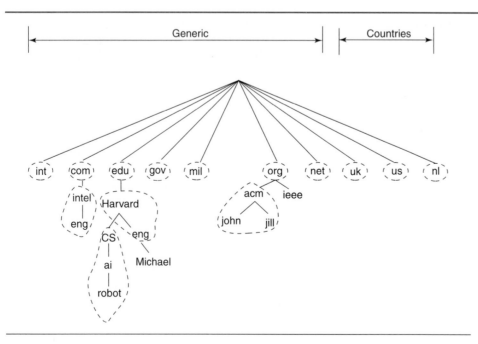

Figure 10–16
DNS name space divided into zones.

Figure 10–17
How a resolver looks up a remote name in eight steps.

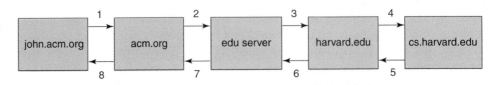

of the name server, such as *ai.cs.harvard.edu* (artificial intelligence, computer science at Harvard) falling under *cs.harvard.edu*, it returns the authoritative resource records. An authoritative record is one that comes from the authority that manages the record and is thus always correct. Authoritative records are in contrast to cached records, which may be out of date.

If, however, the domain is remote and no information about the requested domain is available locally, the name server sends a query message to the top-level server for the domain requested. Let us consider the example of Figure 10–17. Here, a resolver on *john.acm.org* wants to know the IP address of the host *michael.cs.harvard.edu*. In step 1, it sends a query to the local name server *acm.org*. Let us suppose the local name server has never had a query for this domain before and knows nothing about it. It may ask a few nearby name servers, but if none of them know, it sends a UDP packet to the server for *edu* given in its database *edu.server.net* (see Figure 10–17). It is unlikely that this server knows the address of *michael.cs.harvard.edu* either, but it must know all of its children, so it forwards the request to the name server for harvard.edu (step 3). In turn, this server forwards the request to cs.harvard.edu (step 4), which must have the authoritative records. Hence, the requested resource record works its way back in steps 5 through 8.

Once the records get back to the acm.org name server, they will be entered into a cache there in case they are needed later. However, this information is not authoritative, since changes made to cs.harvard.edu will not be propagated to all the caches in the whole wide world that may know about it.

It is worth mentioning that the query method described in this example is known as a *recursive query*, since each server that does not have the requested information finds it somewhere then reports back. An alternate form is also possible. In this form, when a query cannot be satisfied locally, the query fails, but the name of the next server to try along the line is returned. This procedure gives the client more control over the search process. Some servers do not implement recursive queries and always return the name of the next server to try.

It is also worth noting that when a DNS client fails to get a request before its timer goes off, it normally will try another server next time. The assumption here is that the server is probably down, rather than the request or reply got lost.

10.9 IPV6

CIDR may buy us a few more years, but everyone realizes that the days (years) of the IP we described in this chapter are numbered. This protocol is known as *IPv4*. Until the mid-1990s, most Internet users were members of universities, high-tech industries, and the government. With the explosion of interest in the Internet, it is likely that in the next millenium it will be used by a much larger group of people with different requirements. *IPv6* is a new Internet protocol that should never run out of address space and should solve many of the problems in the older IPv4. Its major goals are to: IPv4 IPv6

- Provide billions of hosts.
- Reduce the size of the routing tables.
- Simplify the protocol, thus allowing routers to process packets faster.
- Pay attention to real-time data.
- Allow the protocol to evolve in the future.
- Permit the old and new protocols to coexist.

IPv6 meets these goals fairly well. It maintains the good features of IP, discards or deemphasizes the bad ones, and adds new ones where needed. In general, IPv6 is not compatible with IPv4, but it is compatible with all other protocols used in the Internet, including TCP, UDP, ICMP, and DNS, sometimes with small modifications being required (mostly to accommodate longer addresses).

First, IPv6 has longer addresses than IPv4. They are 16 bytes long, which solves the problem that IPv6 was set out to remedy by providing an effectively unlimited supply of Internet addresses. We will have more to say about addresses shortly.

The second major improvement of IPv6 is the simplification of the header. It contains only seven fields (versus 13 in IPv4). We will discuss the header shortly, too.

The third improvement is better support for options. This change is essential with the new header, because fields that previously were required are now optional. In addition, the way options are represented is different, making it simple for routers to skip over options not intended for them. This feature speeds up packet processing time.

A fourth area in which IPv6 represents a big advance is in security. Authentication and privacy are key features of the new IP.

Finally, more attention has been paid to type of service than in the past. IPv4 actually has an 8-bit field devoted to this matter, but with the expected growth in multimedia traffic in the future, much more is needed.

10.9.1 The Main IPv6 Header

The IPv6 header is shown in Figure 10–18. The *version* field is always 6 for IPv6 (and 4 for IPv4). During the transition period from IPv4, which will probably take a number of years, routers will be able to examine this field to determine what kind of packet they have.

Figure 10–18
IPv6 header.

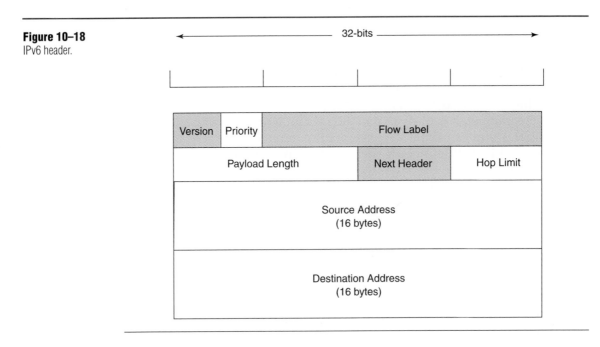

The *priority* field is used to distinguish between packets whose sources can be flow controlled and those that cannot. Values 0 through 7 are for transmissions that are capable of slowing down in the event of congestion. Values 8 through 15 are for real-time traffic whose sending rate is constant, even if all the packets are being lost. Audio and video packets fall into the latter category. This distinction allows routers to better deal with packets in the event of congestion. Within each of the priority groups, lower-number packets are less important than higher-numbered ones. The IPv6 standards suggest, for example, to use one for news, four for file transfer protocol (FTP), and six for Telnet connections, since delaying a news packet for a few seconds is not noticeable, but delaying a Telnet packet certainly is.

The *flow label* is still experimental but will be used to allow a source and destination to set up a pseudo-connection with particular properties and requirements. For example, a stream of packets from one software application on a certain source host to a software application on a certain destination host might have stringent delay requirements and thus need reserved bandwidth. The flow can be set up in advance and given an identifier. When a packet with a nonzero flow label shows up, all routers can look it up in internal tables to see what kind of special treatment it requires. In effect, flows are an attempt to have it both ways—the flexibility of a datagram subnet and the guarantees of a virtual-circuit subnet.

The source address, destination address, and flow number designate each flow. Thus, multiple flows may be active at the same time between a given pair of IP addresses. Also, in this way, even if two flows coming from different hosts but with the same flow number pass through the same router, the router will be able to tell them apart using the source and destination address.

The *payload length* field tells how many bytes follow the 40-byte header shown in Figure 10–18. The name was changed from the IPv4 *total length* field because the meaning was changed slightly. The 40 header bytes are no longer counted as part of the length as they used to be.

The *next header* field provides a means to simplify the header structure. The reason the header could be simplified is that there can be additional (optional) extension headers. This field tells which of the (currently) six extension headers, if any, follows this one. If this header is the last IP header, the *next header* field tells which transport protocol header is the last IP header and which transport protocol handlers (e.g., TCP, UDP) to pass the packet to.

The *hop limit* field is used to keep packets from living forever. It is, in practice, the same as the *time to live* field in the IPv4, normally a field that is decremented on each hop. In theory, in IPv4, it was a time in seconds, but no router used it that way, so the name was changed to reflect the way it is actually used.

Next come the *source address* and *destination address* fields. This is a fixed 16-byte address field used to convey the source and destination address. The IPv6 address space is divided up as shown in Table 10–5. Addresses beginning with 8 zeros are reserved for IPv4 addresses. Two variants are supported, distinguished by the next 16 bits. These variants relate to IPv6 packets that will be tunneled over the existing IPv4 infrastructure. The use of separate prefixes for provider-based and geographic-based addresses is a compromise between two different ways in which the Internet may operate in the future. Provider-based addresses are necessary for companies such as AT&T, MCI, Sprint, British Telecom, and so on, that are currently providing telephone services. Each of these companies will be given some fraction of the address space. The first 5 bits following the 010 prefix are used to indicate in which registry the provider is located. Three registries are currently in operation: North America, Europe, and Asia. Later on, up to 29 additional registries might be added. Each registry can further divide the remaining 15 bytes as necessary. It is expected that many of them will use a 3-byte provider number, giving about 16 million providers. This system will allow large companies to act as their own provider. One other possibility is to use 1 byte to indicate national providers and let them do further allocation. In this manner, additional levels of structure can be introduced as needed.

The geographic model is the same as the current Internet, in which providers do not play a large role. In this way, IPv6 can handle both kinds of addresses.

The link and site local addresses have only a local significance. They can be reused at each organization without conflict. They cannot be propagated outside organizational boundaries. This regulation makes the link and site local addresses well suited to organizations that currently use "firewalls" to enclose themselves off from the rest of the Internet.

Multicast addresses have a 4-bit flag field and a 4-bit scope field following the prefix, then a 112-bit group identifier. One of the flag bits distinguishes

Table 10–5
IPv6 addresses.

Prefix	Usage	Fraction
0000 0000	Reserved	1/256
0000 0001	Unassigned	1/256
0000 001	Open system interconnection (OSI) NSAP addresses	1/128
0000 010	Novell NetWare IPX addresses	1/128
0000 011	Unassigned	1/128
0000 1	Unassigned	1/128
0001	Unassigned	1/16
001	Unassigned	1/8
010	Provider-based addresses	1/8
011	Unassigned	1/8
100	Geographic-based addresses	1/8
101	Unassigned	1/8
110	Unassigned	1/8
1110	Unassigned	1/16
1111 0	Unassigned	1/32
1111 10	Unassigned	1/64
1111 110	Unassigned	1/128
1111 1110 0	Unassigned	1/512
1111 1110 10	Link local use addresses	1/1024
1111 1110 11	Site local use addresses	1/1024
1111 1111	Multicast	1/256

permanent from transient groups (recall the definition of permanent and temporary groups earlier in this chapter). The scope field allows a multicast to be limited to the current link, site, organization, or planet. The values of these four scopes are chosen such that more scopes can be added later. For example, the planet scope is 14, so code 15 is available to allow future expansion of the Internet to other planets, solar systems, and galaxies.

IPv6 also supports a new type of address known as *anycasting addresses.* Anycasting is similar to multicasting in that the destination is a group of addresses, but instead of trying to deliver the packet to all of them, it tries to deliver to just one, usually the nearest one. For example, when contacting a group of cooperating file servers, a client can use anycasting to reach the nearest one, without having to know which one that is.

A notation for the 16-byte addresses has been developed. They are written as eight groups of four hexadecimal digits with colons between the groups as follows:

```
8000:0000:0000:0000:0134:6781:89AB:CDEF
```

Since many addresses will have many zeros inside them, the leading zeros in front of the digits can be omitted, and the 16 zeroes can be replaced by 2 colons. Thus, the above address can read like this:

```
8000::134:6781:89AB:CDEF
```

Finally, IPv4 addresses can be written as a pair of colons and dotted decimal numbers, for example:

```
::192.31.20.46
```

It should be noted that there are 2^{128} addresses in IPv6. IPv6 will allow 7×10^{23} IP addresses per square meter, if the entire earth were covered with computers.

At this point we should briefly compare the IPv4 header and the IPv6 header (Figure 10–18) and explain what has been left out in IPv6. The *IHL* is gone because the IPv6 header has a fixed length. The *protocol* field was taken out because the *next header* field tells what follows the last IP header (e.g., UDP or TCP segment). All the fields related to fragmentation were removed because IPv6 takes a different approach to fragmentation. IPv6 attempts to send packets of the right size at first transmission. If the packet is too large, the router sends back an error message to the sending host, which tries to break the packet into smaller pieces and resume transmission.

Finally, the *checksum* field is gone because calculating it greatly reduces performance. With the reliable networks now used, combined with the fact that the data link layer and the transport layer normally have their own checksums, the value of having a checksum in the IP header is not worth the trouble considering the price performance has to pay.

10.10   SOCKETS AND TRANSPORT SERVICE PRIMITIVES

Transport service primitives are implementation-dependent access functions that allow application programs to access the transport service. Each transport service has its own access primitives (let us call them software functions or software routines for now). In this section we will first examine a simple (hypothetical) transport service and then look at a "real world" example—sockets.

To get an idea of what a transport service might be like, consider the five primitives listed in Table 10–6. This transport interface is truly bare bones, but it gives the essential flavor of what a connection-oriented transport interface has to do. It allows software application programs to establish, use, and drop connections, which is sufficient for many applications.

To see how the primitives (functions) listed in Table 10–6 might be used, consider a configuration with a server and a number of remote clients. The application on the server executes a LISTEN primitive, typically by calling a library procedure that makes a system call to make that piece of software in the server application block until a client turns up. *Block* is a term used in multitasking

Table 10–6
The primitives
(functions) of a
simple transport
protocol.

Primitive (Function)	(Message) sent	Explanation
LISTEN	(none)	Go to sleep (do nothing) until some remote program tries to connect with us
CONNECT	Connection request	Try to establish connection with the far end application program
SEND	Data	Send information to the far end
RECEIVE	(none)	Go to sleep (do nothing) until a transport protocol data unit (message) arrives
DISCONNECT	Disconnection request	I want to disconnect (drop) the connection

systems like UNIX in which a process (piece of software) is moved from the running state to the idle state. The process stays in the "not running" state waiting for some event to happen. If an event occurs, the process is rescheduled to run at the point where it left off. When the software running on the client wants to talk to the server, it executes a CONNECT primitive. The system sends a CONNECT REQUEST message to the server, and the client waits for confirmation from the server that it accepted the request.

Recall from our discussion in Chapter 7 that a protocol data unit (PDU) is the data exchange between two corresponding layers on two different machines. That data is enclosed in the payload of the layer below, and so forth. The same principle applies here to the transport layer. Let us call the messages exchanged between two transport entities on two different machines (a server and a client) a *transport protocol data unit* (TPDU). TPDUs will be used to denote the messages sent from a transport entity (software) to a transport entity. Thus, based on our discussion in Chapter 7, TPDUs (exchanged by the transport layer) are contained in packets (exchanged by the network layer). In turn, packets are contained in frames (exchanged by the data link layer). When a frame arrives, the data link layer processes the frame header and passes the contents of the frame payload field up to the network software. The network layer software processes the packet header and passes the contents of the packet payload up to the transport software. This nesting is illustrated in Figure 10–19.

Let us get back to our client-server example. The client's CONNECT call causes a CONNECTION REQUEST TPDU to be sent to the server. When it arrives, the transport software checks to see that the server is blocked (idling) on a LISTEN (i.e., waiting to handle requests). When this TPDU arrives, the server software is unblocked (rescheduled for running) and the connection is established. Data can now be exchanged using the SEND and RECEIVE primitives.

A state diagram for connection establishment and release for these simple primitives is given in Figure 10–20. Each transition is triggered by some event, either a primitive executed by the local transport user or an incoming packet. For simplicity, we assume here that each TPDU is acknowledged separately.

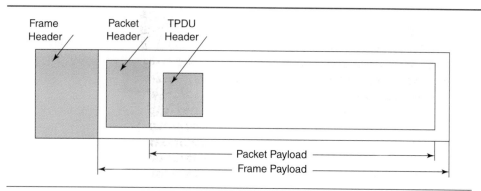

Figure 10–19
Nesting of TPDUs, packets, and frames.

Frame Header / Packet Header / TPDU Header /

Packet Payload

Frame Payload

We also assume that a symmetric disconnection model is used, with the client going first. Please note that this model is unsophisticated. We have seen a more sophisticated model in our discussion of the TCP protocol section earlier in this chapter.

10.10.1 Berkeley UNIX Sockets

Let us now examine how the transport primitives described in Section 10.9 are used in practice. In this section we will present an overview of Berkeley sockets, which are application programming interfaces to the TCP/IP protocol stack as shown in Figure 10–21. In this figure we see that sockets are set between the application (layer 5–7 of the OSI reference model) and layer 4 (TCP/UDP).

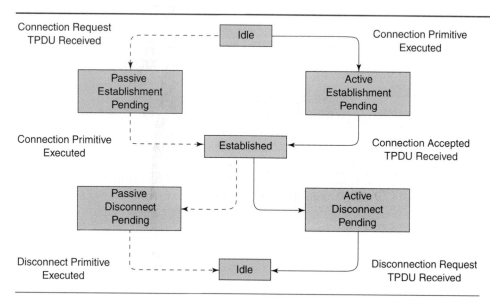

Figure 10–20
A state diagram for a simple connection management scheme. Transitions labeled in italic are caused by packet arrivals. The solid lines show the client's state sequence. The dashed lines show the server's state sequence.

Figure 10–21
TCP/IP protocol
stack with sockets.

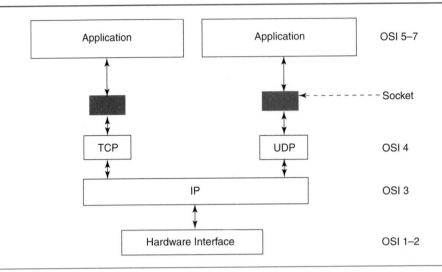

To further understand the role of sockets, consider a timeline of the typical scenario that takes place between a TCP client and server. First the server is started, then some time later a client is started that connects to the server. We assume that the client sends a request to the server, and the server processes the request and sends back a reply to the client. This process continues until the client closes its end of the connection, which sends an end-of-file notification to the server. The server then closes its end of the connection and either terminates or waits for a new client connection. This scenario is shown in Figure 10–22.

To invoke the functions afforded via the socket interface, the application running on the server or the client must invoke a UNIX C function call to one of the following functions, depending on the state of the connection according to Figure 10–22:

- `socket()`: Create a new communication end point.
- `bind()`: Attach a local address to a socket.
- `listen()`: Announce willingness to accept connections.
- `accept()`: Block the caller (the application software) until a connection attempts to arrive.
- `connect()`: Actively attempt to establish a connection.
- `write()`/`send()`: Send some data over the connection.
- `read()`/`receive()`: Receive some data from the connection.
- `close()`: Release the connection.

Protocols supported by the socket implementation are not limited to TCP/IP. In fact, the call to the socket function includes a field that specifies in which protocol the application wishes to establish the session. The protocol suite supported consists of TCP/IP, X.25/OSI, ATM, SPX and IPX, etc.

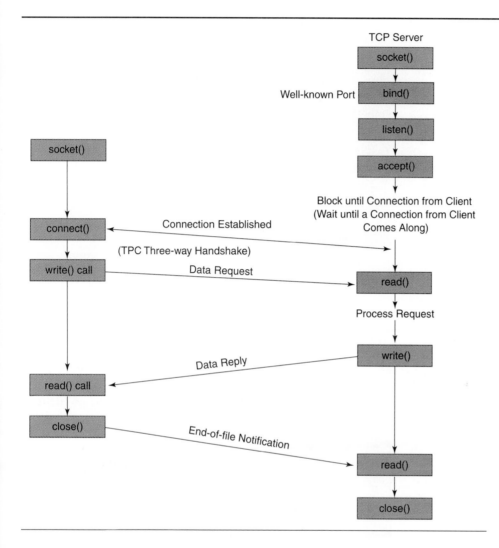

Figure 10–22
Socket functions for
elementary TCP
client-server.

10.10.2 Windows Sockets

In the late 1980s, Windows had one built-in interface known as NetBIOS. Users who wanted a different protocol (IPX/SPX, DECnet, or TCP/IP) had to purchase add-on software from third-party vendors. Third-party software companies provided TCP/IP protocol stacks and proprietary software development kits. There was a small catch—the IP vendors only sold proprietary solutions. Figure 10–23 illustrates this dilemma.

WINSOCK 1.1 came along to resolve the issues of compatibility among third-party vendors' software and Microsoft solutions. All TCP/IP vendors supported that socket interface developed by Microsoft to replace the NetBIOS solution. Figure 10–24 shows a single TCP/IP protocol stack similar to that

Figure 10–23
Two incompatible
interfaces to the
TCP/IP protocol
stack.

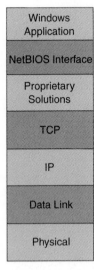

Figure 10–24
WINSOCK 1.1, sup-
ported by all TCP/IP
vendors.

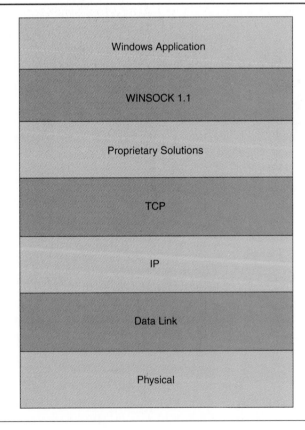

introduced in Section 10.9.1. WINSOCK 1.1 provides a core set of routines derived directly from Berkeley UNIX sockets. It also introduced Windows-specific extensions.

Microsoft then introduced a new version of a Windows socket, WINSOCK 2.0. This newer version allows an application to run over a variety of protocols, thus allowing an application to support more that one protocol at a time. It also implemented quality of service (QoS) aspects that allowed an application to inform the network what bandwidth and latency it needs. WINSOCK 2.0 specifies three levels of QoS services:

- Guaranteed.
- Predictive.
- Best effort.

10.11 SUMMARY

We presented in this chapter the concept of internetworking, known with the popular name of the Internet. We examined the structures and formats of the TCP/IP protocol elements. TCP/IP protocols will be around for years to come. Therefore, it is important to understand their roles in the Internet and their relationship to the OSI reference model.

REVIEW QUESTIONS AND PROBLEMS

1. Describe how connections are managed in TCP/IP. What protocol is used to perform that function?

2. Describe the handshake scenario that takes place between two TCP/IP endpoints.

3. What is fragmentation? Give an example of how it works.

4. List and describe the fields in an IPv4 address.

5. List and describe the fields in an IPv6 address.

6. What are the shortcomings of IPv4?

7. How does IPv6 resolve the shortcomings in IPv4?

8. How does classless interdomain routing (CIDR) work?

9. What is the domain name space?

10. How are domain names resolved in the Internet? Give an example.

11. What is a domain name server (DNS), and how does it resolve names it does not have?

12. Give the format for IPv6 protocols.

13. Give three examples of transport primitives.

14. What are sockets?

15. What are the functions of the Berkeley UNIX sockets?

16. What are the advantages of WINSOCK 2.0 over WINSOCK 1.1?

11

Introduction to Wireless Communications

OBJECTIVES
Topics covered in this chapter include the following:

- Operation of cellular telephony.
- Analog cellular technology.
- Digital cellular technology.
- Frequency allocation and frequency reuse.
- Design and expansion of a cellular system.

11.1 INTRODUCTION

No area of telecommunications has experienced the recent explosive growth like that of wireless communications. Of course, wireless communications is essentially radio communications, which is not all that new. However, the maturation of various technologies (such as very large scale integration of circuitry and advances in signal processing), cost reductions in technologies, new frequency availability, and business pressures for mobility and networking in the past few years have finally made wireless communications efficient and cost effective.

Wireless telecommunications is, unfortunately, a simple and all-encompassing phrase. As the name implies, *wireless telecommunications* means communicating without the use of wires or other physical guides or conduits (such as fiber) through some distance. There are numerous examples of wireless communications, including remote controls for TVs, stereos, garage door openers, pagers, cordless phones, and cellular phones to name a few.

wireless telecommunications

There are several advantages to wireless communications, with the most significant being that the callers need not be tied to a particular location; that is, we can have mobile terminals. Also, the medium, unlike wires, fiber, or coaxial cable, is simply air (for satellite communications it is space) and is free. The primary disadvantage is that, because of the possibility of interference among users, the electromagnetic spectrum must be subdivided, referred to as *frequency allocation*.

Many topics associated with wireless are the same as with other areas of telecommunications, but some functions, such as power and bandwidth, have a different significance. For instance, we wish to minimize power consumption in the mobile units because they are usually run off batteries. We also wish to minimize the broadcast power of the base station to reduce the effect of interference with other base stations. Bandwidth has one set of problems with terrestrial communications, for instance in a twisted pair, and a different set of problems with cellular technology where air is the transmission medium. One interesting problem that has to be handled by a cellular phone system is locations of the communicating parties; that is, the calling party must identify itself and then the system must find the party being called.

In this chapter we intend to explain the technology associated with the major form of wireless communications: cellular telephony.

11.2   PAGERS AND GPS SERVICES

paging

Pagers and paging have been around for a number of years. *Paging* is a comparatively simple and inexpensive means of communications. It is a classic example of simplex transmission in that it is entirely one way without any response from the receiver; however, two-way paging is likely to be the next wave. There are narrow and wide area pagers; some simple pagers operate only within a building, others in a radius of a couple miles, and still others can provide worldwide coverage. The concept of a pager is straightforward: broadcast a message into the air via a transmitter or transmitters and assume that the correct receiver will get the message. Pagers typically require large transmitters (high power) and many transmitters (in wide area paging systems) over which the message is simulcast. Also, since the messages are typically short, the data rates need not be high.

GPS

A similar system to paging is position-locating systems, which also require only a receiver. The modern position-locating systems rely on the *global positioning system (GPS)* satellites that are in medium earth orbit. A GPS receiver will receive signals from multiple satellites and, from the received data, calculate the location within a radius of a few meters. These systems have revolutionized navigation for ships and airplanes and they are now available in automobiles.

11.3   INTRODUCTION TO CELLULAR TELEPHONY

cell, and base station

The fastest growing and most commonly practiced and recognized version of wireless telecommunications is the cellular phone. The cellular concept is remarkable in its apparent simplicity but is rather complicated to implement because of the vast number of calls to be serviced. Think of a large flat plane divided into hexagonal *cells* as shown in Figure 11–1. Assume that a *base station* exists at the center of each cell. Mobile units within a particular cell are served

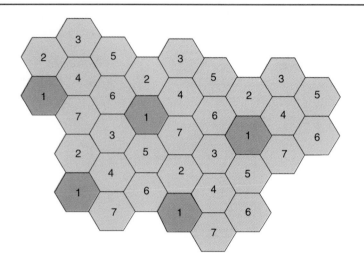

Figure 11–1
The cellular concept.

by that cell's base station, and if a mobile unit moves from one cell to another a *handoff* occurs (to the new cell's base station).

handoff

If each cell had its own unique set of frequencies, we would quickly use up the available bandwidth, but that is not necessary. If the power is adjusted properly and the spacing is appropriate, we can have sets of cells with the same frequencies, with virtually no interference, allowing for frequency reuse. Notice the groups of shaded cells in Figure 11–1. They constitute the set of frequencies used for this system. Within the group of seven cells the frequencies are different, but then these seven cells are reused over and over again. This concept allows the manufacturers of standard telephones (mobile units) to use the same set of channels. These mobiles can then be used throughout the country. Not all systems use sets of seven cells; additionally, some do not reuse frequencies at all, as will be discussed later in this chapter.

If we were dealing with devices such as walkie-talkies, when you talked on one cellular phone the signal would go through the air and be received directly by your friend on another cell phone. But of course that is not practical, especially if you are any distance away from each other and other people are using walkie-talkies in the vicinity. The other people using walkie-talkies would have to have their own frequencies to avoid interference with your conversation—a totally impractical situation. Therefore, each cell phone connects to a base station that makes the connection to another cell phone or a regular phone in the public switched telephone network (PSTN).

The base stations are connected to switches, which are connected to other switches, which are ultimately connected to the PSTN. How are these connections made? There is no need to do this through the air and use up more bandwidth, so typically it is done with land lines, which are usually fiber optic cables, although line-of-sight microwave links are also often used.

Figure 11–2
The architecture of
a cellular phone
connection.

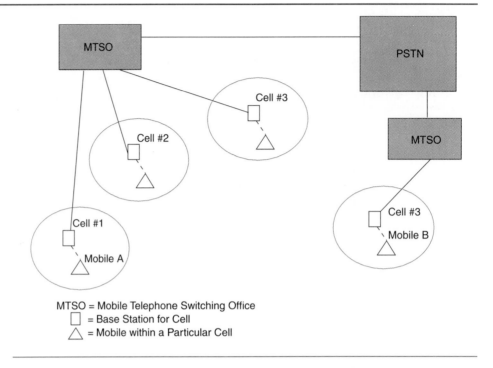

MTSO = Mobile Telephone Switching Office
☐ = Base Station for Cell
△ = Mobile within a Particular Cell

Figure 11–2 illustrates the interconnection of two cellular phones through base stations, switches, and the PSTN. We will discuss the details of how this phone call is set up shortly, but for now the important point is what the path looks like. Notice that there is no direct connection between two wireless phones through the air. Even if they happened to be in the same cell the call would be interconnected through at least a switch and a channel set up for each mobile unit.

11.3.1 Cellular Operation

The cellular concept is straightforward and ingenious. The terrain is divided into cells represented as hexagons or circles (hexagons look neater, but ragged-looking circles are closer to reality), as shown in Figure 11–1. In most systems, seven sets of cells (reuse factor of 7) are repeated. Within the pattern of seven cells, each cell has a unique set of frequencies to use, thereby avoiding interference among cells. There are no adjacent cells where the same frequencies are used. Notice the shaded cells (number 1), which have the same set of frequencies but are never adjacent to each other. Cell 1, in each set of seven cells, is always surrounded by cells 2, 4, 7, 5, 3, and 6 (going around cell 1 in a clockwise direction).

When a terminal within a cell initiates a call, the base station establishes a channel for the call. The channel consists of two carrier frequencies: one frequency is for the base station to transmit to the mobile unit (referred to as the *forward channel*) and another frequency is for the mobile unit to transmit to the base station (referred to as the *reverse channel*). Within a channel, the forward and reverse sets of frequencies are separated to provide full-duplex operation.

forward channel
reverse channel

If a terminal moves from one cell to another, a handoff must occur. The base station currently serving a mobile continuously measures the received signal. If that signal falls below a specific level, then a handoff sequence is initiated. The first step is for the base station to send a message to its switch (*mobile telephone switching office, MTSO,* or mobile switching center, MSC). The switch asks nearby base stations to measure the signal from the mobile unit, and the base station having the strongest signal will then handle the call. This usually requires switching the mobile to another channel (referred to as a *hard handoff*). If all goes well, the switching to another base station is virtually unnoticed by those involved in the conversation.

MTSO

hard handoff

If the unit happens to be adjacent to an area served by another system, then an intersystem handoff is initiated. This type of handoff is more complicated, but the principle is the same. In the end, the mobile will be connected to a new base station, which is part of a different MTSO.

11.3.2   Cell Size and Expansion of Systems

In areas of low population density and thus little cellular traffic, the cells can be up to several kilometers in diameter, depending on the lay of the land—hills, valleys, large buildings, and other obstructions. All of these configurations affect the design and layout of cells. Of course, power restrictions also determine how large a cell can be. As we move into more densely populated areas, including suburbs and cities, the phone traffic expands proportionately as do building obstructions. Another area of increased mobile phone usage is along major highways and at the intersection of two or more major highways. These conditions necessitate having smaller cells.

There is a lot of research and developmental software on the subject of designing cells. Most of the fundamental mathematics and computer science theories have been developed, but many researchers are still studying the nature of the optimum definitions of cells. Many telecommunications engineers and technologists are involved in cell design and antenna design and positioning.

In the recent past, an explosive growth of cellular telephone usage has occurred. Of course, manufacturers and service providers are delighted with the increased business, but there are consequent problems. The biggest problem occurs when all the channels are in use and there is no room for another call. This leads to call *blocking*—not allowing a subscriber to initiate a call. Incidentally, it is a dilemma for service operators to choose between allowing a new call versus allowing a handoff of a call in progress into a busy cell. Typically, companies have decided that customers find it more annoying for a call to be terminated than to have to wait to start a call. The service provider will leave some channels unused even though a caller is waiting to initiate a call. The

blocking

main problem is that the increased traffic can only be accommodated in the following ways:

- Expand bandwidth and reallocate frequencies—this is a problem because only so much spectrum is available (the allocation was expanded from 40 MHz to 50 MHz) and the channels cannot be made narrower without serious interference problems.
- Subdivide the cells so there are fewer users per cell—this is the main way users have been added, but there are practical limitations on how small cells can be.
- Develop and implement new technologies—this is also being done (TDMA and CDMA).

11.3.3   Signal Degeneration in Cellular Systems

A signal moving down a copper wire is attenuated so many dB per foot due to loss in the wire. Also, a light signal in a fiber optic cable is attenuated. We can use amplifiers and regenerators at the appropriate spots to bolster these signals. However, we do not have such a luxury for radio signals in the air, and these signals also degenerate with distance. Of course, this degeneration is another limitation on cell size.

A common unit of power is the dB where dB is defined as $P(dB) = 10\log [P(W)]$. However, this is a fairly large unit for cellular systems and more often the unit dBm is used, which is defined as: $P(dBm) = 10\log[P(mW)] = 10\log[\text{watts}/1\text{mW}]$. For example, a transmitter transmitting 2 watts of power could be expressed as:

(a) $P(dB) = 10 \log[P(W)] = 10 \log(2) = 3.01$ dB or,

(b) $P(dBm) = 10 \log[P(mW)] = 10 \log(2W/1mW) = 10\log(2000) = 33.0$ dBm.

There are four major impediments regarding signal integrity in cellular telephony:

1. The power loss due to distance, assuming free space (no obstructions), can be determined. The received powers, P_R, can be determined from $P_R = kP_T/(4\pi)^2d^2$ where P_T is the transmitted power, d is the distance, and k is a constant that is determined from such variables as transmitter and receiver antenna gains (G_t and G_r), wavelength of the signal (λ), and losses in the system hardware (L). The formula for k is: $k = G_tG_r\lambda^2/L$.

2. Objects such as buildings and hills in the path of the signal cause both signal loss due to *absorption* and additional signal paths due to *reflection*. These multiple paths can cause a single signal to arrive at the receiver at different times, producing phase shifts.

slow fading

3. Motion of the terminal across the terrain causes what is called *slow fading*; that is, at different spots the reception will be stronger or weaker than in other spots. These signal changes are mostly random because of the general unpredictability of the terminal's motion. However, in rural areas the main effect is decreasing power due to distance (approximately a linear decline with the log of the distance from the transmitter). In cities and suburbs the random effect is more noticeable.

4. *Rayleigh fading* (or *fast fading*) is due to the terminal moving quickly through a cell. The various rays reaching the receiver undergo a Doppler shift (the same as the familiar examples of train whistle frequency changes or the red shift of stars). Sometimes the signals add and sometimes they subtract, adversely affecting the quality of the received signal even when the average received signal is strong.

Rayleigh (fast) fading

There are various competing factors in the quality of service relative to power. On the one hand, we would like to boost the power from the base stations and from the mobiles (forward and reverse), but on the other hand increasing power increases the likelihood of interference among cells. Also, since mobiles are typically battery powered, increasing the transmit power of a mobile unit proportionately reduces its battery life. Thus, system designers look for other methods to increase the quality of service.

11.3.4   Roaming, Intersystem Operations, and IS-41

Roaming occurs when a subscriber for a particular service provider moves into an area administered by another service provider. Typically, this has meant an extra charge tacked on to the user's bill; however, competition among the service providers is, if not totally eliminating these expensive roaming charges, at least making the regions where there is no roaming charge larger and larger.

roaming

Aside from charging customers money for roaming, technological problems need to be overcome to actually allow roaming. In the early days of cellular telephony, large and/or nearby cellular carriers made agreements and common procedures to set up and document calls for billing purposes. This system was problematic, however, in that there was no national standard for this rapidly growing industry. In the late 1980s and early 1990s the Telecommunications Industry Association (TIA) developed *interim standard 41 (IS-41)* so a national standard (U.S. and Canada) would provide services to roaming subscribers.

IS-41

IS-41 is actually a whole seven-layer protocol. The standard specifies devices, interfaces, switches, and even databases. IS-41 borrows from and is consistent with the *global system for mobile communications (GSM)*. GSM is the digital cellular system standard used in Europe and is a very comprehensive standard in comparison to most U.S. standards that are mostly concerned with the interface between the base and mobile terminals, referred to as the *air interface*.

GSM

A whole vocabulary of messages is defined by IS-41. These messages are mainly used to manage intersystem handoffs. Roaming is coordinated by two major databases: a *home location register (HLR)* and a *visitor location register (VLR)*. The critical element in making roaming work properly is the *system identifier (SID)* number. Every operating company receives an SID when it receives an FCC license. The SID associates any mobile phone with its service provider and thereby indicates whether it is roaming or not; that is, if the mobile unit is outside its service area its SID will not match the SID of the base station where it is, and it is considered to be a "roamer."

HLR
VLR
SID

Most of the large wireless service providers are phasing out roaming and have instituted one-rate type plans. From a bureaucratic and financial point of view, this is probably the end of roaming, but from a physical, software, and technical point of view, roaming will still exist because the connections have to be made to the various systems. There are other industry plans to be considered such as local number portability (LNP) and calling party pays (CPP). The industry is so new and is growing so fast that it is difficult to predict which systems and which standards will survive.

11.3.5 Multiple Access

Let us consider a single cell with numerous mobile units within that cell. All of these phones are trying to access the base station for the cell. How can the base station service these various phones? There must be a way to separate them and have them operate somewhat, or apparently, simultaneously; that is, we need multiple access. Recall that multiple access was also a problem that had to be resolved in LANs.

FDMA

There are three basic techniques for multiple access. The first generation of phones used a totally analog approach that is referred to as *frequency-division multiple access* or *FDMA*. Similar to radio stations on the AM or FM dial, the channels within a cell are separated by frequency. Each mobile unit in a cell is assigned a set of two frequencies for forward and reverse transmissions, or duplex operation. The available bandwidth determines how many phones can operate within the cell.

TDMA

The second generation of cellular telephony uses *time-division multiple access* or *TDMA*. This system follows interim standards 54 and 136, sometimes referred to as *North American TDMA* or *NA-TDMA*. The TDMA system actually uses a combination of frequency- and time-division multiplexing. Each frequency used is separated into time slots and channels are assigned to time slots within the frequencies. This system increases the amount of available channels compared to standard FDMA and thus provides higher spectrum efficiency.

CDMA

Another second-generation technology uses *code-division multiple access* or *CDMA*. In this scheme of multiple access, all the cellular terminals use the full frequency spectrum but are separated by an individual code for each mobile unit. This code becomes part of the modulation technique before transmission and it is up to the base station to extract the signal for a particular mobile unit from all the other signals by demodulating with the code for that mobile unit.

Figure 11–3 diagrams the three different multiple access techniques. Notice in Figure 11–3(a) that in the FDMA method the channels are defined by their frequencies, and that a full-duplex channel consists of two frequencies with one for forward (base station to mobile) and one for reverse (mobile to base station), for example, "a1" and "a," respectively. Figure 11–3(b) shows the TDMA method in which each frequency is divided into time slots, thus providing more channels per frequency. In Figure 11–3(c) the PN codes define and separate the channels.

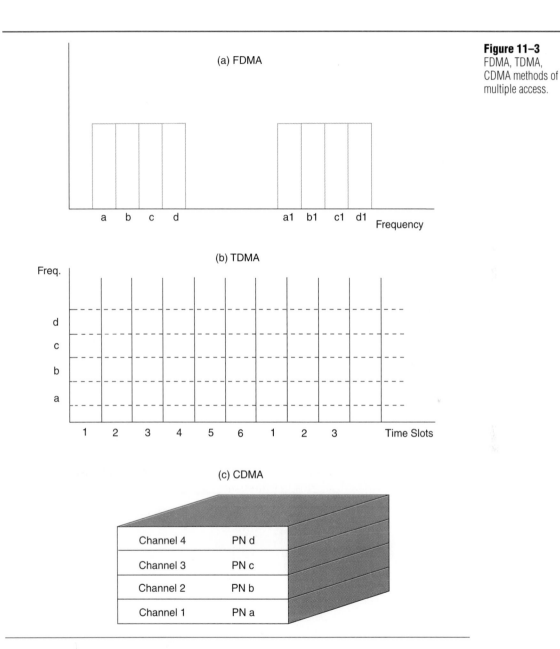

Figure 11–3
FDMA, TDMA,
CDMA methods of
multiple access.

11.3.6 Analog Systems, Advanced Mobile Phone System (AMPS)

The first-generation cellular system is the *advanced mobile phone system (AMPS)*. AMPS
Although digital systems are expanding quickly, millions of cellular phones
still employ the analog system. In fact, digital phones manufactured for use in
the U.S. must also have the dual capability of using AMPS. There are two advantages to dual-mode phones: (1) digital phones have the hardware to roam

into all analog areas, and (2) it prevents manufacturers and service providers from jumping on the digital bandwagon totally and forcing analog phone customers to throw out their analog phones and sign a digital contract.

If an infinite amount of frequency spectrum were available, we could divide the frequencies and allocate each cell phone its own set of forward and reverse frequencies. There is a limited spectrum, however, so each cell has a unique set of frequencies and seven different sets are reused far enough apart so that those cells using the same frequencies do not interfere with each other. Thus, we have frequency-division multiple access.

In any given area, the AMPS frequency allocation is as follows: To promote competition the frequencies are divided so that within each area there are always two spectrum allocations, A and B. One operating company uses the A frequencies and another uses the B frequencies. The band of frequencies for forward transmission (base station to mobile) is 869–894 MHz and the band for reverse transmission is 824–849 MHz. Each channel occupies 30 kHz. Dividing up the frequencies creates 416 channels in the forward band and 416 channels in the reverse band for a total of 416 pairs of 30 kHz channels or $416 \div 7 = 59$ channel pairs per cell. Some of the channels are required for signaling and the remaining channels for traffic.

What type of signal is actually transmitted by the mobile unit? The obvious signal is the carrier frequency that is within the bands just discussed. The carrier is modulated by the voice signal, but actually there is more (see Figure 11–4). Each base station has its own *supervisory audio tone (SAT)*. When the call is set up, a base station and its SAT are established. During the call both the base station and the mobile inject the SAT into the modulator and both continually monitor the SAT to ensure that the mobile unit is receiving the right signal and not that of another base station. The analog signal (usually

SAT

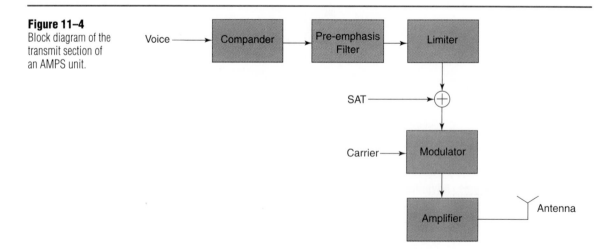

Figure 11–4
Block diagram of the transmit section of an AMPS unit.

voice) is also processed before modulation. Because of the high dynamic range of the speech, the signal is also compressed or companded.

Two channels handle the actual conversations: the *forward voice channel (FVC)*, and the *reverse voice channel (RVC)*. The mobile switching center sets up the call by directing the appropriate base station to dedicate an FVC and RVC pair of channels for the duration of the call. The base station monitors the strength of the RVC and SAT signals to help in handoff decisions. If these signals fall below a certain threshold, then the MSC initiates a handoff.

FVC
RVC

AMPS Signaling and Messages Channels 313–354 are used exclusively for system control information. When a call is in progress, AMPS uses a system called *blank-and-burst*. As the name implies, the system interrupts the conversation by blanking the sound and inserts a control message. This process takes only 100 ms or less and is virtually unnoticed by the callers because it only sounds like a click and, as long as the clicks are infrequent, it is not bothersome.

blank-and-burst

Four AMPS logical signaling channels are used to carry messages. The formats for each of the control channels are similar in that they all begin with an alternating binary sequence for synchronization. This sequence is produced in AMPS by frequency shifts using Manchester-coded frequency shift keying. These signals are created using the carrier ±8 kHz. Also, every control channel format will follow the synchronization bits with an 11-bit Barker code, which consists of 11100010010. Figure 11–5 shows the general format for these control channels. After the Barker code, comes the actual control message.

1. *The forward control channel (FOCC)* is broadcast by the base station and received by the mobile units in its area. As mentioned earlier, each area is subdivided into two systems, A and B. These phones are numbered such that A phones have even numbers and B phones have odd numbers. Therefore, each frame carries a 28-bit code word for terminals with even phone numbers (A) and a 28-bit code word for terminals with odd phone numbers (B). Among the messages carried over the FOCC are the mobile identifier number (MIN) page for a called mobile and the message for a mobile to move to a specific voice channel. The FOCC also broadcasts global messages (pertinent to all mobiles in the area), such as the frequency terminals should use to transmit registration messages and the power level for messages transmitted by the terminals. The base channel continuously transmits on the FOCC. When there are no actual messages, the system sends filler messages that contain a 3-bit number specifying the transmit power level for messages transmitted by the mobile units.

FOCC

Bit Sync	Baker Code	Control Information
1010101....	11100010010	Binary Code Words

Figure 11–5
General format for AMPS control channels.

RECC

2. The *reverse control channel (RECC)* is random access from the mobile unit to the base station. It is random access because any mobile could be transmitting on the channel and there must be some means of dealing with a number of mobiles contending for the base station's attention. The mobile observes the busy/idle bit on the FOCC for an available window to transmit its message. The format follows the general format shown in Figure 11–5. After the Barker code is a 7-bit digital color code, which exists for each base station and allows the mobile unit to confirm it is part of this base station. Following the digital color code is the actual message. The types of messages sent over the RECC *logical channel* are acknowledgment of receipt of page, sending the mobile's electronic serial number (ESN), initiation of a call request along with MIN, and the called party's number.

logical channel

3. The FVC occupies the same band as the conversation and is thus in-band signaling. Since this is a dedicated channel, it is actually a one-to-one message with no need to identify the receiving mobile unit. Each message is repeated 11 times to increase the probability of proper reception at the mobile unit. The FVC carries such critical messages as an alert of an incoming call, handoff, and power level change. An example of a 28-bit handoff message is shown in Figure 11–6. This 28-bit code is further encoded into a 40-bit code using the Bose-Chaudhari-Hocquenghem block code before being attached to the sync code and the Barker code and then repeated 11 times and finally transmitted.

4. The RVC is also in the same band as the conversation. It has a similar format to the FVC and carries only messages in response to messages from the base station, such as confirmation of receipt of a power control message.

Two other signals of importance are the SAT and ST signals. The SAT, previously mentioned, is the supervisory audio tone, and is transmitted by both the base station and the mobile unit. There are three different SAT signals: 5970, 6000, and 6030 Hz. SAT signals are used to distinguish nearby base stations from each other. The signaling tone (ST) is a 200-ms burst of alternating 1s and 0s sent by the mobile unit to indicate an end of call.

In the early 1990s, an enhanced analog system, referred to as *N-AMPS*, was developed and implemented by Motorola. The N stands for narrowband. This system divides the 30-kHz AMPS channel into three 10-kHz channels, essentially increasing the capacity by a factor of three, thus making it useful for densely populated areas.

Figure 11–6
28-bit handoff
message format.

No. of Bits 2	2	2	8	3	11
10 (Start Bits)	SAT (New)	SAT (Current)	Unused	New Power Level	New Channel #

11.3.7 Time-Division Multiple Access (TDMA)

With the demand for cellular telephony far exceeding expectations, the various companies examined new technologies to extend or replace the analog AMPS system. Not only is system capacity being approached but telephone users have also come to expect more services than can be provided by a typical analog system; a higher-quality signal; a better and more secure user authentication process; and an improvement in roaming. Time-division multiple access (TDMA) is a response to these demands.

TDMA has approximately three times the capacity of AMPS, but because it uses digital technology, it has many other features and improvements as well. Enhanced versions of this system are often advertised as digital personal communications services (PCS) systems. TDMA is actually a hybrid system that implements multiple access by a combination of time and frequency division. Each frequency is divided into time slots, so the total number of physical channels becomes a product of the number of frequencies in a cell and the number of time slots per frequency times the reuse factor (number of cells before reuse). This combination of frequency- and time-division multiple access is officially referred to as *United States digital cellular (USDC)* or as *digital AMPS (D-AMPS)*.

The specifications for this first generation of digital cellular are laid out in *interim standard 54 (IS-54)*. IS-54 specifies dual mode, allowing for a smooth transition to digital by allotting the same frequency band and spacing as AMPS but with multiple users on each carrier (time division). So a given base station may have some of its channels dedicated to the digital cellular units and other channels dedicated to analog. As more digital phones are being used more channels become dedicated to USDC.

IS-54

The voice channels each occupy 30 kHz of bandwidth as in the AMPS systems. However, each frequency is divided into time slots as was shown in Figure 11–3(b). The six slots together constitute a frame and are transmitted in 40 ms, therefore the rate could be expressed as 25 frames per second. (Research has shown that 40 ms is approximately the upper limit of human tolerance for breaks in speech. So 40 ms is an important number in cellular telephony.) With the whole frame occupying 40 ms, the time for a single slot is $1/6$ of 40 or 6.67 ms. For full-rate transmission, each user has two time slots per frame, but for half-rate speech transmission, each user has one time slot per frame. See Figure 11–7 for an example of full-rate and half-rate use of the time slots.

North American-TDMA, or NA-TDMA, is similar to the previously discussed system that the Europeans developed and implemented (in the mid-80s to early 90s) called the global system for mobile communications (GSM), which is a digital system virtually replacing the various analog systems in operation

	Slot 1	Slot 2	Slot 3	Slot 4	Slot 5	Slot 6
Full-rate	User # 1	User # 2	User # 3	User # 1	User # 2	User # 3
Half-rate	User # 1	User # 2	User # 3	User # 4	User # 5	User # 6

Figure 11–7
Example of full and half rate in TDMA.

Figure 11–8
IS-136 digital traffic
channel (DTCH) for
one slot.

(a) The Six Data Fields of One Slot for Forward (Base to Mobile)

Bits	28	12	130	12	130	11	1
Field	SYNC	SACCH	DATA	DVCC	DATA	DL	RES'D

(b) The Five Data Fields of One Slot for Reverse (Mobile to Base)

Bits	6	6	16	28	122	12	12	122
Field	G	R	DATA	SYNC	DATA	SACCH	DVCC	DATA

throughout the European continent. Unlike the European system, North American service providers are required to make dual systems available; that is, every cell phone with digital technology must also be able to work with analog technology, AMPS. The applicable standard for NA-TDMA is *interim standard 136 (IS-136).*

IS-136

Duplex transmission is accomplished with different frequencies as in the AMPS system, so both sets of frequencies are time multiplexed into time slots. Each slot contains a total of 324 bits. There are six slots per frame, therefore each frame transmits a total of $6 \times 324 = 1944$ bits. Under IS-136 there are two basic logical channels: the *digital traffic channel (DTCH)*, and the *digital control channel (DCCH)*. Figure 11–8 shows a time slot in the digital traffic channel and how the 324 bits of the slot are allocated.

DTCH
DCCH

SACCH

Note that both forward and reverse transmissions contain the *slow associated control channel (SACCH)*, which is used to carry control messages between the mobile and the base station spread out over several time slots. Among the information carried by the SACCH are power measurements and requests from the mobile for a handoff. Unlike the AMPS where handoffs are totally out of the control of the mobile unit, IS-54 and IS-136 provide for mobile-assisted handoff (MAHO).

Notice in Figure 11–8(b) that guard (G) and ramp (R) time is indicated for the reverse direction (mobile to base station), but no such data fields are needed in the forward direction. The reason for this is that the mobile actually shuts down between transmissions (remember it is time multiplexed so it should not transmit continuously, which has the added advantage of conserving power), but then needs the *ramp time*, 6 bits or .123 ms, to come up to full power. The *guard time* is used to allow some space between the end of the previous slot and the beginning of the present time slot.

ramp time
guard time

Figure 11–9 shows the contents of one slot of a DCCH. The total number of bits is, as in the DTCH, 324 bits. In actuality the slots are each subdivided into two blocks. Thirty-two blocks are used to form a superframe and two superframes are combined into a hyperframe. You will also notice that the format and some of the fields for a slot are the same as for the DTCH. Figure 11–9(a) shows the forward and 11–9(b) shows the reverse formats for a slot.

(a) The Six Data Fields of One Slot for Forward (Base to Mobile)

Bits	28	12	130	12	130	10	2
Field	SYNC	SCF	DATA	SFP	DATA	SCF	RES'D

(b) The Five Data Fields of One Slot for Reverse (Mobile to Base)

Bits	6	6	16	28	122	24	12	122
Field	G	R	PREAM	SYNC	DATA	SYNC+	DVCC	DATA

Figure 11–9
IS-136 digital control channel (DCCH) for one slot.

In Figure 11–9(a) and (b) we see the data, sync, G, and R fields just as in the DTCH. In Figure 11–9(a), the forward DCCH, the shared channel feedback (SCF) provides feedback information to mobile units in regard to various requests and information sent by the mobile units to the base station. The superframe phase (SFP) indicates what block number the current block is (1 of 32).

In Figure 11–9(b), the pream and sync+ fields carry additional synchronization information. The last data field is sometimes replaced with an abbreviated slot containing 78 bits of data and 44 bits of ramp and guard time.

TDMA requires the storing of information. Thus, although it could be implemented with analog technology, it is more practical with digital technology. On the transmit side, the information is first stored, compressed, and then transmitted during the appropriate time slot. On the receiver side, the information must be stored as received because of the high rate of speed and then played back at the normal rate of speed such that the gaps (between the allocated time slots) are not noticed.

Among the similarities with AMPS are the 30-kHz carrier spacing and the maximum power for an NA-TDMA terminal at 4 watts; however, it has 11 power levels (3 more than AMPS), and modulation is different. In AMPS systems the modulation is FM whereas in TDMA it is $\pi/4$ DQPSK, which is essentially a sine wave with one of eight possible phase angles. Note that $360° = 2\pi$ and $2\pi/8 = \pi/4$, thus the spacing between phase angles is $\pi/4$ and hence its name.

The 28 bits of sync in both the forward and reverse directions allow for frame synchronization; the sync is also used for training an *adaptive equalizer*. An equalizer works by first sending a known signal, which the receiver is trained on, and then by using this information (on how the known signal is affected by the transmission) to modify the received message and thereby improve it.

adaptive equalizer

Aside from the equalizer, another innovation used in TDMA systems is interleaving. One problem that arises in digital encoding of speech is that if a relatively long fade in the signal or a long burst of noise occurs, important bits could be lost. The approach is to interleave data from two different speech frames. In a block interleaver the data is in the form of a matrix with alternating data by rows; for example, row 1 might have encoded speech from frame A, then row 2 would have speech from frame B, then row 3 would be back to frame A, and so on.

This technique would allow a noise burst or signal fade to affect only the discontinuous part from each frame. The deinterleaver at the receiver will then produce speech without long strings of corrupted speech. As mentioned before, 40 ms is a critical unit of time in cellular telephony. Thus, the interleavers and deinterleavers are all designed to introduce no more than 40 ms of delay.

11.3.8   Code-Division Multiple Access (CDMA)

Code-division multiple access (CDMA) is a different approach to multiple access when compared to FDMA and TDMA. CDMA depends on spreading the used spectrum, in fact to the point where a channel requires not 30 kHz but actually 1.25 MHz. The concept of spread spectrum has been around since World War II, when it was used by the military to implement secure and relatively noise-resistant communications.

CDMA had been proposed as a means of multiple access for cellular telephony for several years, but it was not practically developed until the early 1990s by Qualcomm. Qualcomm, a company with headquarters in San Diego, holds most of the patents on CDMA and licenses other companies to use the technology. CDMA systems follow the interim standard 95 (IS-95). IS-95, similar to IS-54, requires the ability of dual-mode operation; that is, all equipment must implement AMPS as well as CDMA.

Some early optimistic predictions were made regarding the improvement in system capacity relative to AMPS—up to 30 or 40 times the capacity—but it looks like the actual improvement will be around 10 times the capacity. CDMA spectrum licenses have been bought up very quickly and expensively, and as of 1998 CDMA is available virtually across the entire U.S.

The basic idea of spread spectrum, as the name implies, is to spread the information to be transmitted across the whole available spectrum, then to extract that information at the receiver. At the expense of heavy processing at the transmitter and receiver, CDMA provides a robust, highly spectrum efficient, and highly secure transmission system.

The two most common spread spectrum techniques are: (1) frequency hopping, which is implemented by changing the carrier frequency with every frame and sequencing through a number of frequencies, thereby creating a large bandwidth of frequencies; and (2) using a binary spreading sequence to modulate a digital carrier before this new signal modulates a radio frequency (RF) carrier. The latter technique is used in CDMA systems.

A simplified view of CDMA would be that each mobile unit has a unique binary code. That code is used to tag its transmissions to the base station and from the base station to the mobile. On receipt of a transmission, the receiver uses the unique code to extract the desired signal from all the other received transmissions.

Therefore, in a CDMA system, signal strength, frequency, and time are essentially irrelevant in distinguishing one channel from another; the code actually defines a channel. The transmission process requires two steps: first, the digitized signal (usually voice) is modulated with a digital carrier to create the spread spectrum signal, and second, the spread spectrum signal is used to modulate an RF sinewave to create the transmit signal.

The spreading procedure is different for forward (base to mobile) and reverse (mobile to base) transmissions. In the forward link the user data is encoded, interleaved, and then spread by 1 of 64 orthogonal Walsh functions before modulating the carrier. Orthogonality ensures that the signals will not interfere with each other. In the reverse link, the data is again interleaved and spread with a Walsh function, but the data is further spread to increase its resistance to interference.

Because of the use of the Walsh codes, 64 physical channels are available to the base stations. Two channels are always used: one channel for pilot and the other for sync. The *pilot channel* continuously transmits a sequence of 0s at 1.2288 Mchips/sec. Terminals can use this signal for phase and timing references as well as signal strength measurements. (A *chip* consists of small data bits that are formed from the original data by multiplying each original bit by a pseudo-noise code, thus dividing the original signal into smaller bits and increasing or spreading its bandwidth.)

The *sync channel* continuously transmits information to the terminals such as the base station's identifier number and system time. The sync and pilot are of course only forward channels. Two other forward logical channels are paging and traffic. The paging channel is sent to terminals not actively involved in calls. The forward traffic channel carries user information (voice), power control messages, and other signaling messages.

There are two basic logical reverse channels: the *access channel* (used to initiate a call, respond to a page, or register its location) and the *traffic channel* (user information and signaling messages).

One of the additional innovations of CDMA is in *soft handoffs*. Recall that with AMPS MTSO totally controlled handoffs and TDMA instituted mobile-assisted handoffs. In CDMA, handoffs are *soft* meaning that instead of breaking contact with the current base station before connecting to another one, there is a period of time when the mobile unit is communicating with both base stations. This process can occur because each mobile has its own code that defines its channel, so there is no possibility that another mobile will be using that channel in either cell.

Power control of the mobile units in a cell is critical to proper operation in that cell. The so-called near–far problem is directly related to power. For CDMA to work properly, the base station must receive signals from all the terminals in the cell (or nearby cells that are within range) at virtually the same power level, within a relatively small tolerance. Therefore, mobiles close to the base station must actually reduce their power so they do not overload signals received from mobile units not as close. The base station sends commands to the transmitting terminals by multiplexing with speech and other control information on a forward traffic channel. The power control is twofold in that there is open loop power control where the terminals themselves make adjustments based on their own measurements. The mobile unit uses information received from the base station along with its own measurements to perform

pilot channel

chip

sync channel

access channel
traffic channel

soft handoffs

closed loop power control. This is a frequent operation—the base station makes a decision as to each terminal's power every 1.25 ms.

11.3.9   Identifiers in Cellular Telephony

A critical issue in cellular telephony is the different entities identifying themselves. By this we mean such things as base stations having their own IDs, each cell phone having its own ID, and networks also having their own IDs. Unfortunately, most entities need more than one ID. See Table 11–1 for a list and description of the identifiers commonly used in the three systems we have discussed: AMPS, NA-TDMA, and CDMA. Notice there is some commonality, but be careful since some of the IDs may be the same but have a different number of bits.

11.3.10   PCS—Cellular Telephony Gets Smarter

PCS

Personal communications services (PCS) has achieved the ultimate goal of all the wireless service providers. Regardless of the technology used, a PCS phone would have several additional services beyond basic cellular telephony, including voice mail, call forwarding, caller ID, call waiting, conference calls, and short message services. In other words, voice communications in PCS is the starting point to access all information services.

True PCS, as defined by the FCC, operates in the frequency range of 1850 to 1990 MHz. Other systems may call themselves PCS and offer similar services, but operate at different frequencies. For instance, TDMA and CDMA systems operating at 800 MHz by companies such as AT&T and Bell Atlantic advertise their services as PCS and do provide PCS services. Omnipoint and Sprint PCS operate at 1900 MHZ, with GSM and CDMA technologies, respectively.

The advantage for companies operating in the 800 MHz range is that they were able to use their existing antenna infrastructure because that frequency range is the same as has been used by the analog cellular system. Also, the phones for these systems have been dual mode, allowing users to tap into AMPS systems in areas where PCS is not yet available. Those companies using the higher-frequency range have had to build new antennas at considerable cost in time and money.

PCS phones, at this time, are considerably more expensive to buy, at 5 to 10 times the cost of a regular cell phone. The call quality is better, with the CDMA systems having the best sound quality. The coverage areas for PCS are gradually increasing, with the areas in and around cities increasing at a faster rate. Security is better, with eavesdropping virtually impossible except that, of course, a dual-mode phone roaming in a non-PCS area is susceptible.

11.3.11   Frequency Allocations

The actual frequencies used by the carriers for the various systems are all in the high MHz range. For the AMPS system there are two sets of frequencies: reverse (mobile to base station or uplink) is 824–849 MHz and forward (base station to mobile or downlink) is 869–894 MHz. The carriers are separated by 30 kHz. Recall that the frequencies are allocated to an A and B provider so that

Table 11–1
Identifier numbers and signals used by AMPS, NA-TDMA, and CDMA.

Name	Description	AMPS	NA-TDMA	CDMA
MIN	Mobile Identifier No.: subscriber's number from operating company	34 bits	34 bits	34 bits
ESN	Electronic Serial No.: phone's number from manufacturer	32 bits	32 bits	32 bits
SID	System Identifier: local area number from regulators	15 bits	15 bits	15 bits
SCM	Station Class Mark: indicates phone's capabilities	4 bits	5 bits	8 bits
SAT	Supervisory Audio Tone: different frequency assigned to each base station by company	Different frequencies in audio range	Different frequencies in audio range	***
DCC	Digital Color Code: used to help distinguish base stations	2bits	2 bits	***
A	Authentication Key: "secret number" *only* stored in phone and secure database of provider.	***	64 bits	64 bits
IMSI	International Mobile Subscriber Identifier	***	50 bits	***
PV	Protocol Version: gives capabilities of phone or base station	***	4 bits	***
SOC	System Operator's Code	***	12 bits	***
BSMC	Base Station Manufacturer's Code	***	8 bits	***
LOCAID	Location Area ID: indicates location of base station	***	12 bits	***
DVCC	Digital Verification Color Code: serves same purpose as SAT	***	12 bits	***
MOB_X	Mobile IDs where X is manufacturer code, model #, firmware revision, or protocol revision	***	***	8 or 16 bits
BASE_X	Base Station IDs where X is identifier, class, latitude, or longitude	***	***	4 to 23 bits
REG_ZONE	Registration Zone: replaces LOCAID	***	***	12 bits
NID	Network ID: group of base stations forming a network	***	***	16
PN_OFFSET	Pseudo-Noise Code Offset: defines delay	***	***	9 bits

*** Not used in this system

Table 11–2
Frequency
allocations.

	AMPS	FDMA/TDMA	CDMA
(Cellular)			
Reverse	824–849 MHz	824–849 MHz	824–849 MHz
Forward	869–894	869–894	869–894
(PCS)			
Reverse		1850–1910 MHz	1850–1910 MHz
Forward		1930–1990	1930–1990
Carrier Spacing	30 kHz	30 kHz	1.25 MHz
Number of Physical Channels/Carrier	1	3	Varies

the total number of channels available is halved. Taking 25 MHz (of the forward or reverse direction) and dividing by 30 kHz yields 833 physical channel pairs, but then dividing it by 2 (2 providers) yields 416 pairs of 30 kHz channels. If we now take the 416 pairs and divide by the reuse factor of 7, we end up with 59 channels per cell.

The TDMA system is assigned to the same set of frequencies as AMPS; however, because of time multiplexing into slots, each carrier handles three physical channels. Thus, TDMA has three times the capacity of AMPS. Also, PCS is allotted frequencies of 1850–1910 MHz for reverse and 1930–1990 MHz for forward channels.

CDMA has the same set of frequencies as TDMA, but these frequencies are used differently. The carrier spacing per channel is approximately 1.25 MHz, thus full duplex requires approximately 2.5 MHz. Of course, this is due to the spread spectrum method of CDMA. The number of physical channels cannot be specified exactly but is soft in that it depends on the signal-to-noise ratio. Table 11–2 summarizes the frequency allocations for the three systems.

The one thing that all three systems (AMPS, TDMA, and CDMA) have in common is the method of duplex operation; that is, the forward versus reverse transmissions. This process is accomplished by frequency separation—as noted above, a set of frequencies is always assigned for forward or downlink transmissions and a different set, at least 45 MHz away, is assigned for reverse or uplink transmissions. TDMA not only separates by frequency but also by time—there is an offset in time of approximately 2 ms so that the mobile does not have to receive and transmit at the same time, thereby simplifying the unit's circuitry.

11.4   WIRELESS NETWORKING

WLAN

Wireless local area networks (WLANs) are becoming more common in businesses, convention centers, college campuses, and many other similar facilities. They are generally most useful in confined areas with short distances (a few hundred feet). Also, they are typically operated on a noninterference basis and are thus unlicensed.

There is a huge infrastructure of wiring in businesses and other similar facilities. WLANs are used where it is difficult to add or change wiring or where the employees are very mobile. One of the factors holding back a large growth in the area of WLANs has been the difficulty of increasing the data rates. Also, such data communications present many problems in that short E-mail messages would usually be acceptable, but any extended data communications run the risk of errors and lost data due to all the obstacles in radio communications. However, recent hardware breakthroughs and the introduction of a new standard by the IEEE, 802.11, are expected to enable the industry to expand.

One important use of WLANs is with portable computers. Most desktop computers (at companies anyway) are connected to a network but portable PCs are not. Connecting laptops through WLANs would allow for mobility and interconnectedness.

11.5   SATELLITE COMMUNICATIONS

Satellite communications are becoming more prevalent. Aside from television broadcasts, which everyone is familiar with, satellites play a role in a variety of other communications applications.

Satellites have revolutionized navigation throughout the globe. Airplanes and ships depend heavily on satellites for accurate measurements of their locations. Individuals in their cars or with hand-held devices can also make use of the global positioning system.

Cellular telephone systems can also make use of satellites. The features and ranges all depend on the number and orbit height of the various satellite systems. Of course, satellites have the advantage of being able to cover wide areas and are particularly useful in hostile or inaccessible areas such as oceans and deserts.

One system, the iridium satellite-based cellular system, has satellites in low earth orbit or LEO (700–1000 km), with 12 orbits and 6 satellites per orbit. This system uses a frequency reuse plan similar to AMPS, has 400 beams per satellite, and covers about 50 miles in diameter per beam.

11.6   SUMMARY

In this chapter we have examined wireless communications with special emphasis on cellular telephony. We have looked at the first generation system, AMPS, which is based on frequency-division multiplexing (FDMA). AMPS is still very popular, and with the presence of dual-mode phones, it will be around for a while. The newer digital techniques of multiple access are time division and code division, TDMA and CDMA, and are rapidly being deployed virtually everywhere.

These digital systems allow for and naturally lead into the latest in services, personal communications systems, PCS. PCS uses the same digital technology but provides for additional data services not previously available with the voice-only systems.

REVIEW QUESTIONS AND PROBLEMS

1. List at least five examples of wireless communication systems.

2. Is a microwave oven a wireless device? How about a telescope?

3. **a.** Define a *handoff*.
   b. How does an AMPS handoff differ from a CDMA handoff?

4. List at least three advantages of CDMA over AMPS. Are there any disadvantages?

5. Explain the difference between forward and reverse transmissions.

6. Explain what is meant by *roaming* from the point of view of (a) the user and (b) the technology.

7. Cellular phones typically have an indicator for roaming. Why is this needed and why is it becoming less needed?

8. What is meant by *dual-mode* mobile terminals?

9. Why is the carrier spacing in CDMA systems 1.25 MHz but in AMPS and TDMA systems it is only 30 kHz?

10. Distinguish between the general term *paging* and paging devices.

11. **a.** What are the specific modulation methods used for AMPS, NA-TDMA, and CDMA?
    b. What are the two stages of CDMA modulation?

12. **a.** How can you expand an AMPS system to incorporate more users?
    b. What are the limitations of AMPS expansion?

13. Repeat question 12 for CDMA.

14. Explain why power control is even more important for CDMA systems than for other access systems.

15. Compare PCS systems with standard cellular telephony in terms of features and frequencies.

16. Using Table 11–1, determine:
    a. Which identifiers are common to all three systems.
    b. Which identifiers are base station identifiers and which are mobile identifiers.

Multimedia
Communications

12.1   INTRODUCTION

Multimedia is the holy grail of telecommunications. When it is mentioned, both the eggheads and the suits begin to salivate. Eggheads see immense technical challenges in providing (interactive) video on demand to every home. Suits see equally immense profit in it. No book on telecommunications would be complete without at least an introduction to the subject.

Multimedia is probably one of the most overused terms of the 1990s. The field is at the crossroads of five major industries: computing, telecommunications, publishing, consumer audio-video electronics, and television/movies/broadcasting. The emergence in the 1970s of a new technological field, computer networking, led to a first phase of active collaboration between computing and telecommunications.

Literally, *multimedia* is two or more media. When most people refer to multimedia, however, they mean the combination of two or more *continuous media*; that is, media that have to be played during some well-defined time interval, usually with some user interaction. From a practical point of view, the two media are normally audio and video—that is, sound plus moving pictures.

multimedia

Transmitting multimedia over telecommunications networks requires a lot of bandwidth. In fact, transmitting live video over a telecommunication network requires up to hundreds of Mbps pipes. Still images

Figure 12–1
Process of delivering
compressed media
over the telecommu-
nication networks.

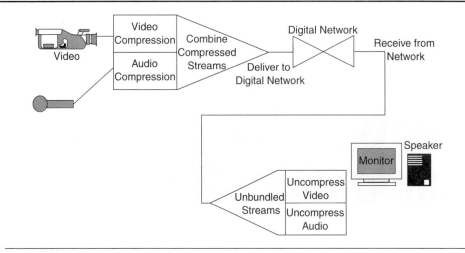

as well can require huge portions of computer memory. For this reason, this type of information representation undergoes some form of compression prior to being delivered over a telecommunication network. Figure 12–1 shows the block diagram of this process. On the receiving end, the compressed information goes through a process of uncompressing to recover, if not all then most of, the origin information intact. In this chapter we discuss transmission of digital video and audio, and the various techniques used in compressing multimedia information.

12.2   OVERVIEW OF TV AND VIDEO

An important phenomenon related to human vision is that if an image is flashed on the retina, it is retained for some period of time (few milliseconds) before decaying. Based on this property, if the eye is exposed to a sequence of images, 50 or more images per second, the eye does not notice that it is looking at discrete images. In fact, the eye perceives the sequence of images as continuous motion. Video systems are designed based on this principle. Video systems fall into two categories: analog and digital video systems.

12.2.1   Analog Video

We begin our discussion of analog video systems with the simple black-and-white television. In this system, a camera scans an electronic beam horizontally and vertically across the image, recording the light intensity as it goes. At the end of the scan, called a *frame*, the beam retraces. The intensity as a function of time is broadcast. The receiving end (the black-and-white television in this case) repeats the scanning process to reconstruct the image. The scanning process used by both the camera and the receiver is shown in Figure 12–2.

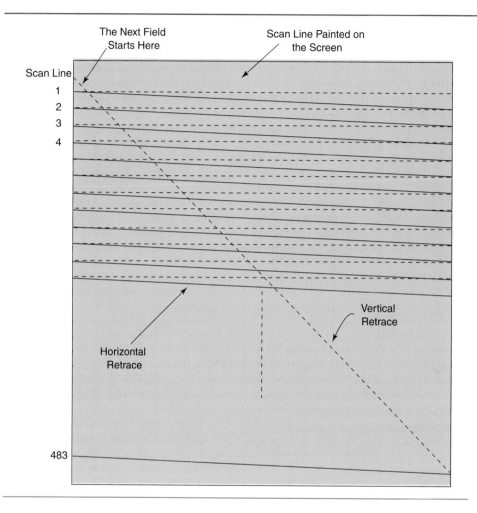

Figure 12–2
Scanning pattern for NTSC video and television.

Scanning parameters vary from country to country. The system used in North America and Japan has 525 scan lines, a horizontal-to-vertical aspect of 4:3 (the ratio of the height to the width of the rectangle shown in Figure 12–2) and 30 frames per second. A frame is the complete scan of the rectangle and thus 30 frames per second is translated to 30 images painted on the screen in a second. The European system has 625 scan lines, with the same ratio of 4:3 and 25 frames per second. In both systems, the top and bottom few lines are not displayed (to approximate a rectangular image). Only 483 of the National Television Standard Committee (NTSC) scan lines and 576 of the 625 PAL/SECAM scan lines are displayed. The beam is turned off during the vertical retrace, so many stations, especially in Europe, use this interval to broadcast TelText (text page containing news, weather, sports, stock prices, etc.).

Color video uses the same scanning pattern as monochrome (black and white), except that instead of displaying the image with one moving beam, three beams moving in union are used. Analog video cameras produce three distinct continuous signals, one for each color component. They capture the intensity of the red, green, and blue component that roughly corresponds to the definition of the primary colors. Each of the primary colors, red, green, and blue *(RGB)*, uses a beam.

Representing a color with an appropriate proportion of red, blue, and green is what video cameras do and what most displays use. It is possible, however, to transform these three values into three other values. Why do so? One reason is to transmit these signals more easily. The important idea is:

> The three green, red, and blue signals may be transformed into three other signals: the luminance that carries information on lightness and brightness, and two color signals. As the human visual system is less sensitive to color than luminance, the color signal may be transmitted or represented with less accuracy.

Figure 12–3 illustrates the function of a TV camera.

Luminance and Color Difference in Analog Broadcast TV   The principle of transforming the RGB signals into luminance and chrominance signals is as old as color TV. The reasons for this process were twofold. First, it allows backward compatibility by allowing old black-and-white TV sets to display color
luminance signal broadcasts. This is the role of the *luminance signal.* Second, it allows the transport of the two chrominance signals in narrower bandwidth and with lower accuracy than the luminance signal.

In television, the luminance signal is usually noted Y. The two chrominance signals are calculated from what is called *color difference signals.* A color

---

Figure 12–3
RGB to *YUV* conversion in broadcast TV.

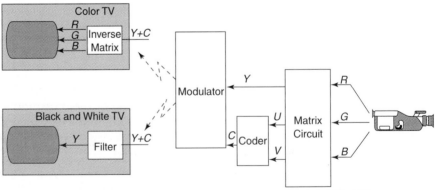

(1) *YUV* are the luminance and chrominance signals in PAL; *YIQ* are the corresponding signals in NTSC.

(2) The *U* and *V* signals are combined into a chroma signal *C*.

difference signal for each of the *red, green,* or *blue* colors is simply obtained by subtracting the luminance signal from the color signal. For example, if R is the red color signal, the red color difference signal is $(R - Y)$. In practice, only two color-difference signals are sufficient if the luminance is available.

Color difference signals are not transmitted as such. Instead, they are combined before transmission into two signals called *chrominance signals.* The transformation is linear by regular electronics. Each of the TV standards defines its own linear transformation:

chrominance
signals

- The NTSC standard refers to the luminance as the Y component and to the two chrominance signals as the I and Q components. I and Q represent in a combined way the hue and the saturation aspect. They are calculated as follows:

$$Y = 0.30R + 0.59G + 0.14B$$
$$I = 0.74(R - Y) - 0.27(B - Y) = 0.60R + 0.28G + 0.23B$$
$$Q = 0.48(R - Y) + 0.41(B - Y) = 0.21R + 0.52G + 0.31B.$$

- The PAL standard calls the luminance Y and the two chrominance signals U and V:

$$Y = 0.30R + 0.59G + 0.11B$$
$$U = 0.493(B - Y) = -0.15R - 0.29G + 0.44B$$
$$V = 0.877(R - Y) = 0.62R - 0.52G - 0.10B.$$

Figure 12–3 illustrates the *RGB* to *YUV* conversion in broadcast TV.

In the past few years, there has been considerable interest in high-definition television (HDTV), which produces sharper images by roughly doubling the number of scan lines. The U.S.A., Europe, and Japan have all developed HDTV systems, all differing and all mutually incompatible. The basic principles of HDTV in terms of scanning, luminance, chrominance, and so on, are similar to the existing systems. However, all three formats have a common aspect ratio of 16:9 instead of 4:3 to match them better to the formats used for movies (which are recorded on 35mm films).

12.2.2 Digital Video

Digital video refers to the representation of the images transmitted for viewing over the digital and public networks. A simple representation is a sequence of frames, each consisting of a rectangular grid of picture elements, or *pixels.* A pixel is sometimes referred to as a *pel.* Each pixel can be a single bit, to represent either black or white. The quality of such a system is similar to what you get by sending a color photograph by fax.

The next step up is to use 8 bits per pixel to represent 256 gray levels. This scheme gives high-quality black-and-white video. For color video, good systems use 8 bits for each of the RGB colors, although nearly all systems mix these into composite video for transmission. Using 24 bits per pixel extends the number of colors to about 16 million, but the human eye cannot even distinguish this many colors, let alone more. Digital color images are produced using three scanning beams, one per color. The geometry is the same as for the analog system of Figure 12–2 except that continuous scan lines are now replaced by neat rows of discrete pixels.

To produce smooth motion, digital video, similar to analog video, must display at least 30 frames per second (the American system). Smoothness of motion is determined by the number of *different* images per second. Whereas flicker is determined by the number of times the screen is painted per second. These two parameters are different. A still image painted at 20 frames per second will not show jerky motion, but it will flicker because one frame starts to decay from the retina before the next one appears. A movie with 20 different frames per second, each of which is painted four times in a row, will not flicker, but the motion will appear jerky.

The significance of these two parameters becomes clear when we consider the bandwidth required for transmitting digital video over a network. Current computer monitors all use a 4:3 aspect ratio so they can use inexpensive, mass-produced picture tubes designed for the television consumer market. Common configurations are 640×480 (VGA), 800×600 (SVGA), and 1024×768 (XGA). An XGA display with 24 bits per pixel and 25 frames per second needs to be fed at 566.231040 Mbps. Even OC-9 is not quite good enough, and running an OC-9 SONET carrier into everyone's house is not exactly what we have in mind. Doubling the rate to avoid flicker is even less attractive. A better solution is to transmit 30 frames per second and have the computer store each one and paint it twice. Broadcast television does not use this strategy because television sets do not have memory. In addition, analog signals cannot be stored in RAM without first converting them to digital forms, which requires extra hardware. As a consequence, interlacing is needed for broadcast television but not for digital video. At this point, we should explain some basic ideas that relate to digital video

Luminance and Color Difference in Digital TV   The analog broadcast TV principles of transforming the three primary color signals into luminance and color differences apply equally to digital TV. The reasons for the process are also of a similar nature: economize the bandwidth and storage capacity by reducing the bit rate. The technique used is called *subsampling* of the color components. This means that fewer samples per line, and sometimes fewer lines per frame are taken. However, the degree of subsampling has to follow certain rules for scaling and compatibility between the various modes. One of these rules is that the ratio between the sampling frequency of the luminance and the sampling frequency of the color difference signals has to be an integer. As a result, all components are sampled at locations extracted from a single grid. The subsampling ratio of the color difference C_d for example— that is, Y sampling frequency: C_d sampling frequency—may be represented by two integers, such as 4:2 for a ratio of 2.

There is a standard notation to indicate the sampling ratio of all three components in a compact form of the type

Y sampling frequency: C_{d1} sampling frequency: C_{d2} sampling frequency.

subsampling

Let us illustrate this by describing how the subsampling is done in certain digital standards.

Subsampling in Studio-Quality TV (4 : 2 : 2)   The ITU-R 601 recommendation refers to the luminance signal as Y and to the color difference signals as C_r and C_b. The luminance is calculated as in NTSC, but C_r and C_b are the strict difference between the value of the red or the blue signal and the luminance ($C_r = R - Y$; $C_b = B - Y$). The standard defines the samples per active line as equal to 720 for the luminance signal. However, the number of active lines per frame differs depending on the analog standard: 486 in NTSC and 576 in PAL/SECAM.

For the two color difference signals, the number of samples per line is half that of the luminance, but the number of lines per frame is unchanged. As a result, the reduction of the bit rate due to subsampling of the chrominance signals is 50 percent on each color stream. As the luminance stream is unchanged, the overall reduction of the bit stream is 33 percent. This bit rate reduction is achieved without noticeable degradation in the quality of the image. Figure 12–4 illustrates subsampling of the color difference signals in studio-quality TV.

Subsampling in Standard Video Conferencing (4 : 1 : 1)   The ITU-TS H.261 recommendation defines a format called *common intermediate format (CIF)* with the following characteristics. The frame size for the luminance is 352 samples per line and 288 lines per frame.

As in studio-quality sampling, the number of samples per line of color difference signals is halved (176). The difference lies in the fact that the number of

CIF

Note: In ITU-R 601 digital TV, the frame is sent in two successive fields each contouring half the lines.

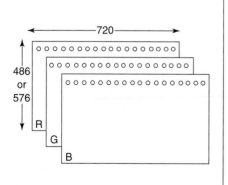
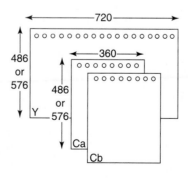

Figure 12–4
Subsampling of the color difference signals in studio-quality TV (ITU-R 601 recommendation).

lines per frame (144) is also halved. As a result, the bit rate of each color stream is a quarter that of the luminance. Subsampling provides an economy of 50 percent on the bit stream. The CIF subsampling structure is also referred to as 4 : 1 : 1—each color difference has one quarter the spatial resolution of the luminance component.

For lower quality, the ITU H.261 standard also defines the quarter-common intermediate format (QCIF), which has only 144 lines per frame and 176 samples per line for the luminance signal. The spatial resolution of the QCIF is of course one quarter of the CIF resolution, but the subsampling ratio (4 : 1 : 1) is the same.

The H.261 recommendation has also defined a super-CIF—704 samples per line and 576 lines per frame that has a resolution close to that of the studio-quality standard. In practice, CIF is the format generally used in video conferencing when the bit rate devoted to the video channel ranges from 100 Kbps to 300 Kbps. We will return to the H.261 standards later in this chapter.

12.3 OVERVIEW OF COMPRESSION TECHNIQUES

All compression systems require two algorithms: one for compressing the data at the source and another for decompressing it at the destination. These algorithms are referred to as the *encoding* and *decoding* algorithms, respectively. These algorithms have certain asymmetries that are important to understand. First, for many applications, a multimedia document, say, a movie will be encoded only once (when it is stored on the multimedia server), but it will be decoded thousands, if not millions, of times (when it is viewed by customers). This *asymmetric connection* means that it is acceptable for the encoding algorithm to be slow and to require expensive hardware, provided that the decoding algorithm is fast and does not require expensive hardware. After all, we may be willing to rent a superfast parallel computer for a few weeks to encode an entire video library, but requiring customers to rent a supercomputer for two hours to view a video is doomed to fail miserably. Many practical compression systems go to great length to make decoding fast and simple, even at the price of making encoding slow and complicated.

On the other hand, for real-time multimedia, such as video conferencing, slow encoding is unacceptable. Encoding must happen on-the-fly, in real time. Consequently, real-time multimedia uses different algorithms or parameters than storing video on disk, often with appreciably less compression.

Second, the encoding/decoding processes do not need to be inevitable. That is, when compressing a file, transmitting it, and then decompressing it, the user expects to get the original back, accurate down to the last bit, as discussed earlier. With multimedia, this requirement does not exist. It is usually acceptable for the video signal after encoding and then decoding to be slightly different than the original. When the decoded output is not exactly equal to the original input, the system is said to be *lossy*. If the input and output are identical, the system is *lossless*. Lossy systems are important because accepting a small amount of

asymmetric connection

information loss can give a huge payoff in terms of the compression ratio possible. Compression schemes can be divided into two general categories: entropy encoding and source encoding. We will now discuss each in turn.

12.3.1   Entropy Encoding

Entropy encoding is used when we want to manipulate the bit streams without regard to what the bit means. This process is generally a lossless, fully reversible technique, applicable to all data. Let us consider some examples.

    Our first example of entropy encoding is the *run-length encoding* we discussed in Chapter 6. In many types of data, it is common to have repeated symbols: strings, bits, numbers, etc. The repeated symbols can be replaced with a special marker not otherwise allowed in the data, followed by the symbol comprising the run, followed by how many times it was repeated consecutively. For example, consider the following string of decimal digits:

```
3150000000000000845871111111111111119786543000000000000000000000000000000000065
```

If we now introduce S as the marker and use two-digit numbers for the repetition count, we can encode the above digit string as

```
315S01284587S1149786543S02965
```

Here the run-length encoding cuts the digit string in more than half.

    Runs are common in multimedia. In audio, silence is often represented by runs of 0s. In video, runs of the same color occur in shots of the ocean water, walls, and many other surfaces. All of these runs can be greatly compressed.

    A second example we consider uses *statistical encoding,* by which we mean using short codes to represent common symbols and long ones to represent infrequent ones. Huffman coding discussed previously in detail in Chapter 6 uses statistical encoding.

    Our third example of entropy encoding is *color look-up table (CLUT)* encoding. Consider an image using *RGB* encoding with 3 bytes per pixel. In theory, the image contains as many as 2^{24} different color values. In practice, it will normally contain many fewer values, especially if the image is a cartoon or a computer-generated drawing, rather than a photograph. Suppose that only 256 color values are actually used. A factor of almost three compression can be achieved by building a 768-byte table listing the *RGB* values of the 256 colors actually used, and then by representing each pixel by the index of its *RGB* value in the table. We transmit the index to the value in the table and not the value itself. Here we see a clear example where encoding is slower than decoding because encoding requires searching the table, whereas decoding can be done with a single indexing operation. The encoder searches the table to find the value in the table that matches the color being encoded and then determines the index in the table.

12.3.2   Source Encoding

Source encoding takes advantage of the properties of the data to produce compressed images. Compression resulting from these techniques is generally

entropy encoding

source encoding

differential
encoding

lossy. Here, too, we will illustrate the concept of source encoding by examining three different techniques used in multimedia compression. Our first technique is *differential encoding,* in which a sequence of values (e.g., a portion of a video image) are encoded by representing each one as the difference from the previous set of values. *Differential pulse code modulation (DPCM),* which we saw in Chapter 2, is an example of this technique. It is lossy because the signal might jump so much between two consecutive values that the difference does not fit in the field provided for expressing differences, so at least one incorrect value will be recorded and some information will be lost. The reason we consider differential encoding to be a kind of source encoding is that takes advantage of the property that large jumps between consecutive data points are unlikely. For example, when transmitting images of an anchor person reading the news, the change in the image is usually the slight movement of the body, the movement of lips and eyes, and other facial expressions while the background is fixed.

Our second technique of source encoding consists of *transformations.* By transforming signals from one domain to another, compression may become much easier. Let us consider the Fourier transformation of Figure 12–5. Here we have a function of time that is represented as a list of coefficients (amplitudes). Given the exact value of the coefficients, the original function can be reconstructed perfectly. However, given only the values of the first, say, eight coefficients rounded off to two decimal places, it may be possible to reconstruct the signal so well that the listener cannot tell that some coefficients have been lost. The gain is that transmitting eight coefficients requires many fewer bits than transmitting the sampled waveform.

Transformations are also applicable to two-dimensional data. Suppose that the 4×4 matrix shown in Figure 12–6(a) represents a gray-scale value of a monochrome image. We can use variable-length encoding to transform these data by subtracting the value in the upper left-hand corner from all elements except itself, as shown in Figure 12–6(b). For example, values between –7 and +7 could be encoded with 4-bit numbers, and values between 0 and 255 could be encoded as a special 4-bit code (–8) followed by an 8-bit number. At the receiving end, all of the values in the matrix will be recovered by simply adding the element in the upper left-hand corner to all of the other elements except itself, as shown in Figure 12–6.

Figure 12–5
Fourier transform of a time varying signal.

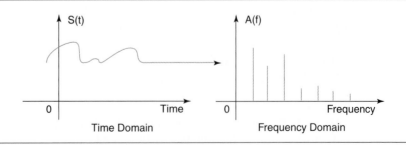

Time Domain   Frequency Domain

160	160	161	160
161	165	166	158
160	167	165	161
159	160	160	160

(a)

160	0	1	0
1	5	6	-2
0	7	5	1
-1	0	1	0

(b)

Figure 12–6
(a) Pixel values for part of an image, and (b) a transformation: the upper left-hand element is subtracted from all elements except itself.

Although this example of transformation is lossless, other more useful ones are lossy. An especially important two-dimensional spatial transformation is the *discrete cosine transform (DCT)*. This transformation has the property that, for images without extreme discontinuities, most of the spatial power is in the first few terms. Thus most of the consequent terms are ignored without much information loss. We will revisit DCT shortly.

DCT

Our third technique of source encoding is *vector quantization*, which is also directly applicable to image data. The principles of vector quantization, slightly simplified, are as follows: The data stream is divided into blocks called *vectors*. For example, when vector quantization is applied to an image, a vector is usually a small rectangular or square block of pixels. Let us assume that all vectors have the same size, and are composed of v octets. A table exists, which contains a set of patterns of v bytes each, as shown in Figure 12–7. This table, called the *codebook*, is available at both the coding and the decoding ends. The codebook may be either predefined or dynamically constructed. For each vector, the codebook is consulted to select the best matching pattern. Once the best matching pattern has been found, its reference—that is, its entry number in the codebook—is transferred. In this technique, instead of transmitting the actual data, the index of the best matching pattern found in a codebook is sent over the network.

vector quantization

What happens if the mismatch between the actual data and the pattern is significant? Then a distortion may be noticed at the receiving end. To remedy this problem, the technique provides for calculating the difference between the actual data and the pattern. The difference is then transmitted together with the index of the pattern, as shown in Figure 12–8. The coding of this index may itself be quantized. So, depending on whether and how the difference is transmitted, vector quantization may be a lossy or lossless scheme.

Vector quantization is particularly well suited to the encoding of data types whose characteristics are well known. Thus, vector tables that approximate a wide range of actual data vectors with sufficient accuracy can be constructed.

Figure 12–7
An example of vector quantization: (a) an image divided into squares, (b) a codebook for the image, and (c) the encoded image.

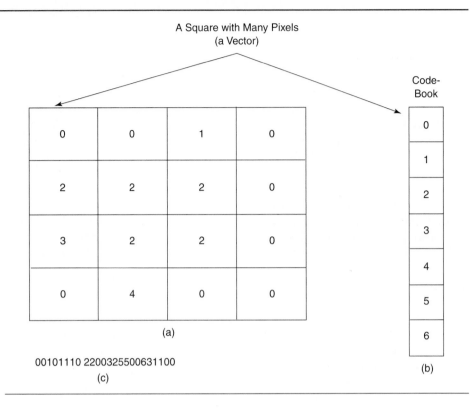

A Square with Many Pixels
(a Vector)

Code-Book

0	0	1	0
2	2	2	0
3	2	2	0
0	4	0	0

(a)

00101110 2200325500631100

(c)

(b)

Figure 12–8
Principle of vector quantization with transmission of the error term.

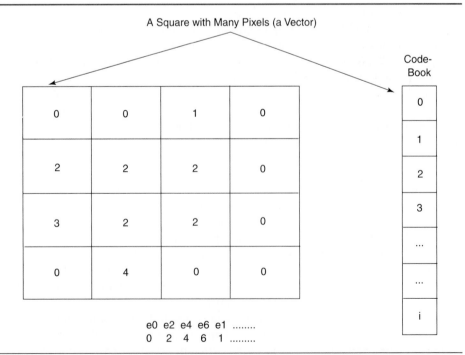

A Square with Many Pixels (a Vector)

Code-Book

0	0	1	0
2	2	2	0
3	2	2	0
0	4	0	0

e0 e2 e4 e6 e1 ....
 0 2 4 6 1 ....

12.4 IMAGE COMPRESSION

Now let us turn our attention to describing the mechanisms of compressing still and motion images. In this section we focus on two particular standards, which are typical examples of how things work. In particular, they will help to show the successive steps and the resulting complexity of image compression. These two standards are JPEG for still images and MPEG-video for moving images.

12.4.1 The JPEG Standard—Compression of Still Images

Joint Photographic Expert Group (JPEG) standards are for compressing continuous-tone gray-scale or color images (e.g., photographs). Photographic experts developed these standards working under the auspices of the ITU, ISO, IEC, and other standards bodies. JPEG has four modes and many options. It is more like a shopping bag full of goodies than an algorithm. It uses a combination of discrete cosine transform, quantization, run-length, and Huffman encoding techniques and supports several modes of operation, including lossless and variable types of lossy modes. Illustrated in Figure 12–9 are the steps for a lossy JPEG compression. We will concentrate on the lossy operations of JPEG and the way it encodes 24-bit RGB.

JPEG standards

Step 1 of encoding an image with JPEG is block preparation. For example, let us say that the input to JPEG is a 640×480 *RGB* image used in NTSC with 24 bits per pixel as shown in Figure 12–10. Since using luminance and chrominance gives better compression, we first compute the luminance, *Y*, and the two chrominances, *I* and *Q* (for NTSC), according to the formulas given in Section 12.3. Using the luminance-to-color differences ratio of $4 : 2 : 2$, separate matrices are constructed for *Y*, *I*, and *Q*, each with elements in the range 0 to 255. Next, blocks of four pixels are averaged in the *I* and *Q* matrices to reduce them to 320×240. This reduction is lossy, but it is not noticeable to the naked eye, since, as stated earlier, the eye responds to luminance more than to chrominance. Finally, each matrix is divided into 8×8 blocks. The *Y* matrix has 4800 blocks, and the *I* and *Q* matrices have 1200 blocks each, as shown in Figure 12–10.

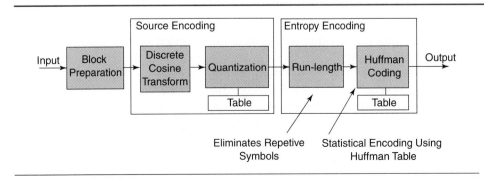

Figure 12–9
The operation of JPEG in lossy sequential mode.

Figure 12–10
(a) *RGB* input data and (b) result after block preparation for JPEG.

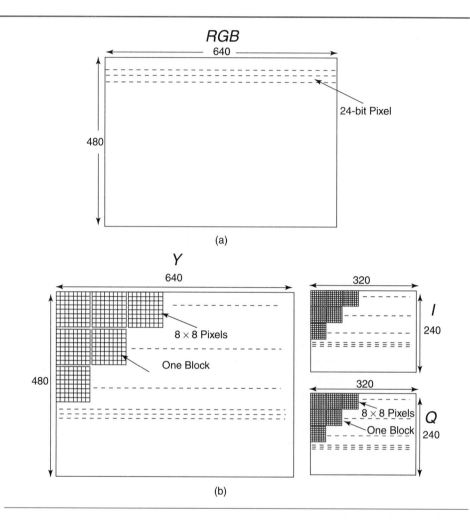

Step 2 of the JPEG process is to separately apply a discrete cosine transformation to each of the 7200 blocks. The output of each DCT is an 8 x 8 matrix of DCT coefficients. For the math lovers, we will get back to DCT later in this chapter. Element (0, 0) of the discrete coefficient block contains the average value for the block. The other elements tell how much spatial power is present at each spatial frequency. In theory, a DCT transformation is lossless. DCT compacts the power of the block near the origin. Prior to DCT transformation, blocks with pixels that vary moderately from the adjacent pixels will result in DCT coefficient blocks with values of elements close to the origin much higher than elements farther from the origin, as seen in Figure 12–11. This property is important, because if we neglect the smaller values in the DCT blocks we will end up with long strings of 0s that we can easily compress using, say, a

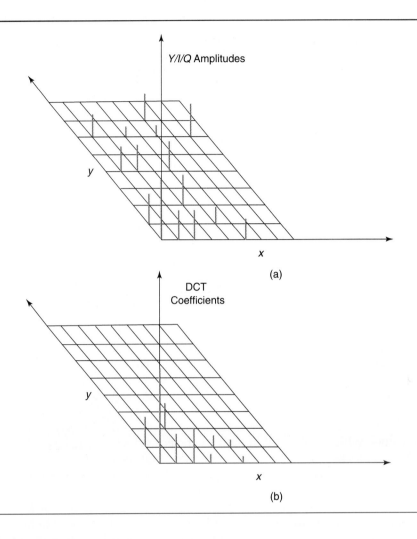

Figure 12–11
(a) One block of the
Y matrix and (b) the
DCT coefficients.

run-length technique. Note that DCT has to be performed on the blocks in each of the three matrices: *Y*, *I*, and *Q*.

What are the benefits of the spatial-to-frequency domain transformation? We already partially gave the answer. In a block representing an image, sampled values usually vary slightly from point to point. Thus, the coefficients of the lowest frequencies will be high, but the medium and high frequencies will have a very small or zero value. They may be neglected. The energy of the signal is concentrated in the lowest spatial frequencies, as shown in Figure 12–12.

Let us illustrate this idea of lower frequencies. Imagine you are a photographer for one day. You were assigned the monotonous task of taking pictures of flat, solid color walls. You take a picture of a white wall and divide it into 8 × 8 blocks. In each block, the amplitude of the signal will be nearly constant

Figure 12–12
Effect of performing DCT on the image.

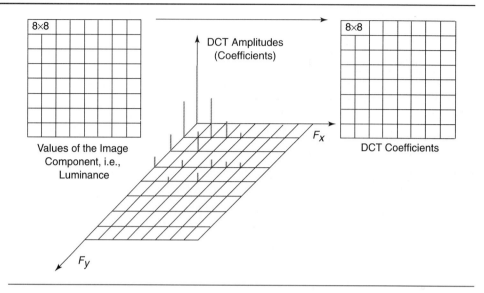

from pixel to pixel. As the value does not change much in each direction, the zero frequency will be high—as the frequency measures the rate of change in each direction—whereas the other frequencies will be nearly zero.

Now you are really bored. You are going to draw a sharp black line on the white wall, so the image becomes more complex. Some of the blocks will be affected. In the affected blocks there will be sudden transition between two consecutive values, which will result in high coefficient values for the highest frequencies. Not all blocks will be affected—only those that the line runs through.

In practice, continuous-tone images such as photographs do not have too many sharp lines or zones. The transitions between areas are usually smooth. Thus, the information is usually contained in the low frequencies. This is the assumption that JPEG makes anyway. Of course walls busy with detailed patterns and colors will result in somewhat different distribution of the DCT coefficients among the frequencies. You can see that JPEG is not designed for complex images, in particular images that look like bitonal images.

Step 3 is called *quantization,* in which the less important DCT coefficients are eliminated. In this lossy transformation we divide each of the coefficients by a weight taken from a table (called the *quantization table*). If the weights are 1, we do nothing to the coefficient. In the quantization table, the weights increase sharply going from the origin (low frequencies) toward the outer parameters of the table. Element (8, 8) in the table equals 64 and element (0, 0) in the table equals 1. This way the higher spatial frequencies are dropped quickly when we divide the DCT coefficients by the elements of the table. Figure 12–13 gives an example of the quantization step and shows values of a quantization table. The objective of the quantization step is to achieve further compression

DCT Coefficients

153	60	40	14	4	2	1	1
92	76	36	10	6	1	0	0
52	38	26	8	7	4	0	0
12	8	6	4	2	1	0	0
4	3	2	0	0	0	0	0
2	2	1	1	0	0	0	0
1	1	0	0	0	0	0	0
0	0	0	0	0	0	0	0

$$c(i,j) = DCT\,(i,j)\,/\,Q(i,j)$$
(Quantization Step)

Quantization Coefficients

153	60	20	3	0	0	0	0
92	76	18	2	0	0	0	0
26	19	13	2	0	0	0	0
3	2	1	1	0	0	0	0
0	0	0	0	0	0	0	0
0	0	0	0	0	0	0	0
0	0	0	0	0	0	0	0
0	0	0	0	0	0	0	0

Figure 12–13
The quantization step in the JPEG compression process.

Quantization Table

1	1	2	4	8	16	32	64
1	1	2	4	8	16	32	64
2	2	2	4	8	16	32	64
4	4	4	4	8	16	32	64
8	8	8	8	8	16	32	64
16	16	16	16	16	16	32	64
32	32	32	32	32	32	32	64
64	64	64	64	64	64	64	64

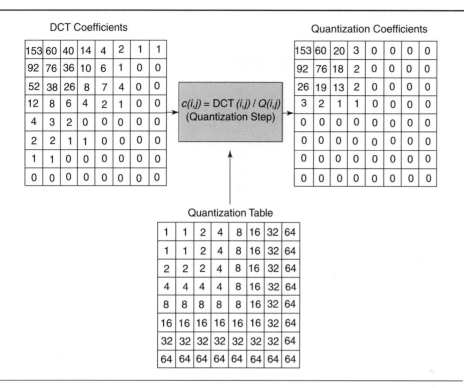

by representing DCT coefficients with no greater accuracy than necessary. The quantization table can be considered as a loss-tuning tool. Increasing the value of the coefficients will increase the compression ratio and reduce the resulting quality of the image.

In video conferencing over the digital networks, it is always a struggle to tune the compression parameters. On the one hand, increasing the values of the quantization table achieves higher compression, and thus a smoother motion, because you can ship more frames per second, say, up to 15 frames per second in most cases, but the transmitted image is grainy. On the other hand, if you decrease the quantization values, the image is sharper but is jerky. You lose the smooth motion, because you achieve less compression, and thus the frames per second transmitted are less, say, 7 frames per second. The result of quantization is a block that contains many 0s. Up to this point we have not really achieved any compression whatsoever by applying the DCT except to get rid of some values that are now represented by 0s. Even 0s take up space in memory and over the transmission facilities. For that reason we are ready to move to step 4.

Step 4 is the application of the run-length (lossless) compression algorithm to the block resulting from the quantization step. After quantization, only certain coefficients have survived. The others have taken the value 0. All these coefficients now have to be sent, but, there is a final step before this occurs. In which order should they be sent? It would be useless to send them line after line, left to right and stop at the bottom. Why? Because run-length coding, the technique that compresses successive values, is to be used. Thus, we need to adopt an order that will maximize the chances for successive values of 0s to occur.

The optimal order is a zigzag sequence. Indeed, high values will be packed around the top left of the table and lower values (0s) around the bottom right. The purpose of the zig zag ordering, as shown in Figure 12–14 is to maximize the possibility of having a long string of 0s, so using run-length results in maximum compression.

The *last step* before transmission or storage is to apply a final entropy-encoding scheme. JPEG allows application of Huffman encoding, which implies (as we know from Chapter 6) that we have variable length codes and requires a code table or a more dynamic technique beyond the scope of this book called *arithmetic encoding*. There you are—exhausted—but at last the image is compressed and ready for storage or transmission over the network.

Figure 12–14
The order in which the run-length algorithm is applied on the blocks.

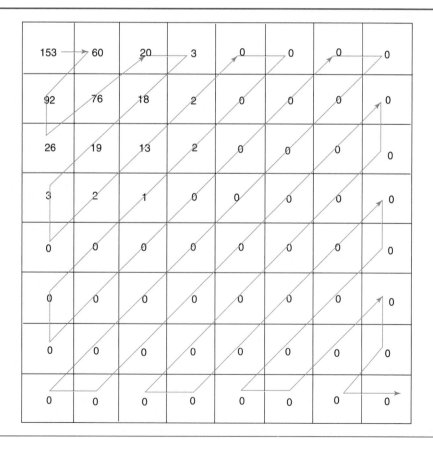

12.4.2   MPEG—Compression of Moving Images

Finally we come to the heart of multimedia communication: The *Motion Picture Expert Group (MPEG) standards.* They contain the main algorithms used to compress videos and have been international standards since 1993. Since movies contain both images and sounds, MPEG is concerned with the compression of video and audio.

MPEG1 was the first standard to be finalized (international standard 11172). It is intended to produce video recorder-quality output (352×240) for NTSC using a bit rate of 1.2 Mbps. Since we saw earlier that this uncompressed video alone can run up to 566 Mbps, getting it down to 1.2 Mbps is not entirely trivial, even at this lower resolution. MPEG1 can be transmitted over twisted-pair transmission lines over a modest distance. MPEG1 is also used for storing movies on CD-ROM.

MPEG2 is the next standard in the MPEG collection (international standard 13818), which was originally designed for compressing broadcast-quality video into 4 to 6 Mbps so it could fit into an NTSC or PAL/SECAM channel. MPEG2 was expanded later to support higher resolution, including HDTV. MPEG4 came along later for use on medium-resolution videoconferencing with low frame rates (10 frames per second) and at low bandwidth (64 kbps). This permits videoconferencing to be held over a single N-ISDN B, also known as N x 64 kbps channel (i.e., 64, 128, 256, 384, 768 kbps granulates). Following this numbering scheme, you may think the next MPEG standard is MPEG8. Actually, this standard does not exist. The standard committee names the standards sequentially, not exponentially. There was a standard called MPEG3 that was supposed to be HDTV, but it was later abandoned and its functions were added to MPEG2. That was the reason the numbers jumped from MPEG2 to MPEG4. Let us now examine the principles of motion picture compression.

Compression consists of eliminating redundant information. In moving images, there are two types of redundancies, or correlations, which can be eliminated or reduced. These redundancies are spatial and temporal correlations.

Spatial Correlation   The compression process addressing spatial correlation deals with individual images. The redundancies within each frame are eliminated or reduced. The techniques used are similar or nearly identical to those we described for the JPEG standard for still images.

Temporal Correlation   *Temporal correlation* is also called *motion compensation.* Temporal correlation consists of identifying redundancies between successive frames. The reduction of temporal correlation usually employs DPCM techniques, which means that only the difference between frames is encoded. Motion compensation is based on the fact that over a certain time interval, frames are similar and certain frames may be partially constructed from other frames. Figure 12–15 illustrates the processing sequence for both spatial and temporal redundancies used in H.261 videoconferencing standard.

MPEG standards

temporal correlation

Figure 12–15
Practical implementation of motion video compression performed in H.261 video conferencing.

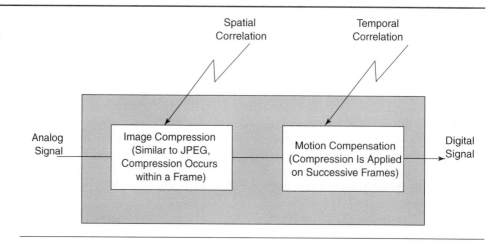

Introduction to Motion Compensation Algorithms   We saw in the section on JPEG how a typical algorithm for reducing spatial correlation works. In this section we study a typical motion compression algorithm. We have selected the MPEG1 motion compression scheme. Our objective is to give you an idea of the types of issues addressed via a real, though simplified, case. We already introduced how MPEG1 fits into the MPEG standards suite.

Reference and Intracoded Frames   Consider three successive frames as shown in Figure 12–16. Look at the first two frames, F1 and F2. F2 may be approximated by pieces of the F1 frame. If F1 serves as a reference to construct another frame like F2, F1 is called a *reference frame,* as shown in Figure 12–17.

reference frame

        Let us assume now that F1 is not reconstructed from any other frames. All the information necessary to display it is contained in F1. Then F1 is called an *intracoded frame (I-frame).* Thus, we have the following definitions:

intracoded frame

- A reference frame is a frame from which other frames are constructed.
- An intracoded frame (I-frame) is a frame not built from any other frames.

        Usually, intracoded frames will also serve as a reference, but not all reference frames need to be intracoded, as we shall soon see.

Motion Vector and Macroblocks   If we are not careful in our compression of, say, a block, which we have already transmitted in F1, we may end up transmitting it again when F2 is sent. For this reason, it is more efficient to transmit only the vectors that indicate the spatial translation of the block from F1 to F2. This vector is called a *motion vector.*

motion vector

        The blocks on which the motion vector are applied are called *matching blocks.* Their actual size depends on the image component. In MPEG1, an image

F1

F2

F3

Figure 12–16
Three successive frames.

Figure 12–17
Reference frame and intracoded frame.

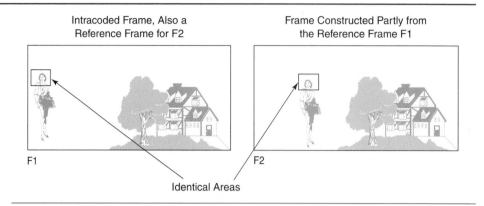

Intracoded Frame, Also a Reference Frame for F2

Frame Constructed Partly from the Reference Frame F1

F1

F2

Identical Areas

is formed from three components; the luminance component and two color difference components are subsampled. Recall that subsampling means that we skip some of the pixels when we encode the color differences, because the eye usually does not notice the degradation in quality. Thus, in practice, matching blocks are squares of 16 × 16 pixels for the luminance component and squares of 8 × 8 pixels of each of the difference components (number of pixels in the block is halved because of subsampling). The combination of the 16 × 16 square and the two 8 × 8 squares is called a *macroblock* as shown in Figure 12–18. The term *macroblock* should not be confused with the 8 x 8

macroblock

Figure 12–18
Macroblocks in MPEG1.

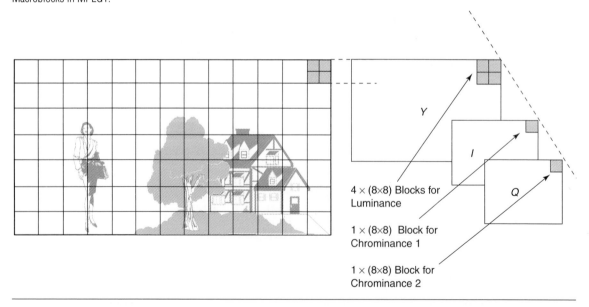

Y

I

Q

4 × (8×8) Blocks for Luminance

1 × (8×8) Block for Chrominance 1

1 × (8×8) Block for Chrominance 2

blocks used in JPEG or in MPEG1 to eliminate the spatial redundancy with DCT techniques. Thus far we have the following:

- A macroblock is one 16 × 16 pixel area. In practice, it is composed of 8 × 8 blocks: four blocks for the luminance and two blocks for each of the color difference components.
- A motion vector indicates the spatial translation of a macroblock between two frames, as shown in Figure 12–19.

Predicted and Bidirectional Frame   Let us go back to the F3 frame in Figure 12–16; it has macroblocks in common with the frame F1. Now let us assume that F3 is constructed from F1 only. F3 is then called a *predicted frame (P-frame)*. It is predicted from a reference frame F1, which also turned out to be an intracoded frame (I-frame). A predicted frame is only constructed from a preceding frame, as shown in Figure 12–20.

predicted frame

Now consider frame F2; again, F2 has macroblocks in common with F1, but on the right side of the image, it also has macroblocks in common with F3. Therefore, conceptually, F2 can be reconstructed by using pieces of F1 and F3. How is it possible to know, at the time F2 is coded, that there will be a matching macroblock in F3? This is possible if F3 is also available at the time F2 is encoded. To

Macroblock

Figure 12–19
Motion vector and macroblocks.

F1

F2

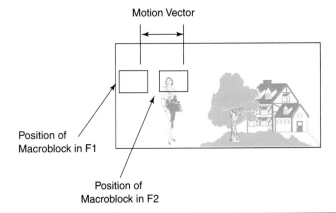

Motion Vector

Position of
Macroblock in F1

Position of
Macroblock in F2

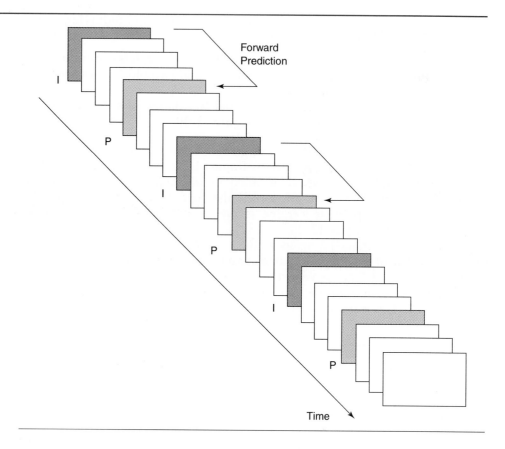

Figure 12–20
Sequence of I and
P frames.

this end, frames F1, F2, and F3 have to be buffered, and F2 only sent when it has been interpolated from F1 and F3. F2 is called a *bidirectional frame (B-frame)* as reconstructed from two directions: from a forward and from a backward frame. (See Figure 12–21.) Up to this point, in addition to the I-frame, we have introduced two additional MPEG frames:

bidirectional frame

- Predicted frames (P-frames) are only reconstructed from preceding frames.
- Bidirectional frames (B-frames) are interpolated from a backward and a forward frame.

Based on this, the sequence is as follows:

1. F1 is encoded without a reference to any other frames.
2. F3 is predicted from F1.
3. F2 is interpolated from an intracoded frame.

In our oversimplified example we assumed that for each macroblock in a predicted or interpolated frame, there is a macroblock that matches it in the refer-

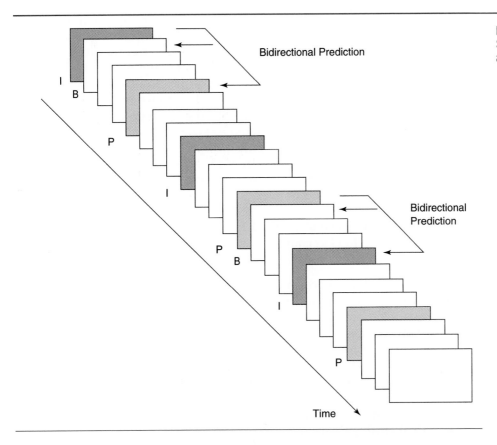

Figure 12–21
Sequence of I and P and B frames.

ence (I) frame(s). There are many reasons, however, why the match may not be exact. Thus, an algorithm that searches for the *best-matching macroblock* within a certain area is needed. If a match is not exact, the arithmetic difference between the actual block and the best-matching block is calculated. As in vector quantization, this difference, which is also a macroblock, is called the *error term*. We will always have the case where no satisfactory matching blocks are found in the area searched. They will then have to be coded in the same way as macroblocks in I-frames.

In the MPEG1 scheme of compression, I-frames are self-contained but less compressed, and P-frames and B-frames are the most compressed. You can think of a highly compressed mode in which the number of B-frames is increased in a sequence of frame transmission. There is, however, a tradeoff to respect. Multiplying B-frames decreases the correlation between them and between the reference frames.

There are typical sequences of I-, P-, and B-frames. One is the sequence *IBBBPBBBI*. Another frequent sequence is *IBBPBBPBBI* for PAL/SECAM, and *IBBPBBPBBPBBI* for NTSC.

The Role of I-Frames   Let us say you are watching an MPEG movie using a VCR. You decide to go to the kitchen to get a drink. You want to stop the movie until you come back. In other words, you want to resume from a given frame or freeze a particular image. If this frame is a P-frame, the player needs to read back the preceding I-frame to build the P-frame. If it is a B-frame you want to start from, the player needs to read backward to retrieve the I-frame and then forward to get the P-frame, as both are involved in its interpolation.

Thus, the I-frames can be considered as synchronization points. It is estimated that the maximum delay between the occurrence of two consecutive I-frames should not exceed 400 ms. In applications where MPEG sequences are used in the VCR mode, it is estimated that one reference frame, be it a P- or an I-frame, should at least occur every 150 ms.

Coding of the I-frame   The steps used in coding I-frames are very similar to those steps used in JPEG in coding continuous-tone images in the lossy mode. The steps are as follows:

- Each of the luminance and chrominances are divided into 8×8 blocks.
- DCT is used to transform each block.
- A quantization table is used to apply the quantization step. As a result, certain coefficients are wiped out.
- The series of most significant coefficients in each block, the DC coefficients, are coded using the DPCM technique—only the difference between two adjacent DC values is coded.
- The coefficients in each block are diagonally ordered, as in JPEG, and the run-length encoding is applied.
- A final Huffman-like encoding step is applied.

Thus, the I-frame has been compressed almost as a still image using JPEG would have been, as shown in Figure 12–22.

Figure 12–22
Processing steps for
I-frame encoding.

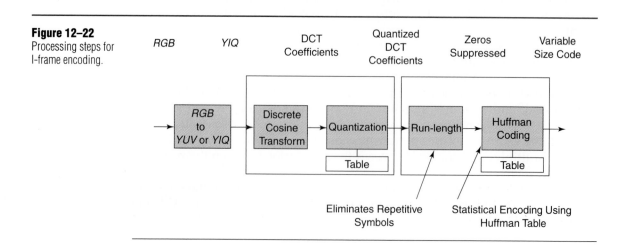

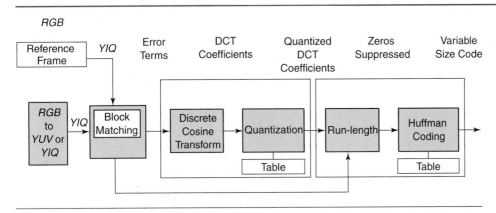

Figure 12–23
Processing steps for P- and B-frames encoding.

Coding of P- and B-frames   By now you should know the coding of the P- and B-frames will be different from the way the I-frames are coded. The following are the steps used in encoding the P-, and B-frames:

- For each macroblock, the reference is searched to find a best match.
- The difference between the actual macroblock and the best-matching macroblock is computed as the motion vector.
- The error term, which is also a macroblock, is transformed using the discrete cosine transform.
- Then a quantization step, diagonal ordering, run-length, and final statistical (Huffman-like) entropy encoding are carried out. Note that the quantization table used is different from that used for I-frames. Note also that unlike with JPEG and I-frames, the DC coefficients are coded in the same way as other coefficients.
- The motion vector of each block is coded using the DPCM technique as adjacent motion vectors often differ only slightly. The resulting sequence of values is finally submitted to a Huffman-like coding step. Figure 12–23 shows the steps used in coding the P- and B-frames.

12.5   AUDIO COMPRESSION REVISITED

In Chapter 2 we described the main technique employed to digitize sounds. In this section, in addition to the techniques introduced in Chapter 2, we present an overview of other techniques and standards used to reduce the bit rate of voice channels.

12.5.1   ADPCM

Adaptive differential PCM (ADPCM) provides nearly equal quality voice but at only 32 kbps. In essence it cuts in half the bandwidth required to transmit the standard PCM voice. Recall that a PCM channel requires 64 kbps for voice

Figure 12–24
An ADPCM
codec: (a) coder,
(b) decoder.

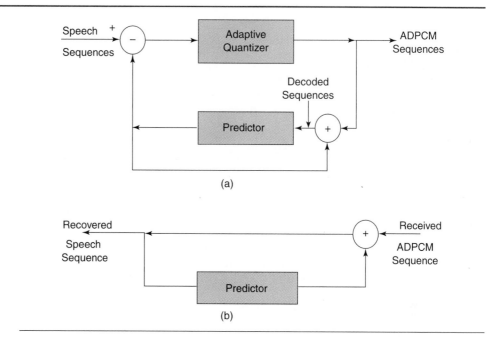

transmission. Key to the operation of ADPCM is the predictor, as shown in Figure 12–24. Predictions are based on the decoded sequence (identical to the sequence at the output of the receive decoder in the absence of transmission errors), which consists of the input speech sequence contained by quantization noise. With ADPCM the predictors have adaptive coefficients.

12.5.2 Vocoders

Vocoder operation is based on an algorithm that attempts to describe the speech production mechanism in terms of a few independent parameters serving as the information-bearing signals. Thus, what are really transmitted over the channel are parameters describing the speech instead of the encoded speech itself. The design considers that the speech is produced from a source-filter arrangement. It is modeled after the human generation of speech. Voice speech is the result of exciting the vocal tracts (which is really a filter) with a series of quasi-periodic glottal pulses, generated by the vocal chords, which is considered the *source*. As its name implies, a vocoder codes speech. It uses an analysis process based on a speech production model and extracts a set of source-filter parameters that are encoded and transmitted. At the far-end decoder, the parameters are decoded and are used to control a speech synthesizer based on the speech production model at the transmit end. The synthesized signal at the receiver resembles the original speech signal. Again, the idea here is not to send the encoded speech, but to send some parameters for the receiving end to control the operations of a number of filters that are used in producing the synthesized speech.

12.5.3 CELP Technique

One of the most popular voice-coding techniques used in multimedia communication is the *codebook excitation linear predictive (CELP) encoder.* This type of encoder provides good voice quality at low bit rates (as low as 4.8 kbps). Eight-kbps CELP encoders provide near-toll-quality speech. CELP encoders employ a vocal tract linear predictor (LP)-based model, a codebook-based excitation model, and an error criterion, which serves to select an appropriate sequence using an optimization process. As a result, a CELP system selects that excitation sequence, which minimizes a "perceptually" weighted square error formed between the input and the "locally" decoded signal. Figure 12–25 shows the process of the CELP synthesis model and some basic ideas on how CELP operates.

CELP encoder

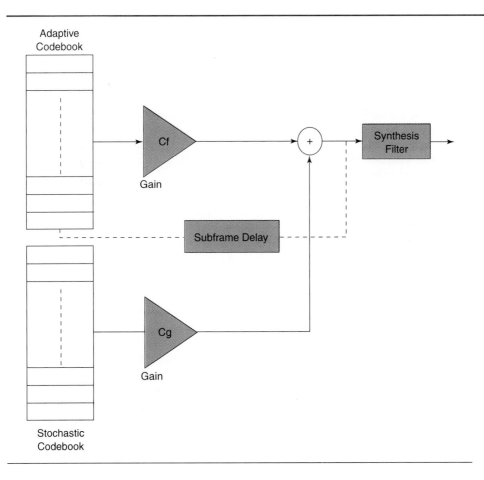

Figure 12–25
CELP synthesis model.

12.5.4 ITU-TS G.721, 722, 723, 728 Standards for Speech

The ITU-TS has recommended a series of schemes for compressing speech. In this section we outline these recommendations and the corresponding compression mechanism.

G.721

This is a standard for converting a 64-kbps stream into a 32-kbps stream. It is based on the ADPCM technique we described in Section 12.5.1. Each value difference is coded with 4 bits. As in G.711, the sampling rate is 8 kHz.

G.722

The objective of the G.722 standard is to provide better sound quality than the conventional G.711 PCM scheme or the G.721 compression technique. G.722 is based on a subband ADPCM method (SB-ADPCM). The speech signal is filtered into a relatively small number of subbands (2 to 16), and each subband signal is translated to zero frequency, sampled at the Nyquist rate, and adaptively encoded. The number of bits used in the encoding process differs for each subband signal. By encoding each subband individually, the quantizing noise is confined within the subband. Subband encoders produce near-toll-quality voice at 16 kbps rate. The bandwidth of a G.722 compressed signal ranges from 50 kHz to 7 kHz, whereas that of G.711 signals is limited to 3.4 kHz. The resulting rate is 48, 56, or 64 kbps—in the standard mode, the sampling rate is 16 kHz and the amplitude depth is 14 bits.

G.723

This is another lossy ADPCM-based compression standard that operates at 24 kbps. The resulting sound quality is inferior to that of the uncompressed G.711 PCM standard or that of the SB-ADPCM-based G.722 standard. Most of the G.723 use is with multimedia conferencing over LAN as part of the H.323 videoconferencing standards.

G.728

G.728 is a standard targeting low bit rate. It operates at 16 kbps but its bandwidth is limited to 3.4 kHz. The resulting sound quality is significantly inferior to that of G.711 or G.722. It is based on a vector quantization scheme called *low-delay code excited linear prediction (LD-CELP)*.

12.6 SUMMARY

JPEG is a standard for continuous-tone gray-scale or color images. It uses a combination of discrete cosine transform, quantization, run-length, and Huffman encoding techniques and supports several modes of operation: sequential encoding, the normal lossy mode, progressive encoding, an enhancement of the first one, and finally a hierarchical encoding mode where different resolution levels can be decoded separately. JPEG consists of a number of steps:

- Block preparation, in which the entire image is divided into individual 8 × 8 blocks: one 8 × 8 block for the luminance and two 8 × 8 blocks for the color difference components.

- DCT transformation is applied to the blocks.
- A quantization step is then applied to the blocks of the DCT coefficients.
- Run-length is used to compress the quantized coefficients.
- Huffman encoding is applied

JPEG works well for continuous-tone images. A compression ratio of 25:1 is common. A number of off-the-shelf and public-domain software and hardware implementations exist.

MPEG standards, on the other hand, are used for compressing motion images. MPEG1 is one of the MPEG standards, in which compression consists of eliminating redundant information. In moving images, there are two types of redundancies or correlations, which can be eliminated or reduced: spatial redundancies and temporal redundancies.

There are a number of ITU recommendations that specify schemes for compressing the audio signal prior to transmission to achieve reduction in the bit rate, and in turn reducing the bandwidth required for transmission. G.711 is the basic 64-kbps PCM digitization standard, G.721 is used to reduce the 64-kbps channel to a 32-kbps channel using ADPCM, and G.722 combines subband and ADPCM encoding to achieve further reduction in the bandwidth required.

REVIEW QUESTIONS AND PROBLEMS

1. What is the bandwidth required for a studio-quality TV channel in the PAL/SECAM system? The PAL/SECAM system uses a transmission of 30 frames per second. Assume that 24 bits are used to encode colors.

2. Do you see a problem transmitting such a channel over a digital network, say, using a T3 facility?

3. If the answer is yes to problem 2, what are the remedies?

4. Give the two classifications of compression schemes.

5. Describe schemes other than compressions used in reducing the bit rate in transmitting studio-quality TV channels.

6. What is JPEG used for?

7. Give the steps used in JPEG sequential lossy mode of compression.

8. Can we use JPEG for compressing motion video? Why or why not?

9. Describe the idea behind MPEG1.

10. What is a reference frame?

11. What is a predicted frame?

12. What is an intracoded frame?

13. Give the sequence of steps used in MPEG1 to compress motion images.

14. Why is the I-frame so important to the operations of MPEG1?

15. What is the idea behind ADPCM?

16. Describe the idea behind subband audio compression.

17. What is CELP?

18. What is LD-CELP?

19. What are G.722 and G.723?

SELECTED BIBLIOGRAPHY

Accredited Standards Committee on Financial Services. *Financial Institution Key Management (Wholesale)*. ANS X9.17. Washington, D.C.: American Bankers Association, 1985.

Ash, G. "Design and Control of Networks with Dynamic Nonhierarchical Routing." *IEEE Communications Magazine* (October 1990).

ATM Forum. "LAN Emulation over ATM Specification—Version 1.0." 1995.

Telecommunications Transmission Engineering. 3 vols. Piscataway, N.J.: Bellcore, 1990.

Bird, D. *Token Ring Network Design*. Reading, Mass.: Addision-Wesley, 1994.

Black, U. *Data Link Protocols*. Englewood Cliffs, N.J.: Prentice Hall, 1993.

———. *Physical Level Interfaces and Protocols*. Los Alamitos, Calif.: IEEE Computer Society Press, 1995.

———. *The V Series Recommendations: Standards for Data Communications over the Telephone Network*. New York: McGraw-Hill, 1995.

Boudec, J. "The Asynchronous Transfer Mode: A Tutorial." *Computer Networks and ISDN Systems* (May 1992).

Carne, E. B. *Telecommunications Primer: Signals, Building Blocks, and Networks*. Upper Saddle River, N.J.: Prentice Hall, 1995.

———. *Telecommunications Topics: Applications of Functions and Probabilities in Electronic Communications*. Upper Saddle River, N.J.: Prentice Hall, 1999.

Chen, K., K. Ho and V. Saksena. "Analysis and Design of a Highly Reliable Transport Architecture for ISDN Frame-Relay Networks." *IEEE Journal on Selected Areas in Communications* (October 1989).

Comer, D. and D. Stevens. *Design, Implementation, and Internals*. Vol. 2 of *Internetworking with TCP/IP*. Englewood Cliffs, N.J.: Prentice Hall, 1994.

———. *Client-Server Programming and Applications*. Vol. 2 of *Internetworking with TCP/IP*. Englewood Cliffs, N.J.: Prentice Hall, 1994.

Comer, D. *Principles, Protocols, and Architecture*. Vol. 1 of *Internetworking with TCP/IP*. Englewood Cliffs, N.J.: Prentice Hall, 1994.

Couch, L. *Modern Communication Systems: Principles and Applications*. Englewood Cliffs, N.J.: Prentice Hall, 1995.

———. *Digital and Analog Communication Systems*. 4th ed. New York: Macmillan, 1993.

DeRose, J. *The Wireless Data Handbook*. Quantum Publishing, 1994.

Goodman, D. *Wireless Personal Communications System*. Reading, Mass.: Addison-Wesley, 1997.

Held, G. *Data Communications Networking Devices*. New York: John Wiley, 1986.

Hioki, W. *Telecommunications*. 2nd ed. Upper Saddle River, N.J.: Prentice Hall, 1994.

Keshav, S. *An Engineering Approach to Computer Networking: ATM Networks, the Internet, and the Telephone Network*. Reading, Mass.: Addison-Wesley, 1997.

Microelectronics Digital/Analog Communications Handbook. Mitel Semiconductor, 1993.

Minoli, D. *Telecommunications Technology Handbook*. Norwood, Mass.: Artech House, 1991.

Pelton, J. *Wireless and Satellite Telecommunications: the Technology, the Market and the Regulations*. Upper Saddle River, N.J.: Prentice Hall, 1996.

Proakis, J., *Digital Communications*. New York: McGraw-Hill, 1989.

Proakis, J. and M. Salehi. *Communications Systems Engineering*. Upper Saddle River, N.J.: Prentice Hall, 1994.

Rappaport, T. *Wireless Communications: Principles and Practices*. Upper Saddle River, N.J.: Prentice Hall, 1996.

Rowe, S. *Telecommunications for Managers*. 4th ed. Upper Saddle River, N.J.: Prentice Hall, 1999.

Stallings, W. *Data and Computer Communications*. 5th ed. Upper Saddle River, N.J.: Prentice Hall, 1996.

GLOSSARY

access channel. In CDMA systems, this logical reverse channel is used to initiate calls, respond to pages, or register the mobile's location.

a-channel or a-bit. The least significant bit in each sixth frame in T1 (superframe and extended superframe format) when in-band signaling is used.

adaptive delta modulation. This technique deals with the slope overload problem in delta modulation systems by varying the step size—step size increases when several 1s are detected and reduces when several 0s are detected.

adaptive equalizer. This system, used to improve signal integrity in mobile communications, works by first sending a known signal, which the receiver uses to improve the actual received message.

address resolution protocol (ARP). The Internet protocol, as part of the TCP/IP suite of protocols, that finds the host being addressed by "asking" which machine has the IP address in question (most hosts are attached to a LAN by an interface board that only understands LAN addresses).

advanced mobile phone system (AMPS). The first-generation cellular system using frequency-division multiple access (FDMA). This system is still in widespread use.

air interface. The interface between the base and mobile terminal.

alarm indication signal (AIS). A special T1 signal, consisting of all 1s, sent downstream from a higher-order system to the lower-order systems indicating that the problem exists in the higher system.

A-law. A logarithmic formula used for compressing an analog signal.

alternate mark inversion (AMI). To facilitate repeaters recognizing 1s, the polarity of a 1 (mark) alternates. For example, if a 1 is + 5V, the next 1 must be –5V, etc. *See also* bipolar violation.

amplitude. Magnitude of the movement of a signal, typically voltage, exhibited as volume.

amplitude modulation (AM). Also known as amplitude shift keying (ASK), AM uses different levels of the signal to indicate different values. Rarely used because of susceptibility to noise.

application layer. Layer 7 (top) of the OSI set of seven layers that constitutes the network architecture. This layer is concerned with providing all services that are directly accessible by the application programs and the ultimate user.

application programming interface (API). These software systems are used to build sophisticated multi-user network-based applications.

architecture. The design of a system including all its components.

asymmetric connection. A circuit in which the bit rate in one direction is different from that of the other direction. An example would be video on demand.

asynchronous transfer mode (ATM). A method of information transfer using fixed-size slots or cells each having 48 octets of information and a 5-octet header. Asynchronous refers to the fact that cells allocated to the same connection may exhibit an irregular recurrence pattern as they are filled due to actual demand. ATM guarantees cell sequence integrity.

asynchronous transmission. Transfer of information such that each character is transmitted as an individual entity without regard to when the previous character was transmitted. *See also* synchronous transmission.

ATM adaptation layer (AAL). This layer maps (and extracts from) the application messages and data units into the information fields essentially isolating the applications and other higher layers from the ATM layer. The AAL consists of two parts: the convergence sublayer (CS) and the segmentation and reassembly sublayer (SAR). There are four protocols (AAL-1, AAL-2, AAL-3/4 and AAL-5), which are used to handle the various types of information.

ATM bearer service. A service offering point-to-point, bidirectional or point-to-multipoint unidirectional virtual connections at either a virtual path and/or a virtual channel level.

ATM layer. A layer in the B-ISDN/ATM model that is concerned with generic flow control, encapsulation, cell VPI/VCI translation, and cell multiplexing/demultiplexing.

attendant services. Automated, computer-controlled handling of calls in a PBX, for example.

attenuation. Signal loss of energy and distortion as it travels through a medium such as air or cable.

automatic call distributor (ACD). A system used to automatically switch large volumes of incoming calls to attendant (answering) positions.

backbone. The main circuit or network in a particular system through which any station may connect to any other station.

balanced line. Both wires of a twisted pair are above ground potential, having equal amplitude but opposite phase. *See* unbalanced line.

bandwidth. The width of the passband; calculated by subtracting the lowest frequency from the highest frequency.

baseband transmission. Signals placed directly onto the transmission medium without modulation by a carrier signal. The signals deteriorate over relatively short distances and must be regenerated.

basic rate interface (BRI). The fundamental or lowest rate for ISDN consisting of 2 B-channels at 64 kbps each and 1 D-channel at 16 kbps, yielding 144 kbps for user data (not including overhead).

base station. A station at or near the center of a cell. Its function is to relay voice and control signals between mobile units and the mobile telephone switching office.

basic system reference frequency (BSRF). An atomic clock located in Hillsboro, Missouri (geographic center of the 48 contiguous states). This stratum 1 clock is used to create a timing signal used for T1 circuits.

baud. A signaling rate defined as the number of signaling elements per second. A signaling element may be one or more bits, therefore baud will be less than or equal to the bit rate.

b-channel or b-bit. The least significant bit in each twelfth frame in T1 (superframe and extended superframe format) when in-band signaling is used.

bidirectional frame (B-frame). A frame interpolated from a backward and a forward frame.

bipolar eight zero substitution (B8ZS). To maintain 1s density, whenever a string of eight 0s is recognized, that string is replaced by 00011011. The receiver recognizes this code and replaces it with the eight 0s.

bipolar three zero substitution (B3ZS). Similar to B8ZS, but because of higher speeds of T3, three sequential 0s cannot be supported.

bit. A binary digit, which can be a 1 or a 0.

bit error rate test (BERT). An accurate measurement of line performance that requires taking the line out of service.

bit rate. The number of bits per second (bps) or (baud) times the number of bits per signaling element.

bit robbing. In a T1, every sixth frame the system "steals" a bit from each PCM word and uses it for signaling without making a noticeable effect on the voice quality. This bit is called, in the superframe, the A-bit in the sixth frame and the B-bit in the twelfth frame.

bit stream. A sequence of electrical pulses representing binary data, usually referring to internal communications within a system.

bit time. The amount of time for one bit.

bipolar violation (BPV). Two consecutive 1s of the same polarity on a T1 line. *See* alternate mark inversion.

blank-and-burst. The system interrupts the conversation by blanking the sound and inserts the control message. The length is 100 ms or less so that it is unnoticed by the callers.

blocking. A system not servicing a call request because of call overload—more calls than can be serviced at that moment.

BORSCHT circuit. The basic functions that every switching system must provide: battery feed, overvoltage protection, ringing signals, supervision, codec, hybrid circuits, and testing.

bridge. A unit that connects one network to another so that terminals on both networks can communicate as if on a single network. Bridges operate on frames, as opposed to packets, at the data link layer, OSI layer 2. *See* brouters *and* routers.

broadband ISDN (B-ISDN). An ISDN system capable of supporting rates greater than the primary rate. Connections support both circuit mode and packet mode services.

broadband transmission. The method of transmission in which the information signal is used to modulate a carrier signal. This type of transmission is well suited to carry voice, data, and video at the same time and is less subject to attenuation than baseband transmission, but is more difficult to configure and modify.

brouter. A device that combines the functions of bridges and routers.

bus topology. All stations are directly connected to the same transmission medium, usually a cable.

busy signal. A composite pulsed (0.5 sec on/0.5 sec off) signal of 480 Hz and 620 Hz sent by the operating company indicating that the called party's line is in use.

byte. A group of eight bits. *See* octet.

call translation. The process, in a software-controlled circuit-switching system, that determines how a call will be handled based on the available input information.

carrier sense multiple access with collision detection (CSMA/CD). In this probabilistic medium-access technique, devices must sense the presence or lack of presence of a carrier on a network before attempting to transmit. Once the device transmits, it will continue to monitor the line to detect collisions with other devices transmitting and act accordingly. This method is problematic as the number of devices on the network increases. This access method has been used by Ethernet and IEEE 802.3 LANs.

cell. (1) A fixed length packet in an ATM system; (2) A geographical area servicing cellular telephones.

cell loss priority (CLP). One bit in an ATM header used to indicate if a cell may be discarded (1) or may not be discarded (0).

central office terminal (COT). The place where communications common carriers terminate and interconnect customer lines.

channel. The conduit or path through which information passes. *See* physical and logical channel.

channel bank. Used to combine 24 or more signal sources (analog) into digital bit streams for T1 circuits.

channel service unit (CSU). The primary (electrical and mechanical) interface between the customer equipment and the digital network; required by the FCC.

chip. In CDMA systems, a binary word formed by multiplying original data by a pseudo-noise code (PN), causing that original data to be broken up into smaller bits and increasing (spreading) the bandwidth.

chrominance signals. A set of two color difference signals, I and Q, formed from the red, green, and blue signals.

circuit switching. The process of establishing a a dedicated physical path between the communicating customers for the duration of a transmission.

clients. Applications and computers that make use of resources of a network controlled by a server.

coaxial cable. A single copper wire covered by insulating material, which in turn is covered by a metallic braid or shield and then covered with insulation, all of which have the same axis; that is, they are coaxial. This construction provides both high bandwidth and excellent noise immunity.

code conversion. The process in which received address information has digits added, deleted, or substituted to process the call over the selected route.

codebook excitation linear predictive (CELP) encoder. One of the most popular voice coding techniques used in multimedia communication.

coding. Replacing quantized values with (usually) 8-bit numbers as part of pulse code modulation.

code-division multiple access (CDMA). A spread spectrum method of multiple-access technique in which the mobile units share the same frequencies but are separated by their codes.

coder/decoder (codec). One device to replace the individual devices formerly used to perform the functions needed in pulse code modulation and demodulation such as sampling, quantizing, encoding, decoding, filtering, and companding.

common channel signaling (CCS). Data communication network used by switches to exchange control information.

common intermediate format (CIF). A format defined by ITU-TS H.261 in which the frame size for the luminance is 352 samples/line and 288 lines/ frame.

companding. Formed from compression/expanding in which the signal is compressed at the transmitter to reduce the dynamic range and then expanded at the receiver to restore the signal to its normal condition.

computer telephony integration (CTI). The automated interaction of people with a computer through the telephone or with a telephone through a computer.

concentration. A switching network, or part of one, with more inputs than outputs. The process of combining calls from many input lines or trunks before the actual switching takes place.

congestion signal. A composite pulsed (0.2 sec on/ 0.3 sec off) signal of 480 Hz and 620 Hz sent by the operating company to indicate a blocked call due to the company having more calls than it can service at the moment.

Consultative Committee on International Telephone and Telegraph (of the International Telecommunication Union) (CCITT). A group that establishes many of the standards and protocols used in the telecommunications industry.

convergence sublayer (CS). A sublayer, within AAL, which attaches headers and trailers to an ATM message and passes the message on to the segmentation and reassembly sublayer, SAR. The CS is concerned with messages while the SAR is concerned with cells.

crosstalk. A form of noise occurring when two wires are electrically coupled such that current in one causes an electromagnetic field to affect the other, thus allowing communication on one path to be heard on another.

customer switching systems. *See* private branch exchange.

cyclic redundancy check (CRC). Method used for error checking of transmitted data. In extended superframe, 6 F-bits are used to carry a 2-kbps CRC-6 channel. The CRC for the current extended superframe occurs in the F-bits of the following superframe.

data circuit terminating equipment (DCE). The equipment between the DTE and the actual transmission line. A modem is considered to be a DCE unit.

data communications. The transfer of digital information principally among computers. *See* voice communications.

data compression. Refers to the ability to remove redundant elements from the transmitted data stream and, at the receiving end, recover the original information. Two general types of data compression are lossy and lossless.

datagram. An implementation of packet switching, the packets are treated independently with all packets containing the full destination address and sequential delivery not guaranteed.

data link layer. Layer 2 of the OSI set of seven layers that constitutes the network architecture. This layer is concerned with providing a relatively error-free path between adjacent stations to higher layers and is not concerned with the content of the information.

data link protocol. Protocol whose purpose is to provide reliable transfer of data and consists of three types: character, byte count, and bit-oriented protocols.

data terminal equipment (DTE). Equipment that converts user information into digital signals for transmission and also receives digital data. A PC could be considered a DTE. See data circuit terminating equipment (DCE).

delta modulation. Instead of sending a binary code representing the step size, as in traditional pulse code modulation, a single bit is sent. This bit indicates whether the current value has increased from previous (1) or decreased from previous (0) sample. Useful for slowly changing signals. *See* slope overload.

demultiplexing. The process of splitting up a multiplexed signal into the original separate signals.

dial tone. A composite continuous signal of 350 Hz and 440 Hz sent by the operating company to acknowledge the subscriber is off-hook and the caller may dial the number.

differential encoding. A sequence of values are encoded by representing each one as the difference from the previous set of values. Used in video compression.

digit translation. Interpreting the collected dialed digits and translating them into systems addresses; performed by call processing software.

digital access crossconnect switch (DACS). A switch system that, by using a time-division multiplexing scheme, switches or crossconnects digital bit streams.

digital control channel (DCCH). One of the two basic logical channels under IS-136. The other is the digital traffic channel (DTCH).

digital loop carrier (DLC). A transport system in the local loop that uses digital multiplexing technology to place 24 analog channels together onto a single DS-1 digital signal (for example, SLC 96), thereby reducing the number of cables to the central office.

digital service unit (DSU). A piece of equipment whose primary function in the digital network is to convert a unipolar signal from the customer's data terminal equipment (DTE) into a bipolar signal for the network and vice versa.

digital traffic channel (DTCH). One of the two basic logical channels under IS-136. The other is the digital control channel (DCCH).

discrete cosine transform (DCT). A two-dimensional lossless source transformation useful in image compression.

discrete multi-tone (DMT). A multilevel encoding scheme achieving a high bit rate by encoding groups of bits in discrete subchannels, referred to as bins.

distribution. The capability of connecting an input to any one of several outputs in a switching network. Same number of inputs as outputs.

domain name system. This is a hierarchical, domain-based naming scheme to implement the mapping of names into network addresses. The top-level domains are either generic (com, edu, org, gov, etc.) or country (uk, au, nz, etc.).

dotted decimal notation. The format in which the 32-bit addresses are written. The lowest IP address is 0.0.0.0 and the highest is 255.255.255.255.

downlink. System or frequency band of the transmission from a satellite to an earth station or base station to a mobile unit. *See* uplink.

DS-0 channel. The basic unit in the digital signaling (DS) hierarchy. DS-0 carries 64 kb/s and accommodates one voice circuit.

DS-1 channel. 24 DS-0 channels, 24×64 kbps + 8 kbps (framing bit) = 1.544 Mbps.

DS-2 channel. 4 DS-1s.

DS-3 channel. 7 DS-2s or 28 DS-1s = 44.7 Mbps.

dual-tone multifrequency (DTMF). Signaling by the use of a simultaneous combination of two frequencies (one from a low group and one from a high group). Used for dialing a number on phone and other signaling applications.

duplex communication. Communication where two parties can both send and receive information. *See* full *and* half duplex.

duration of service. Time for dialing, ringing, and conversation.

802.2. logical link control (IEEE LAN standard).

802.3. Carrier sense multiple access/collision detection, CSMA/CD (IEEE LAN standard).

802.4. Token passing bus (IEEE LAN standard).

802.5. Token passing ring (IEEE LAN standard).

electronic industries association 232 standard (EIA-232). An electrical standard derived from the long-standing standard RS-232. This standard specifies such parameters as open-circuit, short-circuit voltages, slew rate, distances, capacitance, etc.

encapsulation. The generation and extraction of a header.

entity. A hardware and/or software module. Peer entities, those in the same layer whether or not in the same system, may interact with each other. Entities make up a subsystem. Entities consist of primitives.

entropy encoding. A compression scheme, which is generally lossless, fully reversible, and applicable to all data.

ESF data link. Twelve F-bits in the extended superframe, sometimes referred to as the D-channel, data channel, embedded operations channel, maintenance data link, or frame data link. Used for a variety of purposes including red alarm (eight 1s followed by eight 0s).

Ethernet. A popular LAN, which was developed jointly by Xerox, DEC, and Intel. This LAN is not identical but is compatible with the IEEE 802.3 standard. Transmission rate was originally specified as 10 Mbps and is implemented with coaxial cables.

expansion. A switching network, or part of one, with more outputs than inputs.

extended superframe (ESF). A group of 24 frames, each containing 24 PCM words. Additional functions (provided by the F-bits) beyond superframe include error checking with CRC-6 protocol and providing carrying system performance data.

extended test controller (XTC). A unit that performs the same tests as the PGTC plus additional tests such as monitoring the digital channel.

fast fading. *See* Rayleigh fading.

feature transparency. A system designed so that the user is unaware of which component (e.g., PBX) is providing the service.

fiber distributed data interface (FDDI). A fiber optic ring configuration of a LAN operating system at a minimum of 100 Mbps.

fiber optic cable. This cable contains one or more glass or plastic fibers (rather than copper) surrounded by insulation known as cladding. Fiber optic cables are characterized by high transmission rates, electrical immunity because they carry light, and relatively low attenuation but higher cost than copper.

f-bit. Framing bit. Can be used to carry signaling information.

filter. A device allowing only certain frequencies to pass through it while severely attenuating other signals at different frequencies. Filters may be digital or analog. Common types are low pass, high pass, band pass, and band reject. Filtering in LANs has a slightly different meaning—bridges examine destination addresses and determine whether or not to forward to another LAN.

flow control. Control of the rate of data from the source as a means of minimizing congestion.

footprint. The area covered by a transmitter such as a geostationary satellite.

forward channel. A channel in mobile communications that transmits from a base station to a mobile unit(s). Also referred to as a *downlink.*

forward control channel (FOCC). Broadcast by the base station and received by mobile units, this channel carries phone number pages, messages to move to a specific voice channel, frequencies to be used, etc.

forward voice channel (FVC). The voice channel from the base station to the mobile unit.

Fourier series. The representation of a signal or function by decomposing it into sine waves, named for the French mathematician.

frame. In digital transmission systems, a group of synchronous serial bits representing data and/or control information. In a T1 system a frame consists of 24 pulse code modulation words with each frame separated by a single framing bit (see super and extended super frames). Each frame is 193 bits (24 channels × 8 bits/channel + 1 framing bit).

framing pattern sequence (FPS). Six F-bits used, by extended superframe, to maintain terminal synchronization and identify frame boundaries. These bits are always the fourth, eighth, sixteenth, twentieth, and twenty-fourth F-bits and always have the values 001011, respectively.

frequency. The number of cycles per second, measured in Hertz.

frequency-division multiplexing (FDM). The division of the available bandwidth into frequency segments that carry separate signals. All signals are sent at the same time but each occupies a different frequency band.

frequency-division multiple access (FDMA). The channels within a cell are separated by frequency. This is the technique used by the analog mobile phone systems.

frequency modulation (FM). FM, also known as frequency shift keying (FSK), uses different frequencies to represent values. It is used in inexpensive and low-speed modems.

full-duplex communication. Both parties can simultaneously send and receive information. *See* half duplex.

gateway-witching systems. Used to route traffic between national and international calls.

global positioning system (GPS). A GPS receiver receives signals from multiple satellites and then calculates the location within a radius of a few meters.

Global System for Mobile Communications (GSM). Digital cellular system standard in Europe that, unlike U.S. standards, is very comprehensive going beyond the air interface.

guard time. This time is used in TDMA systems to allow space between the end of a previous slot and the beginning of the next.

guided media. Physical media that constrain the signals within boundaries based on the physical construction. Examples include wires, fiber cable, and coaxial cable. *See* unguided media.

half-duplex communication. Two-way alternate communication with each party able to send *and* receive information but not simultaneously.

handoff. The process of a mobile unit changing from one base station to another. *See* hard *and* soft handoff.

hard handoff. A handoff in which a mobile separates from its current base station and then connects to the new base station. *See* soft handoff.

header error control (HEC). An 8-bit field used for error detection and correction in an ATM header that is equal to the sum of the byte 01010101 and the CRC calculated over the rest of the header. The HEC is also used to delineate the cell boundary.

high-level data link control (HDLC). A bit-oriented data link protocol capable of transmitting bit streams that are not necessarily multiples of character length. It is the ISO standard for the data link layer, layer 2, of the OSI.

high-speed Ethernet (100Base-T). This system offers 100 Mbps speeds and follows IEEE Ethernet standards with three variations: 100Base-TX (two pairs of category 5 UTP cable), 100Base-T4 (four pairs of category 3, 4, or 5 UTP), and 100Base-FX (multimode optical fiber).

home location register (HLR). A database that along with the visitor location register (VLR), is used to coordinate roaming mobile phone service.

hub. A device used in star-configured networks to amplify and retransmit signals to and from the various stations.

Huffman code. This data compression technique encodes the most probable, or most commonly used, characters with fewer bits than the least probable characters.

hunt groups. A number of users of a PBX system, for example, configured in a group to answer incoming calls in a round-robin manner based on their availability.

impulse noise. Irregular pulses or noise spikes of short duration and large amplitude, caused by mechanical switches, power changes, lightning, etc.

in-band signaling. Signaling information carried in the same bandwidth as the voice and data messages.

Institute of Electrical and Electronic Engineers (IEEE). An international organization, founded and head-quartered in the U.S., that defines electrical and communications-related standards.

integrated services digital networks (ISDN). A set of standards covering the various types of applications sent through digital channels with the B, Bearer, channel(s) operating at multiples of 64 kbps and a separate signaling channel, D or Demand channel, which is 16 kbps or 64 kbps.

Interim Standard 41 (IS-41). A seven-layer protocol developed by the telecommunications industry association (TIA) for wireless communications, including standards for roaming.

Interim Standard 54 (IS-54). Specifies the first-generation digital cellular system (FDMA-TDMA); also requires dual-mode phones.

Interim Standard 136 (IS-136). The applicable standard for CDMA systems.

Interim Standard 154 (IS-154). The applicable standard for NA-TDMA systems.

interleaving. To minimize the effect of a long fade or long burst of noise, data from two different speech frames are interleaved and then deinterleaved at the receiver.

International Standards Organization (ISO). An organization with representatives from virtually all countries, chartered with the task of developing standards for numerous devices, from automobiles to computers. ISO developed the open system interconnection (OSI) model.

International Telecommunications Union (ITU). An agency established by the United Nations to provide worldwide standards for telecommunications.

International Telecommunications Union-Telecommunications Standardization Sector (ITU-T). That part of the ITU concerned with global telecommunications standards.

Internet Control Message Protocol (ICMP). This protocol, which is part of the suite of TCP/IP protocols, provides feedback from routers to hosts about problems in the communication environment.

intracoded frame (I-frame). A frame not built from any other frames; may serve as a reference frame.

IPv4. Internet protocol version 4 is the version in operation in the 1990s.

IPv6. This new version of the Internet protocol is gradually replacing version 4. Among the major improvements are virtually unlimited address space, simplified protocol, and accommodations for further evolution.

Joint Photographic Expert Group (JPEG) standards. Standards for compressing continuous-tone grayscale or color images.

key telephone systems. A business system in which several lines are leased, with the customer having the control over those lines typically through keys on the telephone sets.

LAN adapter. A hardware board, usually inserted in an expansion slot of a PC, to allow direct connection to a LAN. Also referred to as *network interface adapters (NIAs)* or *network interface cards (NICs)*.

LAN protocol. Protocols used to manage the exchange of information among users connected to a LAN. The most popular LAN protocols include NetBIOS, TCP/IP, XNS, and SPX/IPX.

layer. In the open systems interconnections (OSI) reference model, systems are divided into seven layers. Each layer is composed of subsystems from various open systems that perform similar functions.

line-of-sight transmission. Geographical arrangement of transmitting antennas and receiving towers so that the wave travels in a straight line from transmitter to receiver; the usual configuration for microwave systems.

link control protocol (LCP). This protocol is used to negotiate data link protocol options during the "establish" phase of data communications. There are 11 types of LCP, including 4 configure types.

local exchange switching systems. Switch telephone lines for residential and business users with connections to other local exchanges and trunk exchanges.

local area network (LAN). Devices connected in a geographically small area such as within a room, a building, or nearby buildings. The network is usually owned by the same organization that owns the devices being connected.

local loop. The connection between a telephone and the central office, typically with twisted wire pairs.

LocalTalk. A proprietary LAN system developed by Apple Computer. This is an inexpensive CSMA/CD system requiring only cables because Apple computers come with the required cards.

logical channel. A path for transmission of data. There may be numerous logical channels multiplexed within a single physical channel.

loose source routing. If this option is selected in an Internet transmission, a complete path is specified from the source to destination as a sequence of IP addresses, but the datagram is allowed to pass through other routers on the way. *See* strict source routing.

loopback. A test procedure in which signals are looped back into the same instrument or computer. A simple example would be a computer sending an E-mail to itself.

lossless compression. This method of data compression preserves all the information in the original data, allowing the data to be fully recovered at the receiver. Two basic methods of lossless compression are replacement of long strings of redundant data with a code and using efficient coding as in the Huffman code.

lossy compression. Essentially implemented by discarding information elements that are considered unimportant to the overall content or image. Mostly used to encode still and motion images.

luminance signal. A signal, labeled as Y, which carries information on lightness and brightness and is formed from the red, green, and blue signals.

macroblock. One 16×16 pixel area composed of 8 $\times$ 8 blocks (four blocks for the luminance and two for each of the color difference components).

media-access control (MAC). A technique for determining which of several stations on a LAN can access the network and to ensure that all stations have a chance to access the network.

message switching. A store-and-forward switching technique in which customers exchange information by sending discrete messages. Typically, this is one way with real-time interconnections; for example, electronic mail and telegrams.

mobile telephone switching office (MTSO). This office controls several base stations and handles the management and coordination of calls.

modem. Modulator-demodulator device that converts digital signals into analog signals (modulates) for transmission over the local loop and will demodulate analog signals it receives.

modified Huffman code. A code adopted by the ITU-T for facsimile data compression.

modified relative element address designate (modified READ). A data compression technique for facsimiles based on the position of changing elements (75% of transitions can be defined by plus or minus one pel from the previous line).

Motion Picture Expert Group (MPEG) Standards. The main algorithms used to compress videos that have been international standards since 1993. Since movies contain both images and sounds, MPEG is concerned with compressing both video and audio.

motion vector. Indicates the spatial translation of a macroblock between two frames.

multicasting. The IP uses a 28-bit address to identify groups and make a best-effort attempt to deliver the message to all the members of the group.

multimedia. Two or more media, usually considered to be audio and video.

multimode fiber. A fiber optic cable that allows many waves to travel through it with the rays refracting off the boundaries at different angles. *See* single-mode fiber.

multipath fading. Delayed and out-of-phase waves interfering with direct waves.

multiple access. The means by which base stations service multiple mobile phones in each cell.

multiplexing. The process of combining two or more signals into one.

multipoint link. A connection or link with more than two parties sharing the link. *See* point-to-point link.

multistation access unit (MAU). A device that clamps onto a cable providing a means for a station to physically attach to a LAN.

NetWare. A widely used LAN network operating system developed by Novell. Its later versions are media-independent, so that it runs on all popular network hardware such as Ethernet, Token Ring, Arcnet, and others.

network. Interconnected elements such as transmission links, switches, and interface equipment all forming a communication infrastructure. Examples include the telephone network and computer networks.

network architecture. A set of layers and protocols defining a network with enough information for implementers to write programs or build hardware for each layer.

network access point. Internet protocol (IP) traffic interchange point.

network interface. The access point for a user into a network; allows the internal architecture of the network to be transparent to the user.

network information center. The agency charged with assigning network numbers for the Internet.

network layer. Layer 3 of the OSI set of seven layers that constitutes the network architecture. This layer is concerned with the transfer of network transmission data units (packets) among network layer service users across the subnet. This layer fragments and reassembles messages into packets and packets into messages. Standards such as X.25, Q.931, ISO 8473, and DOD IP apply.

network operating system (NOS). A specialized software that extends the operating system functions to a network environment and supports LAN hardware. Typical NOS functions include print, file and database, remote access, messaging, network management, and communications services.

noise. Any undesired disturbance in a system. *See* crosstalk, impulse *and* thermal.

null suppression. Null or blank suppression compresses data by scanning and replacing strings of blanks with a two-character code.

Nyquist sampling theorem. To attain a valid digital representation of an analog wave, the sampling must be done at a rate that is at least twice the highest frequency component of that signal. For example, in a voice system the sampling could take place at 6800 Hz (2 × 3400 Hz) but practically it is sampled at 8 kHz.

octet. A group of eight bits. *See* byte.

off-hook. The telephone set taken off its cradle or hook, thereby closing a switch and completing a current path to alert the phone company the user is ready to initiate a call.

on-hook. The telephone set placed on its cradle or hook, thereby opening a switch and breaking the current path to alert the phone company the user is terminating a call. *See* off-hook.

open systems interconnection reference model (OSI). A telecommunications seven-layer architecture comprised of the physical (layer 1), data link, network, transport, session, presentation, and application (layer 7) layers.

out-of-band signaling. Signaling that takes place out of the voice or data channel. In time-division systems, this may occur as a specially allocated bit position, time slot, or dedicated channel.

packet switching. The routing of packets (discrete units of messages) through the telephone system to their destination. Messages are fragmented into packets (each about 128 octets long), which have additional information including a destination address, are each forwarded along the best available path through the network, and are reassembled at the destination.

paging. Pagers are a classic example of simplex transmission in that it is entirely (although this is changing) one way without a response from the receiver.

pair gain test controller (PGTC). A device used to isolate a transmission problem to a particular section in the loop.

passband. The range of frequencies that will pass through a communications channel. *See* bandwidth.

personal communications services (PCS). Third-generation mobile communications that offer additional services such as voice mail, call forwarding, caller ID, short messages, etc. Strictly defined, PCS operates just below the 2 GHz band.

phase shift. The time difference between two events. For two sine waves it would be the time between two corresponding points on the sine waves.

phase shift keying (PSK). At the beginning of a bit time the phase of the signal is altered with the change of phase indicating the value being represented. More precisely, this is called differential phase shift keying (DPSK). Used in moderate speed modems.

phase shift modulation (PM). *See* phase shift keying.

physical layer. Layer 1 (bottom) of the OSI set of seven layers that constitutes the network architecture. This layer is concerned with the actual hardware and physical transmission of analog and digital data.

physical medium. Twisted-pair wire, optical fiber, coaxial cable, air, etc., which is used to carry the signal.

pilot channel. In CDMA systems this channel is used to continuously transmit a sequence of 0s at 1.2288 Mchips/sec and is used for phase, timing reference, and signal strength measurements.

pixel. A picture element that contains gray-scale information.

point-to-point-connection. Only two parties share the link. *See* multipoint link.

point-to-point protocol (PPP). A character-oriented (as opposed to the bit-oriented HDLC) protocol developed to overcome the deficiencies of SLIP. It handles error detection, permits authentication, and offers many other improvements over SLIP. PPP provides three functions: framing, link control, and a means of negotiating network layer options independently of the network layer protocol used.

predicted frame (P-frame). A frame constructed from a preceding frame.

presentation layer. Layer 6 of the OSI set of seven layers that constitutes the network architecture. This layer is concerned with the form of user data, including code conversion of, for example, ASCII to EBCDIC.

primary rate interface (PRI). ISDN primary rate circuits operate at 1.544 Mbps providing 23 B-channels at 64 kbps and 1 D-channel at 64 kbps. The international definition for primary rate is 30B + 1D or 2.048 Mbps.

primitive. A basic operation of an entity. The lowest level of a layer.

private branch exchange (PBX). These switching systems are similar to local exchanges, except they are privately owned or leased and serve a company, university, hospital, etc. Newer PBXs provide a direct termination of a T1 into a digital trunk interface.

protocol. A set of rules that governs communications among different entities within a telecommunications network.

protocol data unit (PDU). User data and protocol control information together.

protocol stack. List of protocols; one per layer, used by a system.

public switched telephone network (PSTN). The telephone infrastructure that has developed over the last century.

pulse code modulation (PCM). The process of converting an analog signal into an encoded digital value for transmission. This process includes sampling, quantization, and coding.

quadrature amplitude modulation (QAM). A modulation technique that varies both phase shift and amplitude but holds carrier frequency constant. Used in high-speed modems.

quality of service (QoS). A group of standards or objectives for the quality of a particular system. For example, in an ATM system the QoS parameters include cell loss ratio, cell misinsertion rate, cell error ratio, cell transfer delay, cell delay variation, cell transfer capacity, and skew.

quantization. Representing the samples of an analog signal by rounding each true sample value to a near value from a finite set of discrete quantities as part of pulse code modulation.

quantizing distortion. The difference between the true value and the quantized, or rounded, value. This is the major source of error in pulse code modulation, which can be minimized by increasing the number of steps or companding.

ramp time. In a TDMA system, this time is allotted to allow the mobile to come back up to full power after being shut down between transmissions.

Rayleigh fading. As a mobile unit moves quickly through a cell, a single ray reaches the unit at different times so that the signal adds and subtracts causing the quality to vary. Also referred to as *fast fading*.

red alarm. If the framing bits are not being received properly (lost synchronization) by a T1 system, the terminal equipment recognizing the problem goes into the red alarm state and sends a bit-2 alarm. This is a red alarm and consists of the second bit of each PCM word in the DS-1 stream being forced low. *See* yellow alarm.

reference frame. A frame from which other frames are constructed.

refraction. The bending of a light wave when it passes from one medium into another of different density.

repeater. A unit that receives digital transmissions and amplifies, retimes, and retransmits the signal. Over copper wires T1 signals need repeaters at approximately every mile. Repeaters are also used in LANs, but they are limited to connecting LANs having the same physical characteristics.

reverse address resolution protocol (RARP). This protocol allows a newly booted workstation to broadcast its Ethernet address. The host sees this request, looks up the Ethernet address in its configuration files, and responds with the corresponding IP address.

reverse channel. A channel in mobile communications that transmits from a mobile unit to the base station. Also called an uplink.

reverse control channel (RECC). A random access channel from the mobile unit to base station; used to send acknowledgments, call requests, and the mobile's electronic serial number (ESN).

reverse voice channel (RVC). The voice channel from the mobile unit to the base station.

ring-back signal. A composite pulsed (2 sec on/4 sec off) signal of 440 Hz and 480 Hz sent by the operating company to indicate that the dialed party is being sent a ringing signal (note this and the ringing signal are independent).

ring error monitor (REM). One device on a token ring that can regenerate lost tokens and remove bad frames from the network.

ringing signal. A 20 Hz, 90V$_{RMS}$ signal sent to alert a subscriber that a call is attempting to come through.

ring topology. In this LAN configuration, adjacent stations are connected to each other and information travels around the ring. If one device is disabled the ring may be disabled, if there is no redundancy or other means of correction built in. The ring may not necessarily be physical but may be logical in nature.

roaming. A mobile phone subscriber moving into an area serviced by another provider.

router. Device that connects autonomous networks of compatible architecture, at the network layer, OSI layer 3. Such devices perform packet routing and are controlled by the protocols of the networks they service, TCP/IP or IPX/SPX, for example. Most routers available today can perform the functions of a bridge and router; that is, they are essentially brouters.

routing. Establishing the connection for a talking path, performed by call processing software.

run-length encoding. This method of data compression eliminates any sequence of repeating characters with a three-character code.

sampling. Periodically measuring the values of an analog signal as part of pulse code modulation.

segment. The unit used for sending and receiving messages in the Internet. A segment consists of a fixed 20-byte header followed by 0 or more data bytes.

segmentation and reassembly sublayer (SAR). A sublayer within the AAL; breaks messages up into 44-octet chunks that are inserted into the payload of the cell. This layer is concerned with cells where the convergence sublayer (CS) is concerned with messages.

serial line IP (SLIP). A very simple protocol developed to connect Sun workstations to the Internet over a dial-up line using a modem. It includes no error detection or correction, no authentication, and is not an approved Internet standard.

server. Computers containing and controlling shared resources that are used by other computers (clients) in a network.

session layer. Layer 5 of the OSI set of seven layers that constitutes the network architecture. This layer is concerned with managing the logical connection, or session, between communicating parties. Among the issues resolved are duplex vs. half duplex, who talks when, etc.

Shannon capacity theorem. This theorem defines the relationship between bandwidth and signal-to-noise ratio and is $C = B \times \log_2 (1 + S/N)$ where C is channel capacity in b/s, B is the bandwidth in Hz, and S/N is the signal-to-noise ratio.

shielded twisted-pair. A twisted pair of insulated wires with a metal foil shield around the pair. The shielding not only helps immunize the wire pair from extraneous noise but also protects other wires from its generated noise. There are various categories of twisted-pair wire depending on how tightly the wires are twisted.

signal. The representation of the data external to the machine, either in analog or digital format.

simplex communication. Communication in one direction only.

single mode fiber. A fiber cable with a very small diameter such that light can travel through it only without refraction. More expensive than multimode fiber, but light can travel longer distances without needing regeneration. *See* multimode fiber.

slope overload. If the sampled signal in a delta modulation system changes rapidly, the replicated staircase cannot keep up and there is a wide gap between the original signal and that at the receiver. *See* adaptive delta modulation.

slow associated control channel (SACCH). This channel carries control messages between the mobile unit and the base station such as power measurements and requests from the mobile for a handoff.

slow fading. The change in signal level for a mobile phone due to its motion as it moves slowly through the terrain.

soft handoff. Instead of breaking contact with the current base station before connecting to another, there is a period of time in CDMA systems when the mobile unit is communicating with both the new and old base stations. *See* hard handoff.

source encoding. A lossy compression technique that takes advantage of the properties of the data.

Three methods of source encoding are differential encoding, transformations, and vector quantization.

space-division multiplexing. The process of combining physically separate signals into a bundled cable.

store-and-forward switching. *See* message switching.

star topology. In this LAN configuration, each station is connected to a central piece of equipment commonly called a hub. All communications must go through the hub, which amplifies and retransmits the signals received on one cable to the other cables.

statistical multiplexing. A method of TDM, used in ATM, in which several connections are assigned the same link based on their traffic characteristics. If a large number of connections are very bursty, they all may be assigned the same link under the assumption that they will not all burst at the same time.

strict source routing. If this option is selected in Internet transmission, a complete path is specified from the source to destination as a sequence of IP addresses, and the datagram is required to follow the exact route. *See* loose source routing.

subnet. A portion of a larger network that may act independently.

subsampling. A technique used in television of taking fewer samples per line and sometimes fewer lines per frame.

subsystem. A part of a layer. Subsystems themselves consist of entities.

superframe. A frame containing 12 standard frames, each containing 24 PCM words. Framing bit functions include locating signaling information and checking terminal synchronization.

supervisory audio tone (SAT). Both the base station and the mobile phone insert this audio signal into their modulators to ensure that the mobile is communicating with the correct base station.

sync channel. In a CDMA system, a channel that continuously transmits information to the terminals such as the base station's ID number and system time.

synchronous optical network (SONET). A standard for digital transmission over optical networks. The entire bandwidth of the fiber is devoted to one channel containing time slots for the various subchannels. It is synchronous and controlled by a master clock having an accuracy of one part in one billion.

synchronous transfer mode (STM). A method of packetized transmission using a circuit-switched network

synchronous transmission. All characters of a message are sent contiguously, one after another. *See* asynchronous transmission.

system administration terminal (SAT). An interface to a PBX, for example, to allow the system operator to administer, maintain, troubleshoot, and configure the system.

system identifier (SID). A number assigned to each mobile phone operating company when it receives a license.

T-carrier systems. Telecommunication systems that digitize many voice signals and bundle these digitized signals into a single digital bit stream (by means of time-division multiplexing) and then demultiplex them at the receiver. The basic standard rate is 1.544 Mb/s.

T1 carrier. A time-division multiplexed system that concentrates 24 voice channels onto a single digital transmission link. Each voice channel is PCM 64 kbps, called DS-0, consisting of 8 bits per sample $\times$ 8000 samples/sec. The output signal is 24 $\times$ 64 kbps (+ 8 framing bits) or 1.544 Mbs, called DS-1. Each time slot is 648 ns.

T1 line driver. Part of the T1 carrier system, this unit conditions the electrical characteristics of the T1 bit stream to create uniform shapes and correct voltage levels. Also, the line driver converts the pulses from unipolar to bipolar.

T3 carrier. A TDM system providing 44.736 Mbps of digital service or the multiplexed capacity of eight T1s. Each time slot is 22.5 ns. Requires greater bandwidth than twisted pair and the primary means of delivery are digital microwave, coaxial cables, and fiber optics. T3 frame size is 4760 bits (compared to 193 bits for T1) and uses B3ZS line code compared to B8ZS for T1.

temporal correlation. A form of motion compensation that consists of identifying redundancies between successive frames.

thermal noise. Noise caused by random motion of electrons due to heat, which is present in all electronic devices and media. Thermal noise is characterized by a hissing sound and is referred to as white noise.

thick Ethernet (10Base5). The IEEE label 10Base5 indicates a CSMA/CD 10 Mbps, baseband operation, and 500 meter segments. Thick refers to the size of the coaxial cable.

thin Ethernet (10Base2). The IEEE label 10Base2 indicates CSMA/CD 10 Mbps, baseband operation, and 200 meter segments. Thin refers to the size of the coaxial cable, which is more flexible and cheaper than thick Ethernet.

time-division multiple access (TDMA). A technique of wireless communications that divides up a particular frequency into time slots that are shared among several subscribers. It is actually a combination of frequency and time division.

time-division multiplexing (TDM). The capacity of the transmission facility is divided into time slots and each channel is assigned a time slot. Only one signal occupies the channel at any particular instant.

time-division (multiplexed) switching. A circuit-switching technique in which time slots of an input bit stream of time-multiplexed data are directed to the appropriate output. It is a combination of space division and time multiplexing.

time slot interchanger. A device, part of a time-division switching system, that separates and switches signals from multiplexed calls. A mirror transfer is performed for the other direction of transmission through the TSI.

token passing. A medium-access control technique for a bus or ring in which computers form a physical or logical ring. When a computer has the token (a special type of data frame), it may transmit and then pass the token on to the next computer.

topology. The actual physical layout of a LAN in which the devices are interconnected. The most commonly used topologies are bus, star, and ring.

traffic channel. In CDMA systems, a logical reverse channel used to send user information and signaling messages.

transmission control protocol/Internet protocol (TCP/IP). A collection of protocols governing the Internet. TCP provides end-to-end communicators a reliable virtual circuit, which is then transported across the Internet using the Internet protocol (IP). TCP provides virtual circuit flow and congestion control in layer 4, while IP is in layer 3.

transmitter window. Number of outstanding (unacknowledged) blocks that the transmitter can have at any time, a flow control parameter.

transport layer. Layer 4 of the OSI set of seven layers that constitutes the network architecture. This layer is concerned with the reliable and transparent transfer of messages between endpoints. This layer complies with standards such as DOD transmission control protocol (TCP), X.214, and X.224.

transport service access point (TSAP). A port plus its host's IP address forms a 48-bit unique access point in the Internet.

trellis-coded modulation. A modulation technique that combines error correction, coding, and modulation for improved reliability but also high speed.

trunks. The transmission links that interconnect switching systems providing the overall hierarchical network.

trunk exchange switching systems. Also called tandem switches, transient switches, and toll switches, these systems are used to interconnect trunks and have no end users directly connected.

trunk interface circuits. A circuit associated with the connection of a trunk to a switching system. Examples include central office trunks, direct inward dialing trunks, digital and analog tie trunks, auxiliary trunks, LAN interfaces, and ISDN primary rate interfaces.

twisted-pair wire. A pair of insulated copper wires twisted together to minimize crosstalk with other wires. *See* balanced *and* unbalanced lines.

twisted-pair Ethernet (10Base-T). Networks following this IEEE standard are limited to 100 Mbps and use unshielded twisted pairs with the length of cable to the hub limited to 100 meters. These networks require at least category 3 UTP cable connected with RJ-45 connectors.

μ-Law. An alternative to the A-Law formula for compression of analog signals.

unbalanced line. One wire of a twisted pair at ground potential (called the return) and the other carrying the signal. *See* balanced line.

unguided media. Any medium not having its own physical boundaries, such as the atmosphere. Wireless and satellite communications use unguided media. *See* guided media.

universal asynchronous receiver/transmitter (UART). A programmable device that performs serial-to-parallel and parallel-to-serial conversions of digital data.

universal synchronous receiver/transmitter (USRT). The interface circuit used to transfer data, which is framed by a synchronization (SYNC) character and an end-of-test character.

unshielded twisted-pair (UTP). An inexpensive, flexible, and easy-to-install wiring technique consisting of a pair of insulated wires twisted together. There are various categories of twisted-pair wire depending on how tightly the wires are twisted.

uplink. System or frequency band of the transmission from an earth station to a satellite or a mobile unit to a base station. *See* downlink.

V.235. ITU-T recommendation identical to V.28 for 56-kbps service with a 34-pin (square) connector rather than the DB-25 connector.

V.24. ITU-T recommendation for mechanical specifications for EIA-232-D.

V.28. ITU-T recommendation for electrical specifications for EIA-232-D.

vector quantization. A method of source encoding useful with image data. The data are divided into blocks or vectors and the actual transformation is made using a codebook.

virtual channel identifier (VCI). A 16-bit number used to associate the two endpoints of an ATM network (instead of by a time slot or bucket number as in STM networks). This number is unique to any given link.

virtual-circuit network. In this implementation of packet switching, all packets follow the same route, a virtual circuit. Each packet carries a common identifier, does not need to carry a complete address, and does not arrive in sequence. The virtual circuit is an end-to-end connection between two users over a store-and-forward network.

virtual path identifier (VPI). An 8-bit number used to identify a group of virtual channels.

visitor location register (VLR). A database that, along with the home location register, is used to coordinate roaming mobile phone service.

voice communications. The transfer of sound (usually voice) primarily through analog signals. *See* data communications.

wide area network (WAN). A network encompassing a large geographical area usually requiring the use of a common carrier.

wireless   telecommunications. Communicating without the use of wires or other physical guides or conduits through some distance.

wireless local area network (WLAN). A LAN in which the interconnections are made using infrared or radio frequencies transmitted over the air.

X.25. The ITU-T recommendation defining the interface between a user and a packet-switched public data network. X.25 defines three layers: physical, link, and packet or network layer of OSI.

yellow alarm. If a T1 terminal receives a red alarm, it responds with a yellow alarm indicating it is sending the information properly and something must be wrong in the transmission line.

zero constraint rule. There must be at least one 1 in every 8 bits, because repeaters need 1s to maintain their timing. Thus, there can be no more than seven 0s in a row. *See* bipolar eight zero substitution.

AAL	ATM adaptation layer
ACD	automatic call distributor
ACK	acknowledgment
ADPCM	adaptive differential pulse code modulation
ADU	alarm display unit
AIS	alarm indication signal
AM	amplitude modulation
AMI	alternate mark inversion
AMPS	advanced mobile phone system
ANSI	American National Standards Institute
API	application programming interface
ARP	address resolution protocol
ARQ	automatic repeat request
ASCII	American Standard Code for Information Interchange
ASK	amplitude shift keying
ATM	asynchronous transfer mode
B3ZS	bipolar three zero substitution
B8ZS	bipolar eight zero substitution
BERT	bit error rate test
B-ISDN	broadband ISDN
BORSCHT	battery feed, overvoltage protection, ringing signals, supervision, codec, hybrid circuits, testing
BPV	bipolar violation
BSRF	basic system reference frequency
CCITT	Consultative Committee on International Telephone and Telegraph
CCS	common channel signaling
CDMA	code-division multiple access
CELP	codebook excitation linear predictive
CIF	common intermediate format
CLP	cell loss priority
CLUT	color look up table
COT	central office terminal
CODEC	coder/decoder
CPU	central processor unit
CRC	cyclic redundancy check
CS	convergence sublayer
CSMA/CD	carrier sense multiple access with collision detection
CSU	channel service unit
CTI	computer telephony integration
CTS	clear-to-send

CTU	channel test unit
DACS	digital access crossconnect switch
DCCH	digital control channel
DCT	discrete cosine transform
DLC	digital loop carrier
DMT	discrete multi-tone
DNS	domain name server or system
DPSK	differential phase shift keying
DSP	digital signal processing
DSU	digital service unit
DTCH	digital traffic channel
DTE	data terminal equipment
DTMF	dual-tone multi-frequency
EIA-232	Electronic Industries Association 232 standard
ESF	extended superframe format
ESN	electronic serial number
FDDI	fiber-distributed data interface
FDM	frequency-division multiplexing
FDMA	frequency-division multiple access
FM	frequency modulation
FOCC	forward control channel
FPS	framing pattern sequence
FSK	frequency shift keying
FVC	forward voice channel
GPS	global positioning system
GSM	global system for mobile communications
HDLC	high-level data link controller
HDTV	high definition television
HEC	header error control
HLR	home location register
ICMP	Internet control message protocol
IEEE	Institute of Electrical and Electronic Engineers
ILMI	intermediate local management interface
ISA	industry standard architecture bus
ISDN	integrated services digital networks
ISO	International Standards Organization
ISP	Internet service provider
ITU	International Telecommunications Union
ITU-T	International Telecommunications Union—Telecommunications Standardization Sector
JPEG	Joint Photographic Expert Group
LAN	local area network
LCP	link control protocol
LFC	local function capability
LIU	line interface unit
LSU	line switching unit
MAC	medium-access control
MAHO	mobile-assisted handoff
MAU	multistation access unit
MIN	mobile identifier number

Modem	modulator-demodulator
Modified-READ	modified relative element address designate
MPEG	Motion Picture Expert Group
MSVC	meta-signaling virtual channel
MTSO	mobile telephone switching office
MVIP	multivendor integration protocol bus
NACK	not ACKnowledgment (Reject)
NAP	network access point
NIA	network interface adapter
NIC	network information center
NIC	network interface card
NNI	network-network interface
NOS	network operating system
NTSC	National Television Standard Committee
OAM	operation and administration management
OLU	optical line unit
OSI	open systems interconnection reference model
PBX	private branch exchange
PCM	pulse code modulation
PCS	personal communications services
PCU	power converter unit
PDU	protocol data unit
PGTC	pair gain test controller
PM	phase-shift modulation
PPP	point-to-point protocol
PRI	primary rate interface
PSK	phase-shift keying
PSTN	public switched telephone network
QAM	quadrature amplitude modulation
QoS	quality of service
RARP	reverse address resolution protocol
RECC	reverse control channel
REJ	reject
REM	ring error monitor
RGB	red, green, blue
RNR	receiver not ready
RR	receiver ready
RTS	request-to-send
RVC	reverse voice channel
S/N	signal-to-noise ratio
SABM	set asynchronous balanced mode
SABME	set asynchronous balanced mode, extended
SACCH	slow associated control channel
SAR	segmentation and reassembly
SAT	supervisory audio tone
SAT	system administration terminal
SEAL	simple efficient adaptation layer
SID	system identifier
SLIP	serial line Internet protocol
SNMP	simple network management protocol

SONET	synchronous optical network
SNMP	simple network management protocol
SPU	signal processor unit
SREJ	selective reject
STM	synchronous transfer mode
STP	shielded twisted pair
SVC	signaling virtual channel
TCP/IP	transmission control protocol/Internet protocol
TDM	time-division multiplexing
TDMA	time-division multiple access
TE	terminal equipment
TIA	telecommunications industry association
TRU	transmit/receive unit
TSAP	transport service access point
TSI	time slot interchanger
UART	universal asynchronous receiver/transmitter
UNI	user-network interface
USDC	united state digital cellular
USRT	universal synchronous receiver/transmitter
UTP	unshielded twisted pair
VCI	virtual channel identifier
VLR	visitor location register
VPI	virtual path identifier
WAN	wide area network
WLAN	wireless local area network
XTC	extended test controller

INDEX